GO MATH!
FLORIDA

FLORIDA

Houghton
Mifflin
Harcourt

Printed in the U.S.A.

ISBN 978-0-544-50083-9

4 5 6 7 8 9 10   1468   23 22 21 20 19 18 17

4500652356   ^ B C D E F G

Dear Students and Families,

Welcome to **Go Math!**, Grade 5! In this exciting mathematics program, there are hands-on activities to do and real-world problems to solve. Best of all, you will write your ideas and answers right in your book. In **Go Math!**, writing and drawing on the pages helps you think deeply about what you are learning, and you will really understand math!

By the way, all of the pages in your **Go Math!** book are made using recycled paper. We wanted you to know that you can Go Green with **Go Math!**

Sincerely,

The Authors

Made in the United States
Text printed on 100% recycled paper

GO MATH!

Authors

Juli K. Dixon
Professor of Mathematics Education
University of Central Florida
Orlando, Florida

Matt Larson
Curriculum Specialist for Mathematics
Lincoln Public Schools
Lincoln, Nebraska

Miriam A. Leiva
Founding President, TODOS:
   Mathematics for All
Distinguished Professor
   of Mathematics Emerita
University of North Carolina Charlotte
Charlotte, North Carolina

Thomasenia Lott Adams
Professor of Mathematics Education
University of Florida
Gainesville, Florida

Fluency with Whole Numbers and Decimals

Extending division to 2-digit divisors, integrating decimal fractions into the place value system and developing understanding of operations with decimals to hundredths, and developing fluency with whole number and decimal operations

1 Place Value, Multiplication, and Expressions **3**

Domains Operations and Algebraic Thinking
Number and Operations in Base Ten

DIGITAL PATH
Go online! Your math lessons are interactive. Use *i*Tools, Animated Math Models, the Multimedia *e*Glossary, and more.

Look for these:

Project In the Chef's Kitchen

Higher Order Thinking

Connect to Health
p. 24

Use every day for Standards Practice.

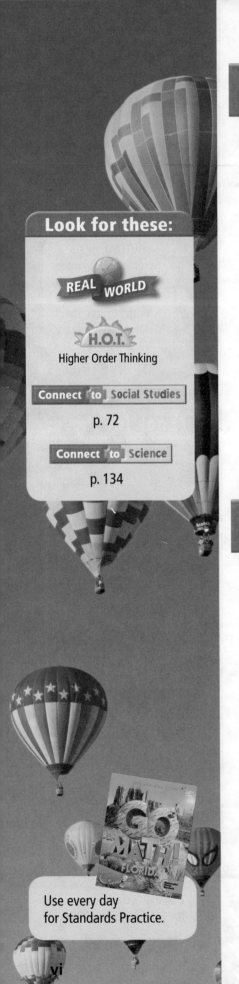

Look for these:

REAL WORLD

H.O.T.
Higher Order Thinking

Connect to Social Studies
p. 72

Connect to Science
p. 134

Use every day
for Standards Practice.

© Houghton Mifflin Harcourt Publishing Company

Look for these:

REAL WORLD

H.O.T.
Higher Order Thinking

Connect to Science

p. 230

Use every day
for Standards Practice.

Operations with Fractions

Project: The Rhythm Track **240**

Developing fluency with addition and subtraction of fractions, and developing understanding of the multiplication of fractions and of division of fractions in limited cases (unit fractions divided by whole numbers and whole numbers divided by unit fractions)

6 Add and Subtract Fractions with Unlike Denominators — **241**

Domain Number and Operations–Fractions

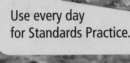

Use every day for Standards Practice.

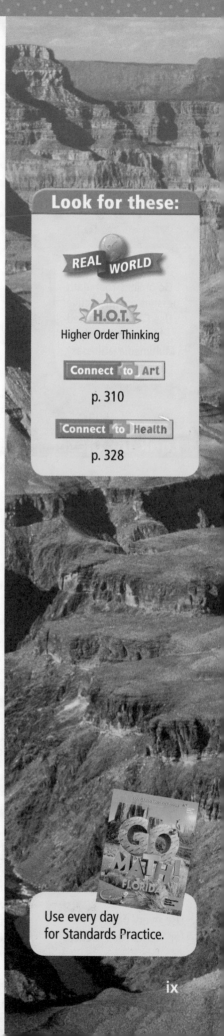

Look for these:

REAL WORLD

H.O.T.
Higher Order Thinking

Connect to Art
p. 310

Connect to Health
p. 328

Use every day for Standards Practice.

Geometry and Measurement

Developing understanding of volume

DIGITAL PATH
Go online! Your math lessons are interactive. Use *i*Tools, Animated Math Models, the Multimedia *e*Glossary, and more.

Look for these:

Project Space Architecture

REAL WORLD

H.O.T.
Higher Order Thinking

Connect to Science
p. 384

Use every day for Standards Practice.

x

10 Convert Units of Measure 403

Domain Measurement and Data

Look for these:

REAL WORLD

H.O.T.
Higher Order Thinking

Connect to Reading
p. 408

Use every day
for Standards Practice.

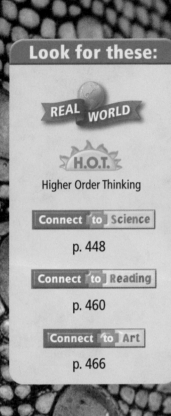

Fluency with Whole Numbers and Decimals

Extending division to 2-digit divisors, integrating decimal fractions into the place value system and developing understanding of operations with decimals to hundredths, and developing fluency with whole number and decimal operations

Chef preparing lunch in a restaurant

Project

In the Chef's Kitchen

Restaurant chefs estimate the amount of food they need to buy based on how many diners they expect. They usually use recipes that make enough to serve large numbers of people.

Get Started

Although apples can grow in any of the 50 states, Pennsylvania is one of the top apple-producing states. The ingredients at the right are needed to make 100 servings of Apple Dumplings. Suppose you and a partner want to make this recipe for 25 friends. Adjust the amount of each ingredient to make just 25 servings.

Important Facts

Apple Dumplings (100 servings)

- 100 baking apples
- 72 tablespoons sugar ($4\frac{1}{2}$ cups)
- 14 cups all-purpose flour
- 6 teaspoons baking powder
- 24 eggs
- 80 tablespoons butter (10 sticks of butter)
- 50 tablespoons chopped walnuts ($3\frac{1}{8}$ cups)

Apple Dumplings (25 servings)

Completed by _____

Place Value, Multiplication, and Expressions

Show What You Know

Check your understanding of important skills.

Name _____

▶ **Place Value** Write the value of each digit for the given number.

1. 2,904

2 _____

9 _____

0 _____

4 _____

2. 6,423

6 _____

4 _____

2 _____

3 _____

▶ **Regroup Through Thousands** Regroup. Write the missing numbers.

3. 40 tens = _____ hundreds

4. 60 hundreds = _____ thousands

5. _____ tens 15 ones = 6 tens 5 ones

6. 18 tens 20 ones = _____ hundreds

▶ **Missing Factors** Find the missing factor.

7. 4 × _____ = 24

8. 6 × _____ = 48

9. _____ × 9 = 63

Be a Math Detective and use the clues at the right to find the 7-digit number. What is the number?

Clues

- This 7-digit number is 8,920,000 when rounded to the nearest ten thousand.
- The digits in the tens and hundreds places are the least and same value.
- The value of the thousands digit is double that of the ten thousands digit.
- The sum of all its digits is 24.

Vocabulary Builder

▶ **Visualize It** ••••••••••••••••••••••••••••••••••••

Sort the review words into the Venn diagram.

Review Words

estimate

factor

multiply

place value

product

quotient

Preview Words

base

Distributive Property

evaluate

exponent

inverse operations

numerical expression

order of operations

period

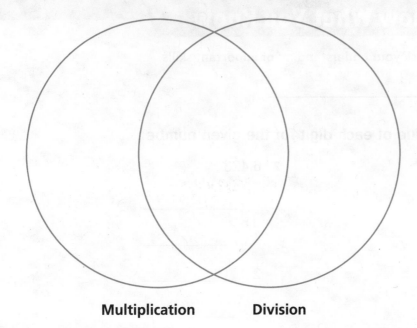

Multiplication   **Division**

▶ **Understand Vocabulary** ••••••••••••••••••••••••••

Write the preview words that answer the question "What am I?"

1. I am a group of 3 digits separated by commas in a multidigit

   number. _____

2. I am a mathematical phrase that has numbers and operation signs

   but no equal sign. _____

3. I am operations that undo each other, like multiplication and division.

   _____

4. I am the property that states that multiplying a sum by a
   number is the same as multiplying each addend in the
   sum by the number and then adding the products.

   _____

5. I am a number that tells how many times the base is used

   as a factor. _____

GO
Online • eStudent Edition • Multimedia eGlossary

Name _____

Place Value and Patterns

Essential Question How can you describe the relationship between two place-value positions?

Investigate

Materials ■ base-ten blocks

You can use base-ten blocks to understand the relationships among place-value positions. Use a large cube for 1,000, a flat for 100, a long for 10, and a small cube for 1.

Number	1,000	100	10	1
Model				
Description	large cube	flat	long	small cube

Complete the comparisons below to describe the relationship from one place-value position to the next place-value position.

A. • Look at the long and compare it to the small cube.

  The long is _____ times as much as the small cube.

• Look at the flat and compare it to the long.

  The flat is _____ times as much as the long.

• Look at the large cube and compare it to the flat.

  The large cube is _____ times as much as the flat.

B. • Look at the flat and compare it to the large cube.

  The flat is _____ of the large cube.

• Look at the long and compare it to the flat.

  The long is _____ of the flat.

• Look at the small cube and compare it to the long.

  The small cube is _____ of the long.

Math Talk MATHEMATICAL PRACTICES
How many times as much is the flat compared to the small cube? the large cube to the small cube? **Explain.**

Draw Conclusions

1. **Describe** the pattern you see when you move from a lesser place-value position to the next greater place-value position.

_____

_____

_____

2. **Describe** the pattern you see when you move from a greater place-value position to the next lesser place-value position.

_____

_____

_____

Make Connections

You can use your understanding of place-value patterns and a place-value chart to write numbers that are 10 times as much as or $\frac{1}{10}$ of any given number.

Hundred Thousands	Ten Thousands	One Thousands	Hundreds	Tens	Ones
		?	300	?	

10 times as much as $\frac{1}{10}$ of

_____ is 10 times as much as 300.

_____ is $\frac{1}{10}$ of 300.

Use the steps below to complete the table.

STEP 1 Write the given number in a place-value chart.

STEP 2 Use the place-value chart to write a number that is 10 times as much as the given number.

STEP 3 Use the place-value chart to write a number that is $\frac{1}{10}$ of the given number.

Number	10 times as much as	$\frac{1}{10}$ of
10		
70		
9,000		

Name _____

Share and Show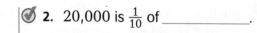

Complete the sentence.

1. 500 is 10 times as much as _____.

☑ 2. 20,000 is $\frac{1}{10}$ of _____.

3. 900 is $\frac{1}{10}$ of _____.

4. 600 is 10 times as much as _____.

Use place-value patterns to complete the table.

Number	10 times as much as	$\frac{1}{10}$ of
☑ 5. 10		
6. 3,000		
7. 800		
8. 50		

Number	10 times as much as	$\frac{1}{10}$ of
9. 400		
10. 90		
11. 6,000		
12. 200		

H.O.T. **Complete the sentence with 100 or 1,000.**

13. 200 is _____ times as much as 2.

14. 4,000 is _____ times as much as 4.

15. 700,000 is _____ times as much as 700.

16. 600 is _____ times as much as 6.

17. 50,000 is _____ times as much as 500.

18. 30,000 is _____ times as much as 30.

19. **Write Math** ► **Explain** how you can use place-value patterns to describe how 50 and 5,000 compare.

_____

_____

Problem Solving ....

H.O.T. Sense or Nonsense?

20. Mark and Robyn used base-ten blocks to show that 300 is
100 times as much as 3. Whose model makes sense? Whose
model is nonsense? **Explain** your reasoning.

Mark's Work

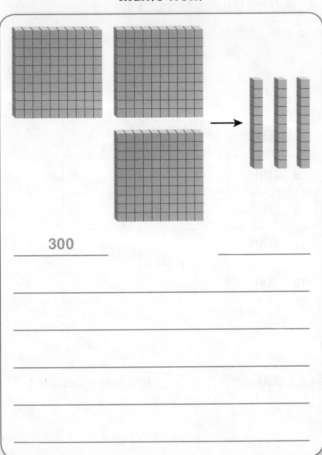

300

Robyn's Work

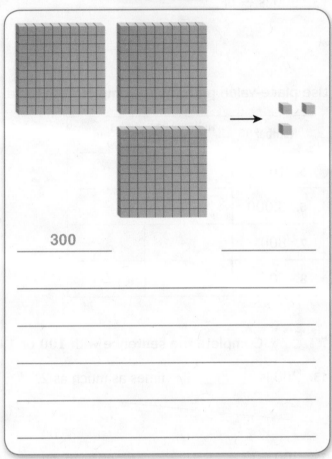

300

• **Explain** how you would help Mark understand why he should have
used small cubes instead of longs.

Place Value of Whole Numbers

Essential Question How do you read, write, and represent whole numbers through hundred millions?

🗝 UNLOCK the Problem REAL WORLD

The diameter of the sun is 1,392,000 kilometers. To understand this distance, you need to understand the place value of each digit in 1,392,000.

A place-value chart contains periods. A **period** is a group of three digits separated by commas in a multidigit number. The millions period is left of the thousands period. One million is 1,000 thousands and is written as 1,000,000.

Periods

MILLIONS			THOUSANDS			ONES		
Hundreds	Tens	Ones	Hundreds	Tens	Ones	Hundreds	Tens	Ones
		1,	3	9	2,	0	0	0
		$1 \times 1,000,000$	$3 \times 100,000$	$9 \times 10,000$	$2 \times 1,000$	0×100	0×10	0×1
		1,000,000	300,000	90,000	2,000	0	0	0

The place value of the digit 1 in 1,392,000 is millions. The value of 1 in 1,392,000 is $1 \times 1,000,000 = 1,000,000$.

Standard Form: 1,392,000

Word Form: one million, three hundred ninety-two thousand

Expanded Form:
$(1 \times 1,000,000) + (3 \times 100,000) + (9 \times 10,000) + (2 \times 1,000)$

Math Idea

When writing a number in expanded form, if no digits appear in a place value, it is not necessary to include them in the expression.

Try This! Use place value to read and write numbers.

Standard Form: 582,030

Word Form: five hundred eighty-two _____ , _____

Expanded Form: $(5 \times 100,000) + ($ _____ $\times$ _____ $) + (2 \times 1,000) + ($ _____ $\times$ _____ $)$

- The average distance from Jupiter to the sun is four hundred eighty-three million, six hundred thousand miles. Write the number that shows this distance. _____

Place-Value Patterns

Canada's land area is about 4,000,000 square miles. Iceland has a land area of about 40,000 square miles. Compare the two areas.

 Example 1 Use a place-value chart.

STEP 1   Write the numbers in a place-value chart.

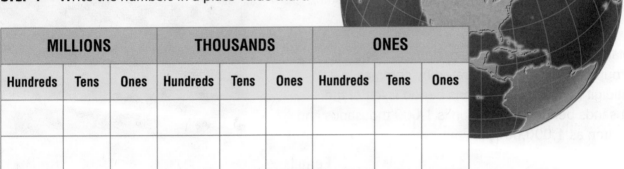

MILLIONS			THOUSANDS			ONES		
Hundreds	Tens	Ones	Hundreds	Tens	Ones	Hundreds	Tens	Ones

STEP 2

Count the number of whole number place-value positions.

4,000,000 has _____ more whole number places than 40,000.

Think: 2 more places is 10 × 10, or 100.

4,000,000 is _____ times as much as 40,000.

So, Canada's estimated land area is _____ times as much as Iceland's estimated land area.

> **Remember**
> The value of each place is 10 times as much as the value of the next place to its right or $\frac{1}{10}$ of the value of the next place to its left.

You can use place-value patterns to rename a number.

 Example 2 Use place-value patterns.

Rename 40,000 using other place values.

40,000	4 ten thousands	4 × 10,000
40,000	_____ thousands	_____ × 1,000
40,000	_____	_____

Name _____

Share and Show

1. Complete the place-value chart to find the value of each digit.

MILLIONS			THOUSANDS			ONES		
Hundreds	Tens	Ones	Hundreds	Tens	Ones	Hundreds	Tens	Ones
		7,	3	3	3,	8	2	0
		7 × 1,000,000	3 × _____	3 × 10,000	_____ × 1,000	8 × 100	_____	0 × 1
		_____	_____	30,000	3,000	_____	20	0

Write the value of the underlined digit.

2. 1,57<u>4</u>,833

_____

3. 598,<u>1</u>02

_____

✓ 4. 7,0<u>9</u>3,455

_____

5. <u>3</u>01,256,878

_____

Write the number in two other forms.

6. (8 × 100,000) + (4 × 1,000) + (6 × 1)

_____

_____

✓ 7. seven million, twenty thousand, thirty-two

_____

_____

On Your Own

Write the value of the underlined digit.

8. 8<u>4</u>9,567,043

_____

9. 9,<u>4</u>22,850

_____

10. <u>9</u>6,283

_____

11. <u>4</u>98,354,021

_____

12. 791,<u>3</u>50

_____

13. 2<u>7</u>,911,534

_____

14. 105,9<u>8</u>0,774

_____

15. 8,26<u>5</u>,178

_____

Write the number in two other forms.

16. 345,000

_____

_____

17. 119,000,003

_____

_____

Problem Solving REAL WORLD

Use the table for 18–19.

Average Distance from the Sun (in thousands of km)			
Mercury	57,910	Jupiter	778,400
Venus	108,200	Saturn	1,427,000
Earth	149,600	Uranus	2,871,000
Mars	227,900	Neptune	4,498,000

18. Which planet is about 10 times as far as Earth is from the sun?

_____

19. Which planet is about $\frac{1}{10}$ of the distance Uranus is from the Sun?

_____

20. **H.O.T.** **What's the Error?** Matt wrote the number four million, three hundred five thousand, seven hundred sixty-two as 4,350,762. **Describe** and correct his error.

_____

_____

21. **Write Math** ► **Explain** how you know that the values of the digit 5 in the numbers 150,000 and 100,500 are not the same.

_____

_____

_____

_____

22. **Test Prep** In the number 869,653,214, which describes how the digit 6 in the ten-millions place compares to the digit 6 in the hundred-thousands place?

Ⓐ 10 times as much as

Ⓑ 100 times as much as

Ⓒ 1,000 times as much as

Ⓓ $\frac{1}{10}$ of

SHOW YOUR WORK

Name _____

Properties

Essential Question How can you use properties of operations to solve problems?

You can use the properties of operations to help you evaluate numerical expressions more easily.

Properties of Addition

Commutative Property of Addition If the order of addends changes, the sum stays the same.	$12 + 7 = 7 + 12$
Associative Property of Addition If the grouping of addends changes, the sum stays the same.	$5 + (8 + 14) = (5 + 8) + 14$
Identity Property of Addition The sum of any number and 0 is that number.	$13 + 0 = 13$

Properties of Multiplication

Commutative Property of Multiplication If the order of factors changes, the product stays the same.	$4 \times 9 = 9 \times 4$
Associative Property of Multiplication If the grouping of factors changes, the product stays the same.	$11 \times (3 \times 6) = (11 \times 3) \times 6$
Identity Property of Multiplication The product of any number and 1 is that number.	$4 \times 1 = 4$

🔑 UNLOCK the Problem REAL WORLD

The table shows the number of bones in several parts of the human body. What is the total number of bones in the ribs, the skull, and the spine?

To find the sum of addends using mental math, you can use the Commutative and Associative Properties.

Part	Number of Bones
Ankle	7
Ribs	24
Skull	28
Spine	26

 Use properties to find 24 + 28 + 26.

$24 + 28 + 26 = 28 + \underline{\hspace{1cm}} + 26$ Use the _____ Property to reorder the addends.

$= 28 + (24 + \underline{\hspace{1cm}})$ Use the _____ Property to group the addends.

$= 28 + \underline{\hspace{1cm}}$ Use mental math to add.

$= \underline{\hspace{1cm}}$

So, there are _____ bones in the ribs, the skull, and the spine.

Math Talk MATHEMATICAL PRACTICES Explain why grouping 24 and 26 makes the problem easier to solve.

Chapter 1 13

Distributive Property

Multiplying a sum by a number is the same as multiplying each addend by the number and then adding the products.

$5 \times (7 + 9) = (5 \times 7) + (5 \times 9)$

The Distributive Property can also be used with multiplication and subtraction. For example, $2 \times (10 - 8) = (2 \times 10) - (2 \times 8)$.

🔑 Example 1 Use the Distributive Property to find the product.

One Way Use addition.

$8 \times 59 = 8 \times ($ _____ $+ 9)$	Use a multiple of 10 to write 59 as a sum.
$= ($ _____ $\times 50) + (8 \times$ _____ $)$	Use the Distributive Property.
$=$ _____ $+$ _____	Use mental math to multiply.
$=$ _____	Use mental math to add.

Another Way Use subtraction.

$8 \times 59 = 8 \times ($ _____ $- 1)$	Use a multiple of 10 to write 59 as a difference.
$= ($ _____ $\times 60) - (8 \times$ _____ $)$	Use the Distributive Property.
$=$ _____ $-$ _____	Use mental math to multiply.
$=$ _____	Use mental math to subtract.

🔑 Example 2 Complete the equation, and tell which property you used.

A $23 \times$ _____ $= 23$

Think: A number times 1 is equal to itself.

Property: _____

_____

B $47 \times 15 = 15 \times$ _____

Think: Changing the order of factors does not change the product.

Property: _____

_____

MATHEMATICAL PRACTICES

Math Talk Explain how you could find the product 3×299 by using mental math.

14

Name _____

Share and Show   . . . .

1. Use properties to find $4 \times 23 \times 25$.

     $23 \times$ _____ $\times 25$ _____ Property of Multiplication

     $23 \times ($ _____ $\times$ _____ $)$ _____ Property of Multiplication

     $23 \times$ _____

      _____

Use properties to find the sum or product.

2. $89 + 27 + 11$ 3. 9×52 ✓ 4. $107 + 0 + 39 + 13$

_____   _____   _____

Complete the equation, and tell which property you used.

5. $9 \times (30 + 7) = (9 \times$ _____ $) + (9 \times 7)$ ✓ 6. $0 +$ _____ $= 47$

_____   _____

Math Talk   MATHEMATICAL PRACTICES
Describe how you can use properties to solve problems more easily.

On Your Own . . . .

Practice: Copy and Solve Use properties to find the sum or product.

7. 3×78 8. $4 \times 60 \times 5$ 9. $21 + 25 + 39 + 5$

Complete the equation, and tell which property you used.

10. $11 + (19 + 6) = (11 +$ _____ $) + 6$ 11. $25 + 14 =$ _____ $+ 25$

_____   _____

12. **H.O.T.** Show how you can use the Distributive Property to rewrite and find $(32 \times 6) + (32 \times 4)$.

_____

Problem Solving REAL WORLD

13. Three friends' meals at a restaurant cost $13, $14, and $11. Use parentheses to write two different expressions to show how much the friends spent in all. Which property does your pair of expressions demonstrate?

_____

_____

14. Jacob is designing an aquarium for a doctor's office. He plans to buy 6 red blond guppies, 1 blue neon guppy, and 1 yellow guppy. The table shows the price list for the guppies. How much will the guppies for the aquarium cost?

_____

Fancy Guppy Prices	
Blue neon	$11
Red blond	$22
Sunrise	$18
Yellow	$19

15. Sylvia bought 8 tickets to a concert. Each ticket costs $18. To find the total cost in dollars, she added the product 8×10 to the product 8×8, for a total of 144. Which property did Sylvia use?

_____

•••••••••••• **SHOW YOUR WORK** ••••••••••

16. **H.O.T. Sense or Nonsense?** Julie wrote $(15 - 6) - 3 = 15 - (6 - 3)$. Is Julie's equation sense or nonsense? Do you think the Associative Property works for subtraction? **Explain**.

_____

_____

_____

17. **Test Prep** Canoes rent for $29 per day. Which expression can be used to find the cost in dollars of renting 6 canoes for a day?

  Ⓐ $(6 + 20) + (6 + 9)$

  Ⓑ $(6 \times 20) + (6 \times 9)$

  Ⓒ $(6 + 20) \times (6 + 9)$

  Ⓓ $(6 \times 20) \times (6 \times 9)$

FOR MORE PRACTICE:
Standards Practice Book, pp. P7–P8

Name _____

Powers of 10 and Exponents

Essential Question How can you use an exponent to show powers of 10?

🔑 UNLOCK the Problem

Expressions with repeated factors, such as $10 \times 10 \times 10$, can be written by using a base with an exponent. The **base** is the number that is used as the repeated factor. The **exponent** is the number that tells how many times the base is used as a factor.

$$\underset{\substack{\text{3 factors}}}{10 \times 10 \times 10} = \overset{\text{exponent}}{10^{\underset{\text{base}}{3}}} = 1,000$$

Word form: the third power of ten

Exponent form: 10^3

🔒 Activity Use base-ten blocks.

Materials ■ base-ten blocks

What is $10 \times 1,000$ written with an exponent?

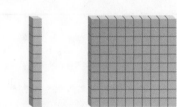  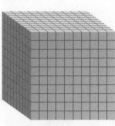

1 one	10 ones	100 ones	1,000 ones
1	1×10	$1 \times 10 \times 10$	$1 \times 10 \times 10 \times 10$
10^0	10^1	10^2	10^3

- How many ones are in 1? _____

- How many ones are in 10? _____

- How many tens are in 100? _____
  Think: 10 groups of 10 or 10×10

- How many hundreds are in 1,000? _____
  Think: 10 groups of 100 or $10 \times (10 \times 10)$

- How many thousands are in 10,000? _____

In the box at the right, draw a quick picture to show 10,000.

So, $10 \times 1,000$ is $10^{\boxed{}}$.

▲ Use $\boxed{T}$ for 1,000.

10,000 ones
$1 \times 10 \times 10 \times 10 \times 10$

$10^{\boxed{}}$

🔲 Example Multiply a whole number by a power of ten.

Hummingbirds beat their wings very fast. The smaller the hummingbird is, the faster its wings beat. The average hummingbird beats its wings about 3×10^3 times a minute. How many times a minute is that, written as a whole number?

Multiply 3 by powers of ten. Look for a pattern.

$3 \times 10^0 = 3 \times 1 = $ _____

$3 \times 10^1 = 3 \times 10 = $ _____

$3 \times 10^2 = 3 \times 10 \times 10 = $ _____

$3 \times 10^3 = 3 \times 10 \times 10 \times 10 = $ _____

So, the average hummingbird beats its wings about _____ times a minute.

Math Talk MATHEMATICAL PRACTICES Explain how using an exponent simplifies an expression.

- What pattern do you see?

_____

_____

Share and Show

Write in exponent form and word form.

1. 10×10

   Exponent form: _____

   Word form: _____

   _____

2. $10 \times 10 \times 10 \times 10$

   Exponent form: _____

   Word form: _____

   _____

Find the value.

3. 10^2

4. 4×10^2

5. 7×10^3

Name _____

On Your Own ·

Write in exponent form and word form.

6. $10 \times 10 \times 10$

exponent form: _____

word form: _____

_____

7. $10 \times 10 \times 10 \times 10 \times 10$

exponent form: _____

word form: _____

_____

Find the value.

8. 10^4

_____

9. 2×10^3

_____

10. 6×10^4

_____

Complete the pattern.

11. $7 \times 10^0 = 7 \times 1 =$ _____

$7 \times 10^1 = 7 \times 10 =$ _____

$7 \times 10^2 = 7 \times 100 =$ _____

$7 \times 10^3 = 7 \times 1,000 =$ _____

$7 \times 10^4 = 7 \times 10,000 =$ _____

12. $9 \times 10^0 =$ _____ $= 9$

$9 \times 10^1 =$ _____ $= 90$

$9 \times 10^2 =$ _____ $= 900$

$9 \times 10^3 =$ _____ $= 9,000$

$9 \times 10^4 =$ _____ $= 90,000$

13. $12 \times 10^0 = 12 \times 1 =$ _____

$12 \times 10^1 = 12 \times 10 =$ _____

$12 \times 10^2 = 12 \times 100 =$ _____

$12 \times 10^3 = 12 \times 1,000 =$ _____

$12 \times 10^4 = 12 \times 10,000 =$ _____

14. **H.O.T.** $10^3 = 10 \times 10^n$ What is the value of n?

Think: $10^3 = 10 \times$ _____ $\times$ _____,

or $10 \times$ _____

The value of n is _____.

15. **Write Math** ▶ **Explain** how to write 50,000 using exponents.

_____

_____

UNLOCK the Problem — REAL WORLD

16. Lake Superior is the largest of the Great Lakes. It covers a surface area of about 30,000 square miles. How can you show the estimated area of Lake Superior as a whole number multiplied by a power of ten?

(A) 3×10^2 sq mi   (C) 3×10^4 sq mi

(B) 3×10^3 sq mi   (D) 3×10^5 sq mi

a. What are you asked to find?

_____

b. How can you use a pattern to find the answer?

_____

c. Write a pattern using the whole number 3 and powers of ten.

$3 \times 10^0 = 3 \times 1 =$ _____

$3 \times 10^1 = 3 \times 10 =$ _____

$3 \times 10^2 =$ _____ $=$ _____

$3 \times 10^3 =$ _____ $=$ _____

$3 \times 10^4 =$ _____ $=$ _____

d. Fill in the correct answer choice above.

17. The Earth's diameter through the equator is about 8,000 miles. What is the Earth's estimated diameter written as a whole number multiplied by a power of ten?

(A) 8×10^1 miles

(B) 8×10^2 miles

(C) 8×10^3 miles

(D) 8×10^4 miles

18. The Earth's circumference around the equator is about 25×10^3 miles. What is the Earth's estimated circumference written as a whole number?

(A) 250,000 miles

(B) 25,000 miles

(C) 2,500 miles

(D) 250 miles

FOR MORE PRACTICE:
Standards Practice Book, pp. P9–P10

Multiplication Patterns

Essential Question How can you use a basic fact and a pattern
to multiply by a 2-digit number?

 UNLOCK the Problem REAL WORLD

How close have you been to a bumblebee?

The actual length of a queen bumblebee is about
20 millimeters. The photograph shows part of a
bee under a microscope, at 10 times its actual size.
What would the length of the bee appear to be at a
magnification of 300 times its actual size?

Use a basic fact and a pattern.

Multiply. 300 × 20

$3 \times 2 = 6$ ← basic fact

$30 \times 2 = (3 \times 2) \times 10^1 = 60$

$300 \times 2 = (3 \times 2) \times 10^2 =$ _____

$300 \times 20 = (3 \times 2) \times (100 \times 10) = 6 \times 10^3 =$ _____

So, the length of the bee would appear to be

about _____ millimeters.

Math Talk MATHEMATICAL PRACTICES
What pattern do you
see in the number sentences and
the exponents?

- What would the length of the bee shown in the photograph appear
  to be if the microscope shows it at 10 times its actual size?

  _____

Example Use mental math and a pattern.

Multiply. 50 × 8,000

$5 \times 8 = 40$ ← basic fact

$5 \times 80 = (5 \times 8) \times 10^1 = 400$

$5 \times 800 = (5 \times 8) \times 10^2 =$ _____

$50 \times 800 = (5 \times 8) \times (10 \times 100) = 40 \times 10^3 =$ _____

$50 \times 8,000 = (5 \times 8) \times (10 \times 1,000) = 40 \times 10^4 =$ _____

Share and Show

Use mental math and a pattern to find the product.

1. $30 \times 4{,}000 =$ _____

   • What basic fact can you use to help you find $30 \times 4{,}000$? _____

Use mental math to complete the pattern.

2. $1 \times 1 = 1$

   $1 \times 10^1 =$ _____

   $1 \times 10^2 =$ _____

   $1 \times 10^3 =$ _____

☑ 3. $7 \times 8 = 56$

   $(7 \times 8) \times 10^1 =$ _____

   $(7 \times 8) \times 10^2 =$ _____

   $(7 \times 8) \times 10^3 =$ _____

☑ 4. $6 \times 5 =$ _____

   $(6 \times 5) \times$ _____ $= 300$

   $(6 \times 5) \times$ _____ $= 3{,}000$

   $(6 \times 5) \times$ _____ $= 30{,}000$

Math Talk MATHEMATICAL PRACTICES

Explain how to find $50 \times 9{,}000$ by using a basic fact and pattern.

On Your Own

Use mental math to complete the pattern.

5. $9 \times 5 = 45$

   $(9 \times 5) \times 10^1 =$ _____

   $(9 \times 5) \times 10^2 =$ _____

   $(9 \times 5) \times 10^3 =$ _____

6. $3 \times 7 = 21$

   $(3 \times 7) \times 10^1 =$ _____

   $(3 \times 7) \times 10^2 =$ _____

   $(3 \times 7) \times 10^3 =$ _____

7. $5 \times 4 =$ _____

   $(5 \times 4) \times$ _____ $= 200$

   $(5 \times 4) \times$ _____ $= 2{,}000$

   $(5 \times 4) \times$ _____ $= 20{,}000$

8. $5 \times 7 =$ _____

   $(5 \times 7) \times$ _____ $= 350$

   $(5 \times 7) \times$ _____ $= 3{,}500$

   $(5 \times 7) \times$ _____ $= 35{,}000$

9. $4 \times 2 = 8$

   $(4 \times 2) \times 10^2 =$ _____

   $(4 \times 2) \times 10^3 =$ _____

   $(4 \times 2) \times 10^4 =$ _____

10. $6 \times 7 = 42$

   $(6 \times 7) \times 10^2 =$ _____

   $(6 \times 7) \times 10^3 =$ _____

   $(6 \times 7) \times 10^4 =$ _____

Use mental math and a pattern to find the product.

11. $(6 \times 6) \times 10^1 =$ _____

12. $(7 \times 4) \times 10^3 =$ _____

13. $(9 \times 8) \times 10^2 =$ _____

14. $(4 \times 3) \times 10^2 =$ _____

15. $(2 \times 5) \times 10^3 =$ _____

16. $(2 \times 8) \times 10^2 =$ _____

17. $(6 \times 5) \times 10^3 =$ _____

18. $(8 \times 8) \times 10^4 =$ _____

19. $(7 \times 8) \times 10^4 =$ _____

22

Name _____

Use mental math to complete the table.

20. 1 roll = 50 dimes   **Think:** 50 dimes per roll × 20 rolls = (5 × 2) × (10 × 10)

Rolls	20	30	40	50	60	70	80	90	100
Dimes	10×10^2								

21. 1 roll = 40 quarters   **Think:** 40 quarters per roll × 20 rolls = (4 × 2) × (10 × 10)

Rolls	20	30	40	50	60	70	80	90	100
Quarters	8×10^2								

	×	6	70	800	9,000
22.	80			64×10^3	
23.	90				81×10^4

Problem Solving REAL WORLD

Use the table for 24–26.

24. **What if** you magnified the image of a cluster fly by 9×10^3? What would the length appear to be?

_____

25. If you magnified the image of a fire ant by 4×10^3 and a tree hopper by 3×10^3, which insect would appear longer? How much longer?

_____

26. ⚡H.O.T.⚡ John wants to magnify the image of a fire ant and a crab spider so they appear to be the same length. How many times their actual sizes would he need to magnify each image?

_____

_____

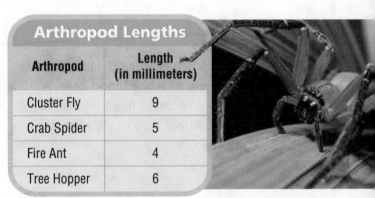

Arthropod Lengths

Arthropod	Length (in millimeters)
Cluster Fly	9
Crab Spider	5
Fire Ant	4
Tree Hopper	6

SHOW YOUR WORK

27. ✹H.O.T.✹ What does the product of any whole-number factor multiplied by 100 always have? **Explain.**

_____

_____

_____

_____

28. Test Prep How many zeros are in the product $(5 \times 4) \times 10^4$?

Ⓐ 3

Ⓑ 4

Ⓒ 5

Ⓓ 6

Connect to Health

Blood Cells

Blood is necessary for all human life. It contains red blood cells and white blood cells that nourish and cleanse the body, and platelets that stop bleeding. The average adult has about 5 liters of blood.

◀ Single red blood cell

▲ Platelet

White blood cell ▶

Use patterns and mental math to solve.

29. A human body has about 30 times as many platelets as white blood cells. A small sample of blood has 8×10^3 white blood cells. About how many platelets are in the sample?

_____

30. Basophils and monocytes are types of white blood cells. A blood sample has about 5 times as many monocytes as basophils. If there are 60 basophils in the sample, about how many monocytes are there?

_____

31. Lymphocytes and eosinophils are types of white blood cells. A blood sample has about 10 times as many lymphocytes as eosinophils. If there are 2×10^2 eosinophils in the sample, about how many lymphocytes are there?

_____

32. ✹H.O.T.✹ An average person has 6×10^2 times as many red bloods cells as white blood cells. A small sample of blood has 7×10^3 white blood cells. About how many red blood cells are in the sample?

_____

FOR MORE PRACTICE:
Standards Practice Book, pp. P11–P12

Name _____

 Mid-Chapter Checkpoint

▶ **Vocabulary**

Choose the best term for the box.

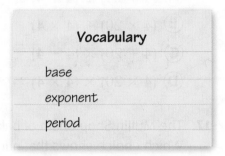

Vocabulary
base
exponent
period

1. A group of three digits separated by commas in a multidigit number is a _____. (p. 9)

2. An _____ is the number that tells how many times a base is used as a factor. (p. 17)

▶ **Concepts and Skills**

Complete the sentence.

3. 7 is $\frac{1}{10}$ of _____.

4. 800 is 10 times as much as _____.

Write the value of the underlined digit.

5. 6,5<u>8</u>1,678

_____

6. 125,<u>6</u>34

_____

7. 34,<u>6</u>34,803

_____

8. 2,<u>7</u>64,835

_____

Complete the equation, and tell which property you used.

9. $8 \times (14 + 7) =$ _____ $+ (8 \times 7)$

_____

10. $7 + (8 + 12) =$ _____ $+ 12$

_____

Find the value.

11. 10^3

_____

12. 6×10^2

_____

13. 4×10^4

_____

Use mental math and a pattern to find the product.

14. $70 \times 300 =$ _____

15. $(3 \times 4) \times 10^3 =$ _____

Fill in the bubble completely to show your answer.

16. DVDs are on sale for $24 each. Which expression can be used to find the cost in dollars of buying 4 DVDs?

   (A) $(4 + 20) + (4 + 4)$

   (B) $(4 \times 20) + (4 \times 4)$

   (C) $(4 + 20) \times (4 + 4)$

   (D) $(4 \times 20) \times (4 \times 4)$

17. The Muffin Shop chain of bakeries sold 745,305 muffins last year. Which choice shows that number in expanded form?

   (A) $(7 \times 100,000) + (45 \times 10,000) + (3 \times 100) + (5 \times 10)$

   (B) $(7 \times 100,000) + (4 \times 10,000) + (5 \times 1,000) + (5 \times 10)$

   (C) $(7 \times 100,000) + (4 \times 10,000) + (5 \times 1,000) + (3 \times 100) + (5 \times 1)$

   (D) $(7 \times 100,000) + (4 \times 10,000) + (3 \times 100) + (5 \times 1)$

18. The soccer field at Mario's school has an area of 6,000 square meters. How can Mario show the area as a whole number multiplied by a power of ten?

   (A) 6×10^4 sq m

   (B) 6×10^3 sq m

   (C) 6×10^2 sq m

   (D) 6×10^1 sq m

19. Ms. Alonzo ordered 4,000 markers for her store. Only $\frac{1}{10}$ of them arrived. How many markers did she receive?

   (A)   4

   (B)   40

   (C)   400

   (D) 1,400

20. Mark wrote the highest score he made on his new video game as the product of $70 \times 6,000$. What was his score?

   (A)   420

   (B)   4,200

   (C)   42,000

   (D) 420,000

26

Multiply by 1-Digit Numbers

Essential Question How do you multiply by 1-digit numbers?

 UNLOCK the Problem REAL WORLD

Each day an airline flies 9 commercial jets from New York to London, England. Each plane holds 293 passengers. If every seat is taken on all flights, how many people fly on this airline from New York to London in 1 day?

 **Use place value and regrouping.**

STEP 1 Estimate: 293 × 9

    Think: 300 × 9 = _____

STEP 2 Multiply the ones.

$$\begin{array}{r} \overset{2}{29}3 \\ \times\ 9 \\ \hline 7 \end{array}$$

9 × 3 ones = _____ ones

Write the ones and the regrouped tens.

▲ The Queen's Guard protects Britain's Royal Family and their residences.

Math Talk MATHEMATICAL PRACTICES
Explain how you record the 27 ones when you multiply 3 by 9 in Step 2.

STEP 3 Multiply the tens.

$$\begin{array}{r} \overset{82}{2}93 \\ \times\ 9 \\ \hline 37 \end{array}$$

9 × 9 tens = _____ tens

Add the regrouped tens.

_____ tens + 2 tens = _____ tens

Write the tens and the regrouped hundreds.

STEP 4 Multiply the hundreds.

$$\begin{array}{r} \overset{82}{2}93 \\ \times\ 9 \\ \hline 2,637 \end{array}$$

9 × 2 hundreds = _____ hundreds

Add the regrouped hundreds.

_____ hundreds + 8 hundreds = _____ hundreds

Write the hundreds.

So, in 1 day, _____ passengers fly from New York to London.

• How can you tell if your answer is reasonable? _____

_____

🔑 Example

A commercial airline makes several flights each week from New York to Paris, France. If the airline serves 1,978 meals on its flights each day, how many meals are served for the entire week?

To multiply a greater number by a 1-digit number, repeat the process of multiplying and regrouping until every place value is multiplied.

STEP 1 Estimate. 1,978 × 7

   Think: 2,000 × 7 = _____

STEP 2 Multiply the ones.

$$\begin{array}{r} 5 \\ 1{,}978 \\ \times\ \ 7 \\ \hline 6 \end{array}$$

7 × 8 ones = _____ ones

Write the ones and the regrouped tens.

STEP 3 Multiply the tens.

$$\begin{array}{r} 55 \\ 1{,}978 \\ \times\ \ 7 \\ \hline 46 \end{array}$$

7 × 7 tens = _____ tens

Add the regrouped tens.

_____ tens + 5 tens = _____ tens

Write the tens and the regrouped hundreds.

STEP 4 Multiply the hundreds.

$$\begin{array}{r} 6\ 55 \\ 1{,}978 \\ \times\ \ 7 \\ \hline 846 \end{array}$$

7 × 9 hundreds = _____ hundreds

Add the regrouped hundreds.

_____ hundreds + 5 hundreds = _____ hundreds

Write the hundreds and the regrouped thousands.

STEP 5 Multiply the thousands.

$$\begin{array}{r} 6\ 55 \\ 1{,}978 \\ \times\ \ 7 \\ \hline 13{,}846 \end{array}$$

7 × 1 thousand = _____ thousands

Add the regrouped thousands.

_____ thousands + 6 thousands = _____ thousands

Write the thousands. Compare your answer to the estimate to see if it is reasonable.

So, in 1 week, _____ meals are served on flights from New York to Paris.

▲ The Eiffel Tower in Paris, France, built for the 1889 World's Fair, was the world's tallest man-made structure for 40 years.

Name _____

Share and Show . . . .

Complete to find the product.

1. 6 × 796   **Estimate:** 6 × _____ = _____

$$\begin{array}{r} 796 \\ \times\ 6 \\ \hline \end{array}$$
Multiply the ones and regroup.

$$\begin{array}{r} 3\ \ \ \\ 796 \\ \times\ 6 \\ \hline 6 \end{array}$$
Multiply the tens and add the regrouped tens. Regroup.

$$\begin{array}{r} 53\ \ \\ 796 \\ \times\ 6 \\ \hline 76 \end{array}$$
Multiply the hundreds and add the regrouped hundreds.

Estimate. Then find the product.

2. Estimate: _____

$$\begin{array}{r} 608 \\ \times\ 8 \\ \hline \end{array}$$

3. Estimate: _____

$$\begin{array}{r} 556 \\ \times\ 4 \\ \hline \end{array}$$

4. Estimate: _____

$$\begin{array}{r} 1,925 \\ \times\ 7 \\ \hline \end{array}$$

On Your Own . . . .

Estimate. Then find the product.

5. Estimate: _____

$$\begin{array}{r} 794 \\ \times\ 3 \\ \hline \end{array}$$

6. Estimate: _____

$$\begin{array}{r} 822 \\ \times\ 6 \\ \hline \end{array}$$

7. Estimate: _____

$$\begin{array}{r} 3,102 \\ \times\ 5 \\ \hline \end{array}$$

 Algebra Solve for the unknown number.

8.
$$\begin{array}{r} 396 \\ \times\ 6 \\ \hline 2,3\ \ 6 \end{array}$$

9.
$$\begin{array}{r} 5,12\ \ \\ \times\ \ \ \ 8 \\ \hline \ \ \ \ 16 \end{array}$$

10.
$$\begin{array}{r} 8,5\ \ 6 \\ \times\ \ \ \ 7 \\ \hline 60,03\ \ \end{array}$$

Practice: Copy and Solve Estimate. Then find the product.

11. 116 × 3   **12.** 338 × 4   **13.** 6 × 219   **14.** 7 × 456

15. 5 × 1,012   **16.** 2,921 × 3   **17.** 8,813 × 4   **18.** 9 × 3,033

Problem Solving REAL WORLD

H.O.T. What's the Error?

19. The Plattsville Glee Club is sending 8 of its members to a singing contest in Cincinnati, Ohio. The cost will be $588 per person. How much will it cost for the entire group of 8 students to attend?

Both Brian and Jermaine solve the problem. Brian says the answer is $40,074. Jermaine's answer is $4,604.

Estimate the cost. A reasonable estimate is _____.

Although Jermaine's answer seems reasonable, neither Brian nor Jermaine solved the problem correctly. Find the errors in Brian's and Jermaine's work. Then, solve the problem correctly.

Brian	**Jermaine**	**Correct Answer**
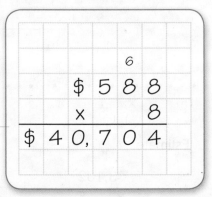	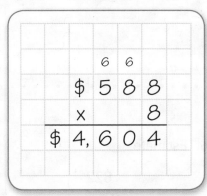	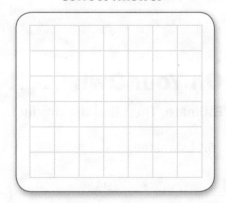

- What error did Brian make? **Explain.** _____

  _____

  _____

- What error did Jermaine make? **Explain.** _____

  _____

- How could you predict that Jermaine's answer might be incorrect

  using your estimate? _____

  _____

  _____

FOR MORE PRACTICE:
Standards Practice Book, pp. P13–P14

Multiply by 2-Digit Numbers

Essential Question How do you multiply by 2-digit numbers?

 UNLOCK the Problem REAL WORLD

A tiger can eat as much as 40 pounds of food at a time but it may go for several days without eating anything. Suppose a Siberian tiger in the wild eats an average of 18 pounds of food per day. How much food will the tiger eat in 28 days if he eats that amount each day?

🔑 **Use place value and regrouping.**

STEP 1 Estimate: 28 × 18

　　　Think: 30 × 20 = _____

_____

STEP 2 Multiply by the ones.

```
   28
 × 18
```

　　　28 × 8 ones = _____ ones

_____

STEP 3 Multiply by the tens.

```
   28
 × 18
```

　　　28 × 1 ten = _____ tens, or _____ ones

_____

STEP 4 Add the partial products.

```
   28
 × 18
```
　　　← 28 × 8

　　　← 28 × 10

　+ _____

Remember

Use patterns of zeros to find the product of multiples of 10.

$$3 \times 4 = 12$$

$$3 \times 40 = 120 \qquad 30 \times 40 = 1{,}200$$

$$3 \times 400 = 1{,}200 \qquad 300 \times 40 = 12{,}000$$

So, on average, a Siberian tiger may eat _____ pounds of food in 28 days.

Example

A Siberian tiger sleeps as much as 18 hours a day, or 126 hours per week. About how many hours does a tiger sleep in a year? There are 52 weeks in one year.

STEP 1 Estimate: 126 × 52

Think: 100 × 50 = _____

_____

STEP 2 Multiply by the ones.

126
× 52
▢▢▢▢   126 × 2 ones = _____ ones

_____

STEP 3 Multiply by the tens.

126
× 52
▢▢▢▢

▢▢▢▢   126 × 5 tens = _____ tens, or _____ ones

_____

STEP 4 Add the partial products.

126
× 52
▢▢▢▢   ← 126 × 2
▢▢▢▢   ← 126 × 50
+ _____
▢▢▢▢

So, a Siberian tiger sleeps about _____ hours in one year.

Math Talk

Are there different numbers you could have used in Step 1 to find an estimate that is closer to the actual answer? **Explain.**

- When you multiply 126 and 5 tens in Step 3, why does its product

  have a zero in the ones place? **Explain.** _____

  _____

Name _____

Share and Show . . . .

Complete to find the product.

1.

$$
\begin{array}{r}
6\ 4. \\
\times\ \ \ \ 4\ 3 \\
\hline
\end{array}
$$

⟵ 64 × _____
⟵ 64 × _____

2.

$$
\begin{array}{r}
5\ 7\ 1 \\
\times\ \ \ \ \ \ 3\ 8 \\
\hline
\end{array}
$$

⟵ 571 × _____
⟵ 571 × _____

Estimate. Then find the product.

3. Estimate: _____

$$
\begin{array}{r}
24 \\
\times\ 15 \\
\hline
120
\end{array}
$$

✓ 4. Estimate: _____

$$
\begin{array}{r}
37 \\
\times\ 63 \\
\hline
\end{array}
$$

✓ 5. Estimate: _____

$$
\begin{array}{r}
384 \\
\times\ 45 \\
\hline
\end{array}
$$

On Your Own . . . .

Estimate. Then find the product.

6. Estimate: _____

$$
\begin{array}{r}
28 \\
\times\ 22 \\
\hline
\end{array}
$$

7. Estimate: _____

$$
\begin{array}{r}
93 \\
\times\ 76 \\
\hline
\end{array}
$$

8. Estimate: _____

$$
\begin{array}{r}
295 \\
\times\ 51 \\
\hline
\end{array}
$$

Practice: Copy and Solve Estimate. Then find the product.

9. 54×31

10. 42×26

11. 38×64

12. 63×16

13. 204×41

14. 534×25

15. 722×39

16. 957×43

Problem Solving REAL WORLD

Use the table for 17–20.

17. How much sleep does a jaguar get in
1 year?

_____

18. In 1 year, how many more hours of sleep
does a giant armadillo get than a platypus?

_____

19. **H.O.T.** Owl monkeys sleep during the
day, waking about 15 minutes after sundown
to find food. At midnight, they rest for an hour
or two, then continue to feed until sunrise.
They live about 27 years. How many hours of
sleep does an owl monkey that lives 27 years
get in its lifetime?

_____

Animal Sleep Amounts

Animal	Amount (usual hours per week)
Jaguar	77
Giant Armadillo	127
Owl Monkey	119
Platypus	98
Three-Toed Sloth	101

SHOW YOUR WORK

20. Three-toed sloths move very slowly, using as
little energy as possible. They sleep, eat, and
even give birth upside down. A baby sloth
may cling to its mother for as much as 36
weeks after being born. How much of that
time is the sloth asleep?

_____

21. **Test Prep** A sloth's maximum speed on the
ground is 15 feet in 1 minute. Even though it
would be unlikely for a sloth to stay in motion
for more than a few moments, how far would
a sloth travel in 45 minutes at that speed?

(A) 60 feet

(B) 270 feet

(C) 675 feet

(D) 6,750 feet

Name _____

Relate Multiplication to Division

Essential Question How is multiplication used to solve a division problem?

You can use the relationship between multiplication and division to solve a division problem. Using the same numbers, multiplication and division are opposite, or **inverse operations.**

$$3 \times 8 = 24 \qquad 24 \div 3 = 8$$

factor   factor   product   dividend   divisor   quotient

UNLOCK the Problem REAL WORLD

Joel and 5 friends collected 126 marbles. They shared the marbles equally. How many marbles will each person get?

- Underline the dividend.
- What is the divisor? _____

One Way Make an array.

- Outline a rectangular array on the grid to model 126 squares arranged in 6 rows of the same length. Shade each row a different color.

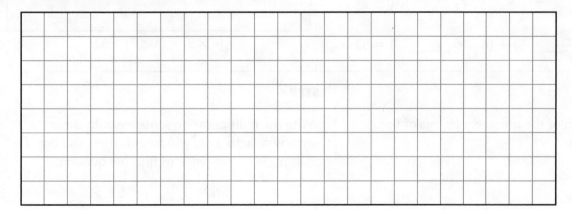

- How many squares are shaded in each row? _____

- Use the array to complete the multiplication sentence. Then, use the multiplication sentence to complete the division sentence.

  $6 \times$ _____ $= 126$ $126 \div 6 =$ _____

So, each of the 6 friends will get _____ marbles.

🔑 Another Way Use the Distributive Property.

Divide. 52 ÷ 4

You can use the Distributive Property and an area model to solve division problems. Remember that the Distributive Property states that multiplying a sum by a number is the same as multiplying each addend in the sum by the number and then adding the products.

STEP 1

Write a related multiplication sentence for the division problem.

Think: Use the divisor as a factor and the dividend as the product. The quotient will be the unknown factor.

$$52 ÷ 4 = \blacksquare$$

$$4 × \blacksquare = 52$$

?

4 | 52

$$4 × ? = 52$$

STEP 2

Use the Distributive Property to break apart the large area into smaller areas for partial products that you know.

(40 + 12) = 52

(4 × _____) + (4 × _____) = 52

? | ?

4 | 40 | 12

$$(4 × ?) + (4 × ?) = 52$$

STEP 3

Find the sum of the unknown factors of the smaller areas.

_____ + _____ = _____

STEP 4

Write the multiplication sentence with the unknown factor that you found. Then, use the multiplication sentence to find the quotient.

$$4 × _____ = 52$$

$$52 ÷ 4 = _____$$

- **Explain** how you can use the Distributive Property to find the quotient of 96 ÷ 8.

_____

_____

_____

Name _____

Share and Show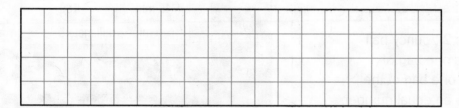

1. Brad has 72 toy cars that he puts into 4 equal groups. How many cars does Brad have in each group? Use the array to show your answer.

$4 \times$ _____ $= 72$ $72 \div 4 =$ _____

<table>
<tr><td></td><td></td><td></td><td></td><td></td><td></td><td></td><td></td><td></td><td></td><td></td><td></td></tr>
<tr><td></td><td></td><td></td><td></td><td></td><td></td><td></td><td></td><td></td><td></td><td></td><td></td></tr>
<tr><td></td><td></td><td></td><td></td><td></td><td></td><td></td><td></td><td></td><td></td><td></td><td></td></tr>
<tr><td></td><td></td><td></td><td></td><td></td><td></td><td></td><td></td><td></td><td></td><td></td><td></td></tr>
</table>

Use multiplication and the Distributive Property to find the quotient.

2. $108 \div 6 =$ _____

3. $84 \div 6 =$ _____

4. $184 \div 8 =$ _____

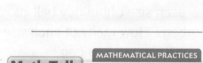

Math Talk MATHEMATICAL PRACTICES
Explain how using multiplication as the inverse operation helps you solve a division problem.

On Your Own

Use multiplication and the Distributive Property to find the quotient.

5. $60 \div 4 =$ _____

6. $144 \div 6 =$ _____

7. $252 \div 9 =$ _____

 Find each quotient. Then compare. Write <, >, or =.

8. $51 \div 3 \bigcirc 68 \div 4$

9. $252 \div 6 \bigcirc 135 \div 3$

10. $110 \div 5 \bigcirc 133 \div 7$

Problem Solving

Use the table to solve 11–13.

11. A group of 6 friends share a bag of the 45-millimeter bouncy balls equally among them. How many does each friend get?

_____

12. **H.O.T.** Mr. Henderson has 2 bouncy-ball vending machines. He buys one bag of the 27-millimeter balls and one bag of the 40-millimeter balls. He puts an equal number of each in the 2 machines. How many bouncy balls does he put in each machine?

_____

13. Lindsey buys a bag of each size of bouncy ball. She wants to put the same number of each size of bouncy ball into 5 party-favor bags. How many of each size of bouncy ball will she put in a bag?

_____

14. **What's the Error?** Sandy writes $(4 \times 30) + (4 \times 2)$ and says the quotient for $128 \div 4$ is 8. Is she correct? **Explain**.

_____

_____

_____

_____

15. **Test Prep** Which of the following can be used to find $150 \div 6$?

  (A) $(6 \times 20) + (6 \times 5)$

  (B) $(6 \times 10) + (6 \times 5)$

  (C) $(2 \times 75) + (2 \times 3)$

  (D) $(6 \times 15) + (6 \times 5)$

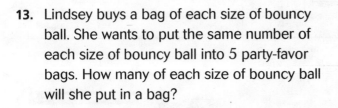

Bouncy Balls

Size	Number in Bag
27 mm	180
40 mm	80
45 mm	180
mm = millimeters	

SHOW YOUR WORK

Problem Solving • Multiplication and Division

Essential Question How can you use the strategy *solve a simpler problem* to help you solve a division problem?

🔑 UNLOCK the Problem · REAL WORLD

Mark works at an animal shelter. To feed 9 dogs, Mark empties eight 18-ounce cans of dog food into a large bowl. If he divides the food equally among the dogs, how many ounces of food will each dog get?

Use the graphic organizer below to help you solve the problem.

Read the Problem	Solve the Problem
What do I need to find? I need to find _____ _____.	• First, multiply to find the total number of ounces of dog food. $8 \times 18 =$ _____
What information do I need to use? I need to use the number of _____ , the number of _____ in each can, and the number of dogs that need to be fed.	• To find the number of ounces each dog gets, I'll need to divide. $144 \div$ _____ $=$ ◼
How will I use the information? I can _____ to find the total number of ounces. Then I can solve a simpler problem to _____ that total by 9.	• To find the quotient, I break 144 into two simpler numbers that are easier to divide. $144 \div 9$ $=$ ◼ $(90 +$ _____ $) \div 9$ $=$ ◼ (_____ $\div 9) + ($ _____ $\div 9) =$ ◼ _____ $+$ 6 $=$ _____

So, each dog gets _____ ounces of food.

Try Another Problem

Michelle is building shelves for her room. She has a plank 137 inches long that she wants to cut into 7 shelves of equal length. The plank has jagged ends, so she will start by cutting 2 inches off each end. How long will each shelf be?

137 inches

Read the Problem	Solve the Problem
What do I need to find?	
What information do I need to use?	
How will I use the information?	

So, each shelf will be _____ inches long.

Math Talk Explain how the strategy you used helped you solve the problem.

Name _____

Share and Show MATH BOARD

❔ UNLOCK the Problem Tips

√ Underline what you need to find.
√ Circle the numbers you need to use.

1. To make concrete mix, Monica pours 34 pounds of cement, 68 pounds of sand, 14 pounds of small pebbles, and 19 pounds of large pebbles into a large wheelbarrow. If she pours the mixture into 9 equal-size bags, how much will each bag weigh?

   First, find the total weight of the mixture.

   _____

   Then, divide the total by the number of bags. Break the total into two simpler numbers to make the division easier, if necessary.

   SHOW YOUR WORK

   Finally, find the quotient and solve the problem.

   So, each bag will weigh _____ pounds.

2. **What if** Monica pours the mixture into 5 equal-size bags? How much will each bag weigh?

   _____

3. Taylor is building doghouses to sell. Each doghouse requires 3 full sheets of plywood which Taylor cuts into new shapes. The plywood is shipped in bundles of 14 full sheets. How many doghouses can Taylor make from 12 bundles of plywood?

   _____

4. Eileen is planting a garden. She has seeds for 60 tomato plants, 55 sweet corn plants, and 21 cucumber plants. She plants them in 8 rows, with the same number of plants in each row. How many seeds are planted in each row?

   _____

On Your Own.................

Choose a
STRATEGY
Act It Out
Draw a Diagram
Make a Table
Solve a Simpler Problem
Work Backward
Guess, Check, and Revise

5. Starting on day 1 with 1 jumping jack, Keith doubles the number of jumping jacks he does every day. How many jumping jacks will Keith do on day 10?

6. **H.O.T.** Starting in the blue square, in how many different ways can you draw a line that passes through every square without picking up your pencil or crossing a line you've already drawn? Show the ways.

7. On April 11, Millie bought a lawn mower with a 50-day guarantee. If the guarantee begins on the date of purchase, what is the first day on which the mower will no longer be guaranteed?

8. **H.O.T.** A classroom bulletin board is 7 feet by 4 feet. If there is a picture of a student every 6 inches along the edge, including one in each corner, how many pictures are on the bulletin board?

9. Dave wants to make a stone walkway. The rectangular walkway is 4 feet wide and 12 feet long. Each 2 foot by 2 foot stone covers an area of 4 square feet. How many stones will Dave need to make his walkway?

10. **Test Prep** Dee has 112 minutes of recording time. How many 4-minute songs can she record?

   (A) 28   (C) 18

   (B) 27   (D) 17

FOR MORE PRACTICE:
Standards Practice Book, pp. P19–P20

Numerical Expressions

Essential Question How can you use a numerical expression
to describe a situation?

 UNLOCK the Problem REAL WORLD

A **numerical expression** is a mathematical phrase that has
numbers and operation signs but does not have an equal sign.

Tyler caught 15 small bass, and his dad caught 12 small bass
in the Memorial Bass Tourney in Tidioute, PA. Write a numerical
expression to represent how many fish they caught in all.

🔑 **Choose which operation to use.**

You need to join groups of different sizes, so use addition.

15 small bass	plus	12 small bass
↓	↓	↓
15	+	12

So, $15 + 12$ represents how many fish they caught in all.

🔑 Example 1 Write an expression to match the words.

A Addition

Emma has 11 fish in her
aquarium. She buys 4 more fish.

fish	plus	more fish
↓	↓	↓
11	+	4

B Subtraction

Lucia has 128 stamps. She uses
38 stamps on party invitations.

stamps	minus	stamps used
↓	↓	↓
128	−	_____

C Multiplication

Karla buys 5 books.
Each book costs $3.

books	multiplied by	cost per book
↓	↓	↓
_____	×	_____

D Division

Four players share 52 cards
equally.

cards	divided by	players
↓	↓	↓
_____	÷	_____

Math Talk MATHEMATICAL PRACTICES
Describe what each
expression represents.

Chapter 1 43

Expressions with Parentheses The meaning of the words in a problem will tell you where to place the parentheses in an expression.

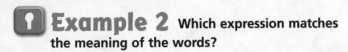 **Example 2** Which expression matches the meaning of the words?

Doug went fishing for 3 days. Each day he put $15 in his pocket. At the end of each day, he had $5 left. How much money did Doug spend by the end of the trip?

- Underline the events for each day.
- Circle the number of days these events happened.

Think: Each day he took $15 and had $5 left. He did this for 3 days.

($15 − $5) ← **Think:** What expression can you write to show how much money Doug spends in one day?

3 × ($15 − $5) ← **Think:** What expression can you write to show how much money Doug spends in three days?

MATHEMATICAL PRACTICES

Math Talk Explain how the expression of what Doug spent in three days compares to the expression of what he spent in one day?

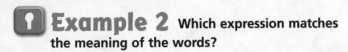 **Example 3** Which problem matches the expression $20 − ($12 + $3)?

Kim has $20 to spend for her fishing trip. She spends $12 on a fishing pole. Then she finds $3. How much money does Kim have now?

List the events in order.

First: Kim has $20.

Next: _____.

Then: _____.

Do these words match the expression? _____

Kim has $20 to spend for her fishing trip. She spends $12 on a fishing pole and $3 on bait. How much money does Kim have now?

List the events in order.

First: Kim has $20.

Next: _____.

Then: _____.

Do these words match the expression? _____

Share and Show ·········

Circle the expression that matches the words.

1. Teri had 18 worms. She gave 4 worms to Susie and 3 worms to Jamie.

(18 − 4) + 3   18 − (4 + 3)

2. Rick had $8. He then worked 4 hours for $5 each hour.

$8 + (4 × $5)   ($8 + 4) × $5

Name _____

Write an expression to match the words.

3. Greg drives 26 miles on Monday and 90 miles on Tuesday.

_____

4. Lynda has 27 fewer fish than Jack. Jack has 80 fish.

_____

Write words to match the expression.

5. $34 - 17$

_____

_____

6. $6 \times (12 - 4)$

_____

_____

Math Talk MATHEMATICAL PRACTICES
Is $4 \times 8 = 32$ an expression? **Explain** why or why not.

On Your Own ....

Write an expression to match the words.

7. José shared 12 party favors equally among 6 friends.

_____

8. Braden has 14 baseball cards. He finds 5 more baseball cards.

_____

9. Isabelle bought 12 bottles of water at $2 each.

_____

10. Monique had $20. She spent $5 on lunch and $10 at the bookstore.

_____

Write words to match the expression.

11. $36 \div 9$

_____

_____

12. $35 - (16 + 11)$

_____

_____

Draw a line to match the expression with the words.

13. Fred catches 25 fish. Then he releases 10 fish and catches 8 more. •

Nick has 25 pens. He gives 10 pens to one friend and 8 pens to another friend. •

Jan catches 15 fish and lets 6 fish go. •

Libby catches 15 fish and lets 6 fish go for three days in a row. •

• $3 \times (15 - 6)$

• $15 - 6$

• $25 - (10 + 8)$

• $(25 - 10) + 8$

Problem Solving REAL WORLD

Use the rule and the table for 14.

14. Write a numerical expression to represent the total number of lemon tetras that could be in a 20-gallon aquarium.

_____

15. **H.O.T.** Write a word problem for an expression that is three times as great as $(15 + 7)$. Then write the expression.

_____

_____

_____

_____

_____

Aquarium Fish	
Type of Fish	Length (in inches)
Lemon Tetra	2
Strawberry Tetra	3
Giant Danio	5
Tiger Barb	3
Swordtail	5

▲ The rule for the number of fish in an aquarium is to allow 1 gallon of water for each inch of length.

16. **What's the Question?** Lu has 3 swordtails in her aquarium. She buys 2 more swordtails.

_____

_____

17. **H.O.T.** Tammy gives 45 stamps to her 9 friends. She shares them equally among her friends. Write an expression to match the words. How many stamps does each friend get?

_____

18. **Test Prep** Josh has 3 fish in each of 5 buckets. Then he releases 4 fish. Which expression matches the words?

   Ⓐ $(3 \times 4) - 5$

   Ⓑ $(5 \times 4) - 3$

   Ⓒ $(5 \times 3) - 4$

   Ⓓ $(5 - 3) \times 4$

SHOW YOUR WORK

FOR MORE PRACTICE:
Standards Practice Book, pp. P21–P22

Evaluate Numerical Expressions

Essential Question In what order must operations be evaluated to find the solution to a problem?

CONNECT Remember that a numerical expression is a mathematical phrase that uses only numbers and operation symbols.

$$(5 - 2) \times 7 \qquad 72 \div 9 + 16 \qquad (24 - 15) + 32$$

To **evaluate**, or find the value of, a numerical expression with more than one operation, you must follow rules called the **order of operations**. The order of operations tells you in what order you should evaluate an expression.

> **Order of Operations**
> 1. Perform operations in parentheses.
> 2. Multiply and divide from left to right.
> 3. Add and subtract from left to right.

🔑 UNLOCK the Problem REAL WORLD

A cake recipe calls for 4 cups of flour and 2 cups of sugar. To triple the recipe, how many cups of flour and sugar are needed in all?

🔑 Evaluate 3 × 4 + 3 × 2 to find the total number of cups.

A Heather did not follow the order of operations correctly.

Heather
○ 3 × 4 + 3 × 2 First, I added.
3 × 7 × 2   Then, I multiplied. ○
42

B Follow the order of operations by multiplying first and then adding.

Name_____
○ 3 × 4 + 3 × 2
○

Explain why Heather's answer is not correct.

_____

So, _____ cups of flour and sugar are needed.

Evaluate Expressions with Parentheses To evaluate an expression with parentheses, follow the order of operations. Perform the operations in parentheses first. Multiply from left to right. Then add and subtract from left to right.

 Example

Each batch of cupcakes Lena makes uses 3 cups of flour, 1 cup of milk, and 2 cups of sugar. Lena wants to make 5 batches of cupcakes. How many cups of flour, milk, and sugar will she need in all?

Write the expression. $5 \times (3 + 1 + 2)$

First, perform the operations in parentheses. $5 \times ($ _____ $)$

Then multiply.   _____

So, Lena will use _____ cups of flour, milk, and sugar in all.

• **H.O.T.** **What if** Lena makes 4 batches? Will this change the numerical expression? **Explain**.

_____

Try This! Rewrite the expression with parentheses to equal the given value.

A $6 + 12 \times 8 - 3$; value: 141

• Evaluate the expression without the parentheses. _____

• Try placing the parentheses in the expression so the value is 141.

  Think: Will the placement of the parentheses increase or decrease the value of the expression?

• Use order of operations to check your work.

  $6 + 12 \times 8 - 3$

B $5 + 28 \div 7 - 4$; value: 11

• Evaluate the expression without the parentheses. _____

• Try placing the parentheses in the expression so that the value is 11.

  Think: Will the placement of the parentheses increase or decrease the value of the expression?

• Use order of operations to check your work.

  $5 + 28 \div 7 - 4$

Name _____

Share and Show

Evaluate the numerical expression.

1. $10 + 36 \div 9$

Think: I need to divide first.

2. $10 + (25 - 10) \div 5$

3. $9 - (3 \times 2) + 8$

MATHEMATICAL PRACTICES

Math Talk Raina evaluated the expression $5 \times 2 + 2$ by adding first and then multiplying. Will her answer be correct? Explain.

On Your Own

Evaluate the numerical expression.

4. $(4 + 49) - 4 \times 10$

5. $5 + 17 - 100 \div 5$

6. $36 - (8 + 5)$

7. $125 - (68 + 7)$

8. $(4 \times 6) - 12$

9. $3 \times (22 - 2)$

10. $23 + (16 - 7)$

11. $(25 - 4) \div 3$

Rewrite the expression with parentheses to equal the given value.

12. $100 - 30 \div 5$
value: 14

13. $12 + 17 - 3 \times 2$
value: 23

14. $9 + 5 \div 5 + 2$
value: 2

UNLOCK the Problem REAL WORLD

15. A movie theater has 4 groups of seats. The largest group of seats, in the middle, has 20 rows, with 20 seats in each row. There are 2 smaller groups of seats on the sides, each with 20 rows and 6 seats in each row. A group of seats in the back has 5 rows, with 30 seats in each row. How many seats are in the movie theater?

back

side   middle   side

a. What do you need to know? _____

_____

b. What operation can you use to find the number of seats in the back

group of seats? Write the expression. _____

c. What operation can you use to find the number of seats in both groups of side seats? Write the expression.

_____

d. What operation can you use to find the number of seats in the middle group? Write the expression.

_____

e. Write an expression to represent the total number of seats in the theater.

_____

_____

f. How many seats are in the theater? Show the steps you use to solve the problem.

_____

_____

16. Test Prep In the wild, an adult giant panda eats about 30 pounds of food each day. Which expression shows how many pounds of food 6 pandas eat in 3 days?

(A) $3 + (30 \times 6)$

(B) $3 \times (30 \times 6)$

(C) $(30 \times 6) \div 3$

(D) $(30 \times 6) - 3$

17. Test Prep Which expression has a value of 6?

(A) $(6 \div 3) \times 4 + 8$

(B) $27 - 9 \div 3 \times (4 + 1)$

(C) $(18 + 12) \times 6 - 4$

(D) $71 - 5 \times (9 + 4)$

Name _____

Grouping Symbols

Essential Question In what order must operations be evaluated to find a solution when there are parentheses within parentheses?

 UNLOCK the Problem REAL WORLD

Mary's weekly allowance is $8 and David's weekly allowance is $5. Every week they each spend $2 on lunch. Write a numerical expression to show how many weeks it will take them together to save enough money to buy a video game for $45.

- Underline Mary's weekly allowance and how much she spends.
- Circle David's weekly allowance and how much he spends.

Use parentheses and brackets to write an expression.

You can use parentheses and brackets to group operations that go together. Operations in parentheses and brackets are performed first.

STEP 1 Write an expression to represent how much Mary and David save each week.

- How much money does Mary save each week?

  Think: Each week Mary gets $8 and spends $2.

  (_____)

- How much money does David save each week?

  Think: Each week David gets $5 and spends $2.

  (_____)

- How much money do Mary and David save together each week? _____

STEP 2 Write an expression to represent how many weeks it will take Mary and David to save enough money for the video game.

- How many weeks will it take Mary and David to save enough for a video game?

  Think: I can use brackets to group operations a second time. $45 is divided by the total amount of money saved each week.

  _____ ÷ [_____]

Math Talk MATHEMATICAL PRACTICES
Explain why brackets are placed around the part of the expression that represents the amount of money Mary and David save each week.

Evaluate Expressions with Grouping Symbols When evaluating an expression with different grouping symbols (parentheses, brackets, and braces), perform the operation in the innermost set of grouping symbols first, evaluating the expression from the inside out.

🔑 Example

John gets $6 for his weekly allowance and spends $4 of it. His sister Tina gets $7 for her weekly allowance and spends $3 of it. Their mother's birthday is in 4 weeks. If they spend the same amount each week, how much money can they save together in that time to buy her a present?

- Write the expression using parentheses and brackets.   4 × [($6 − $4) + ($7 − $3)]

- Perform the operations in the parentheses first.   4 × [_____ + _____]

- Next perform the operations in the brackets.   4 × _____

- Then multiply.   _____

So, John and Tina will be able to save _____ for their mother's birthday present.

- **H.O.T.** **What if** only Tina saves any money? Will this change the numerical expression? **Explain.**

_____

Try This! Follow the order of operations.

A 4 × {[(5 − 2) × 3] + [(2 + 4) × 2]}

- Perform the operations in the parentheses.   4 × {[3 × 3] + [_____ × _____]}

- Perform the operations in the brackets.   4 × {9 + _____}

- Perform the operations in the braces.   4 × _____

- Multiply.   _____

B 32 ÷ {[(3 × 2) + 7] − [(6 − 4) + 7]}

- Perform the operations in the parentheses.   32 ÷ {[_____ + _____] − [_____ + _____]}

- Perform the operations in the brackets.   32 ÷ {_____ − _____}

- Perform the operations in the braces.   32 ÷ _____

- Divide.   _____

Name _____

Share and Show

Evaluate the numerical expression.

1. $12 + [(15 - 5) + (9 - 3)]$

$12 + [10 + \underline{\hphantom{00}}]$

$12 + \underline{\hphantom{000}}$

_____

⊘ 2. $5 \times [(26 - 4) - (4 + 6)]$

_____

⊘ 3. $36 \div [(18 - 10) - (8 - 6)]$

_____

On Your Own

Evaluate the numerical expression.

4. $4 + [(16 - 4) + (12 - 9)]$

_____

5. $24 - [(10 - 7) + (16 - 9)]$

_____

6. $16 \div [(13 + 7) - (12 + 4)]$

_____

7. $5 \times [(7 - 2) + (10 - 8)]$

_____

8. $[(17 + 8) + (29 - 12)] \div 6$

_____

9. $[(6 \times 7) + (3 \times 4)] - 28$

_____

10. $3 \times \{[(12 - 8) \times 2] + [(11 - 9) \times 3]\}$

_____

11. $\{[(3 \times 4) + 18] + [(6 \times 7) - 27]\} \div 5$

_____

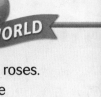

UNLOCK the Problem REAL WORLD

12. Dan has a flower shop. Each day he displays 24 roses. He gives away 10 and sells the rest. Each day he displays 36 carnations. He gives away 12 and sells the rest. What expression can you use to find out how many roses and carnations Dan sells in a week?

a. What information are you given? _____

_____

_____

b. What are you being asked to do? _____

_____

_____

c. What expression shows how many roses Dan sells in one day? _____

d. What expression shows how many carnations Dan sells in one day? _____

e. Write an expression to represent the total number

of roses and carnations Dan sells in one day. _____

f. Write the expression that shows how many

roses and carnations Dan sells in a week. _____

13. Evaluate the expression to find out how many roses and carnations Dan sells in a week.

_____

14. **Test Prep** Which expression has a value of 4?

Ⓐ $[(4 \times 5) + (9 + 7)] + 9$

Ⓑ $[(4 \times 5) + (9 + 7)] \div 9$

Ⓒ $[(4 \times 5) - (9 + 7)] \times 9$

Ⓓ $[(4 + 5) + (9 + 7)] - 9$

FOR MORE PRACTICE:
Standards Practice Book, pp. P25–P26

Name A-J crees

▶ **Vocabulary**

1. The _inverse operations_ states that multiplying a sum by a number is the same as multiplying each addend in the sum by the number and then adding the products. (p.14)

Vocabulary
Distributive Property
inverse operations

▶ **Concepts and Skills**

Complete the sentence.

2. 7,000 is 10 times as much as _700_.

3. 50 is $\frac{1}{10}$ of _50_.

Complete the equation, and tell which property you used.

4. $4 \times (12 + 14) = $ _160_ $ + (4 \times 14)$

5. $45 + 16 = $ _61_ $ + 45$

Find the value.

6. 10^2
_100_

7. 3×10^4
_30,000_

8. 8×10^3
_8,000_

Estimate. Then find the product.

9. Estimate: _3,500_
579
× 6

10. Estimate: _66,000_
7,316
× 9

11. Estimate: _14,000_
436
× 32

Use multiplication and the Distributive Property to find the quotient.

12. $54 \div 3 = $ _18_

13. $90 \div 5 = $ _18_

14. $96 \div 6 = $ _16_

Evaluate the numerical expression.

15. $42 - (9 + 6)$
_30_

16. $15 + (22 - 4) \div 6$
_5 r 3_

17. $6 \times [(5 \times 7) - (7 + 8)]$

© Houghton Mifflin Harcourt Publishing Company

Fill in the bubble completely to show your answer.

18. Erica's high score on her new video game is 30,000 points. Maria's high score is $\frac{1}{10}$ of Erica's. How many points did Maria score?

(A) 30

(B) 300

(C) 3,000

(D) 30,000

19. Rich makes $35 a week mowing lawns in his neighborhood. Which expression can be used to show how much money he makes in 8 weeks?

(A) $(8 + 30) + (8 + 5)$

(B) $(8 \times 30) + (8 \times 5)$

(C) $(8 + 30) \times (8 + 5)$

(D) $(8 \times 30) \times (8 \times 5)$

20. Mr. Rodriguez bought a supply of 20 reams of printer paper. Each ream contains 500 sheets of paper. How many sheets of printer paper are there?

(A) 1,000

(B) 5,000

(C) 10,000

(D) 100,000

21. Harvester ants are common in the southwestern United States. A single harvester ant colony may have as many as 90,000 members. What is that number written as a whole number multiplied by a power of ten?

(A) 9×10^4

(B) 9×10^3

(C) 9×10^2

(D) 9×10^1

Name _____ Storm Rizer

Fill in the bubble completely to show your answer.

22. Megan used the following expression to find the quotient of a division problem.

$$(4 \times 12) + (4 \times 6)$$

What was the division problem and the quotient?

Ⓐ $24 \div 4 = 6$

Ⓑ $48 \div 4 = 12$

Ⓒ $64 \div 4 = 16$

Ⓓ $72 \div 4 = 18$

23. It is 1,325 feet from Kinsey's house to her school. Kinsey walks to school each morning and gets a ride home each afternoon. How many feet does Kinsey walk to school in 5 days?

Ⓐ 6,725 feet

Ⓑ 6,625 feet

Ⓒ 6,525 feet

Ⓓ 5,625 feet

24. An adult elephant eats about 300 pounds of food each day. Which expression shows about how many pounds of food a herd of 12 elephants eats in 5 days?

Ⓐ $5 + (300 \times 12)$

Ⓑ $5 \times (300 \times 12)$

Ⓒ $(300 \times 12) \div 5$

Ⓓ $(300 \times 12) - 5$

25. Carla can type 265 characters a minute on her computer keyboard. At that rate, how many characters can she type in 15 minutes?

Ⓐ 2,975

Ⓑ 3,875

Ⓒ 3,905

Ⓓ 3,975

26. Donavan copied the problem below from the board. He missed one of the numbers needed to show his work. What number is missing in his work? **Explain** how you found the missing number.

$$17 \times 5 = (\blacksquare + 7) \times 5$$

$$= (\blacksquare \times 5) + (7 \times 5)$$

$$= 50 + 35$$

$$= 85$$

_____

_____

_____

► **Performance Task**

27. Drew's weekly allowance is $8.00. His friend Jan's weekly allowance is $10. Drew spends $3 a week and Jan spends $4 a week.

A Write two expressions to show how much money each person has at the end of the week. Use parentheses.

Drew has _____.

Jan has _____.

B Drew and Jan decide that they want to put their money together to buy a video game. Write an expression that shows how much they can save each week. **Explain**.

_____

_____

C The video game Drew and Jan want to buy costs $55. Write an expression to show how many weeks it will take them to save enough to buy the video game. Use parentheses and brackets in your expression. Then evaluate the expression.

_____

_____

Chapter 2

Divide Whole Numbers

Show What You Know ✓

Check your understanding of important skills.

Name _____

▶ **Meaning of Division** Use counters to solve.

1. Divide 18 counters into 3 equal groups. How many counters are in each group?

   _____ counters

2. Divide 21 counters into 7 equal groups. How many counters are in each group?

   _____ counters

▶ **Multiply 3-Digit and 4-Digit Numbers** Multiply.

| 3. | 321
 × 4 | 4. | 518
 × 7 | 5. | 4,092
 × 6 | 6. | 8,264
 × 9 |

▶ **Estimate with 1-Digit Divisors** Estimate the quotient.

7. 2)312   8. 4)189   9. 6)603   10. 3)1,788

The height of the Gateway Arch shown on the Missouri quarter is 630 feet, or 7,560 inches. Be a math detective to find how many 4-inch stacks of quarters make up the height of the Gateway Arch. If there are 58 quarters in a 4-inch stack, how many quarters high is the arch?

Vocabulary Builder

▶ **Visualize It** ••••••••••••••••••••••••••••••••••••

Complete the Flow Map using the words with a ✓.

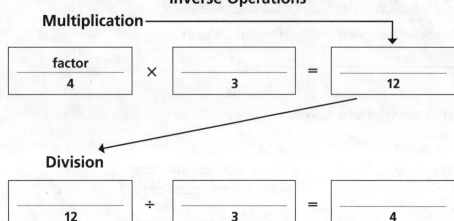

Inverse Operations

Multiplication

factor				
4	×	3	=	12

Division

12	÷	3	=	4

Review Words

compatible numbers
✓dividend
✓divisor
estimate
✓factor
partial quotients
✓product
✓quotient
remainder

▶ **Understand Vocabulary** ••••••••••••••••••••••••••

Use the review words to complete each sentence.

1. You can _____ to find a number that is close to the exact amount.

2. Numbers that are easy to compute with mentally are called

   _____.

3. The _____ is the amount left over when a number cannot be divided evenly.

4. A method of dividing in which multiples of the divisor are subtracted from the dividend and then the quotients are

   added together is called _____.

5. The number that is to be divided in a division problem is the

   _____.

6. The _____ is the number, not including the remainder, that results from dividing.

Name _____

Place the First Digit

Essential Question How can you tell where to place the first digit of a quotient without dividing?

UNLOCK the Problem ❭ REAL WORLD

Tania has 8 purple daisies. In all, she counts 128 petals on her flowers. If each flower has the same number of petals, how many petals are on one flower?

- Underline the sentence that tells you what you are trying to find.
- Circle the numbers you need to use.
- How will you use these numbers to solve the problem?

_____

 Divide. 128 ÷ 8

STEP 1 Use an estimate to place the first digit in the quotient.

Estimate. 160 ÷ _____ = _____

The first digit of the quotient will be in

the _____ place.

STEP 2 Divide the tens.

```
      1
  8)128
  -
```

Divide. 12 tens ÷ 8
Multiply. 8 × 1 ten

Subtract. 12 tens − _____ tens
Check. _____ tens cannot be shared among 8 groups without regrouping.

STEP 3 Regroup any tens left as ones. Then, divide the ones.

```
     16
  8)128
  -8↓

  −
```

Divide. 48 ones ÷ 8
Multiply. 8 × 6 ones

Subtract. 48 ones − _____ ones
Check. _____ ones cannot be shared among 8 groups.

Math Talk MATHEMATICAL PRACTICES
Explain how estimating the quotient helps you at both the beginning and the end of a division problem.

Since 16 is close to the estimate of _____, the answer is reasonable.

So, there are 16 petals on one flower.

© Houghton Mifflin Harcourt Publishing Company

🔒 Example

Divide. Use place value to place the first digit. 4,236 ÷ 5

STEP 1   Use place value to place the first digit.

5)4,236

Look at the thousands.

4 thousands cannot be shared among 5 groups without regrouping.

Look at the hundreds.

_____ hundreds can be shared among 5 groups.

The first digit is in the _____ place.

Remember

Remember to estimate the quotient first.

Estimate: 4,000 ÷ 5 = _____

STEP 2   Divide the hundreds.

```
    8
5)4,236
 -
```

Divide. _____ hundreds ÷ _____

Multiply. _____ × _____ hundreds

Subtract. _____ hundreds − _____ hundreds

Check. _____ hundreds cannot be shared among 5 groups without regrouping.

STEP 3   Divide the tens.

```
   84
5)4,236
 -40↓
   23
  -20
    3
```

Divide. _____

Multiply. _____

Subtract. _____

Check. _____

_____

STEP 4   Divide the ones.

```
   847
5)4,236
 -40
   23
  -20
    36
   -35
     1
```

Divide. _____

Multiply. _____

Subtract. _____

Check. _____

_____

So, 4,236 ÷ 5 is _____ r_____.

Math Talk   MATHEMATICAL PRACTICES
Explain how you know if your answer is reasonable.

Name _____

Share and Show

Divide.

1. 3)579

2. 5)1,035 ✓

3. 8)1,766 ✓

Math Talk

MATHEMATICAL PRACTICES

As you divide, explain how you know when to place a zero in the quotient.

On Your Own

Divide.

4. 8)275

5. 3)468

6. 4)3,220

7. 6)618

8. 4)716

9. 9)1,157

10. 6)6,827

11. 7)8,523

Practice: Copy and Solve   Divide.

12. $645 \div 8$

13. $942 \div 6$

14. $723 \div 7$

15. $3,478 \div 9$

16. $3,214 \div 5$

17. $492 \div 4$

18. $2,403 \div 9$

19. $2,205 \div 6$

20. $2,426 \div 3$

21. $1,592 \div 8$

22. $926 \div 4$

23. $6,033 \div 5$

UNLOCK the Problem REAL WORLD

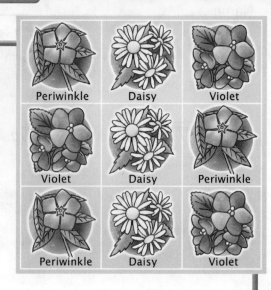

Periwinkle   Daisy   Violet
Violet   Daisy   Periwinkle
Periwinkle   Daisy   Violet

24. Rosa has a garden divided into sections. She has 125 daisy plants. If she plants an equal number of the daisy plants in each section of daisies, will she have any left over? If so, how many daisy plants will be left over?

a. What information will you use to solve the problem? _____

_____

b. How will you use division to find the number of daisy plants left over? _____

_____

c. Show the steps you use to solve the problem. Estimate: $120 \div 3 =$ _____

d. Complete the sentences:

Rosa has _____ daisy plants. She puts an equal number in each

of _____ sections.

Each section has _____ plants.

Rosa has _____ daisy plants left over.

25. H.O.T. One case can hold 3 boxes. Each box can hold 3 binders. How many cases are needed to hold 126 binders?

_____

26. Test Prep In which place is the first digit in the quotient $1,497 \div 5$?

(A) thousands

(B) hundreds

(C) tens

(D) ones

FOR MORE PRACTICE:
Standards Practice Book, pp. P31–P32

Divide by 1-Digit Divisors

Essential Question How do you solve and check division problems?

 UNLOCK the Problem REAL WORLD

Jenna's family is planning a trip to Oceanside, California. They will begin their trip in Scranton, Pennsylvania, and will travel 2,754 miles over 9 days. If the family travels an equal number of miles every day, how far will they travel each day?

- Underline the sentence that tells you what you are trying to find.
- Circle the numbers you need to use.

Divide. 2,754 ÷ 9

STEP 1

Use an estimate to place the first digit in the quotient.

Estimate. 2,700 ÷ 9 = _____

The first digit of the quotient is in

the _____ place.

STEP 2

Divide the hundreds.
_____

STEP 3

Divide the tens.
_____

STEP 4

Divide the ones.

Since _____ is close to the estimate of _____, the answer is reasonable.

So, Jenna's family will travel _____ miles each day.

$$9 \overline{) 2,754}$$

Math Talk MATHEMATICAL PRACTICES
Explain how you know the quotient is 306 and not 36.

© Houghton Mifflin Harcourt Publishing Company

CONNECT Division and multiplication are inverse operations. Inverse operations are opposite operations that undo each other. You can use multiplication to check your answer to a division problem.

🔑 **Example** Divide. Check your answer.

To check your answer to a division problem, multiply the quotient by the divisor. If there is a remainder, add it to the product. The result should equal the dividend.

$$
\begin{array}{r}
102 \ r2 \\
6\overline{)614} \\
-6 \\
\hline
01 \\
-0 \\
\hline
14 \\
-12 \\
\hline
2
\end{array}
$$

$$
\begin{array}{r}
102 \quad \leftarrow \text{quotient} \\
\times \ 6 \quad \leftarrow \text{divisor} \\
\hline
 \\
+ \ 2 \quad \leftarrow \text{remainder} \\
\hline
 \quad \leftarrow \text{dividend}
\end{array}
$$

Since the result of the check is equal to the dividend, the division is correct.

So, $614 \div 6$ is _____.

You can use what you know about checking division to find an unknown value.

Try This! Find the unknown number by finding the value of *n* in the related equation.

Ⓐ
$$
\begin{array}{r}
63 \\
7\overline{)}
\end{array}
$$

$$n = 7 \times 63$$

↑   ↗   ↗

dividend   divisor   quotient

Multiply the divisor and the quotient.

$n =$ _____

Ⓑ
$$
\begin{array}{r}
125 \ r \ \\
6\overline{)752}
\end{array}
$$

$$752 = 6 \times 125 + n$$

↑   ↑   ↑   ↖

dividend   divisor   quotient   remainder

Multiply the divisor and the quotient.

$752 = 750 + n$

Think: What number added to 750 equals 752?

$n =$ _____

Name _____

Share and Show ....

Divide. Check your answer.

1. 8)624 Check.   **2.** 4)3,220 Check.   **3.** 4)1,027 Check.

Math Talk MATHEMATICAL PRACTICES
Explain how multiplication can help you check a quotient.

On Your Own..

Divide.

4. 6)938   **5.** 4)762   **6.** 3)5,654   **7.** 8)475

Practice: Copy and Solve Divide.

8. 4)671   **9.** 9)2,023   **10.** 3)4,685   **11.** 8)948

12. $1,326 \div 4$ **13.** $5,868 \div 6$ **14.** $566 \div 3$ **15.** $3,283 \div 9$

Algebra Find the value of *n* in each equation. Write what *n* represents in the related division problem.

16. $n = 4 \times 58$ **17.** $589 = 7 \times 84 + n$ **18.** $n = 5 \times 67 + 3$

n = _____   *n* = _____   *n* = _____

Problem Solving ⟩REAL WORLD⟩

Use the table to solve 19–20.

19. If the Welcome gold nugget were turned into 3 equal-sized gold bricks, how many troy ounces would each brick weigh?

_____

20. **Pose a Problem** Look back at Problem 19. Write a similar problem by changing the nugget and the number of bricks. Then solve the problem.

_____

_____

_____

_____

_____

Large Gold Nuggets Found

Name	Weight	Location
Welcome Stranger	2,284 troy ounces	Australia
Welcome	2,217 troy ounces	Australia
Willard	788 troy ounces	California

⟩ SHOW YOUR WORK ⟩

21. **H.O.T.** There are 246 students going on a field trip to pan for gold. If they are going in vans that hold 9 students each, how many vans are needed? How many students will ride in the van that isn't full?

_____

22. One crate can hold 8 cases of trading cards. How many crates are needed to hold 128 cases of trading cards?

_____

23. **Test Prep** At a bake sale, a fifth-grade class sold 324 cupcakes in packages of 6. How many packages of cupcakes did the class sell?

Ⓐ 1,944   Ⓒ 64

Ⓑ 108   Ⓓ 54

68

FOR MORE PRACTICE:
Standards Practice Book, pp. P33–P34

Division with 2-Digit Divisors

Essential Question How can you use base-ten blocks to model and
understand division of whole numbers?

Investigate

Materials ■ base-ten blocks

There are 156 students in the Carville Middle School chorus. The
music director wants the students to stand with 12 students in
each row for the next concert. How many rows will there be?

A. Use base-ten blocks to model the dividend, 156.

B. Place 2 tens below the hundred to form a rectangle. How
many groups of 12 does the rectangle show? How much of
the dividend is not shown in this rectangle?

_____

C. Combine the remaining tens and ones into as many groups
of 12 as possible. How many groups of 12 are there?

_____

D. Place these groups of 12 on the right side of the rectangle
to make a larger rectangle.

E. The final rectangle shows _____ groups of 12.

So, there will be _____ rows of 12 students.

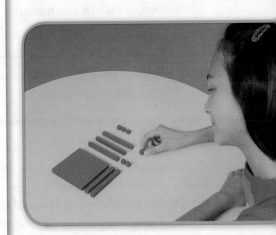

Draw Conclusions . . . .

1. **Explain** why you still need to make groups of 12 after Step B.

_____

_____

2. **Describe** how you can use base-ten blocks to find the quotient 176 ÷ 16.

_____

_____

_____

_____

Make Connections

The two sets of groups of 12 that you found in the Investigate are partial quotients. First you found 10 groups of 12 and then you found 3 more groups of 12. Sometimes you may need to regroup before you can show a partial quotient.

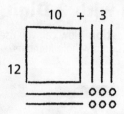

You can use a quick picture to record the partial products.

Divide. 180 ÷ 15

MODEL Use base-ten blocks.

STEP 1 Model the dividend, 180, as 1 hundred 8 tens.

Model the first partial quotient by making a rectangle with the hundred and 5 tens. In the Record, cross out the hundred and tens you use.

The rectangle shows _____ groups of 15.

STEP 2 Additional groups of 15 cannot be made without regrouping.

Regroup 1 ten as 10 ones. In the Record, cross out the regrouped ten.

There are now _____ tens and _____ ones.

STEP 3 Decide how many additional groups of 15 can be made with the remaining tens and ones. The number of groups is the second partial quotient.

Make your rectangle larger by including these groups of 15. In the Record, cross out the tens and ones you use.

There are now _____ groups of 15.

So, 180 ÷ 15 is _____.

RECORD Use quick pictures.

Draw the first partial quotient.

Draw the first and second partial quotients.

> **Math Talk** MATHEMATICAL PRACTICES
> Explain how your model shows the quotient.

Share and Show

Use the quick picture to divide.

1. 143 ÷ 13

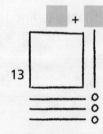

© Houghton Mifflin Harcourt Publishing Company

Name _____

Divide. Use base-ten blocks.

2. 168 ÷ 12

3. 154 ÷ 14

☑ **4.** 187 ÷ 11

Divide. Draw a quick picture.

5. 165 ÷ 11

6. 216 ÷ 18

7. 196 ÷ 14

8. 195 ÷ 15

☑ **9.** 182 ÷ 13

10. 228 ÷ 12

Math Talk MATHEMATICAL PRACTICES
Explain how Exercise 10 is different from Exercises 7-9.

Pony Express

The Pony Express used men riding horses to deliver mail between St. Joseph, Missouri, and Sacramento, California, from April, 1860 to October, 1861. The trail between the cities was approximately 2,000 miles long. The first trip from St. Joseph to Sacramento took 9 days 23 hours. The first trip from Sacramento to St. Joseph took 11 days 12 hours.

Before the Pony Express ended in 1861, there were 100 stations, 80 riders, and 400 to 500 horses. The riders were young men about 20 years old who weighed about 120 pounds. Each rider rode 10 to 15 miles before getting a fresh horse. Riders rode a total of 75 to 100 miles each trip.

Solve.

11. Suppose two Pony Express riders rode a total of 165 miles. If they replaced each horse with a fresh horse every 11 miles, how many horses would they have used?

_____

12. Suppose a Pony Express rider was paid $192 for 12 weeks of work. If he was paid the same amount each week, how much was he paid for each week of work?

_____

13. Suppose three riders rode a total of 240 miles. If they used a total of 16 horses, and rode each horse the same number of miles, how many miles did they ride before replacing each horse?

14. ⭐H.O.T. Suppose it took 19 riders a total of 11 days 21 hours to ride from St. Joseph to Sacramento. If they all rode the same number of hours, how many hours did each rider ride?

Partial Quotients

Essential Question How can you use partial quotients to divide by 2-digit divisors?

🔓 UNLOCK the Problem REAL WORLD

People in the United States eat about 23 pounds of pizza per person every year. If you ate that much pizza each year, how many years would it take you to eat 775 pounds of pizza?

• Rewrite in one sentence the problem you are asked to solve.

_____

_____

🔑 **Divide by using partial quotients.**

$775 \div 23$

STEP 1

Subtract multiples of the divisor from the dividend until the remaining number is less than the multiple. The easiest partial quotients to use are multiples of 10.

_____

STEP 2

Subtract smaller multiples of the divisor until the remaining number is less than the divisor. Then add the partial quotients to find the quotient.

COMPLETE THE DIVISION PROBLEM.

$$23\overline{)775}$$
$$-$$
$$545$$

10×23 | 10

$775 \div 23$ is _____ r _____.

So, it would take you more than 33 years to eat 775 pounds of pizza.

Remember

Depending on the question, a remainder may or may not be used in answering the question. Sometimes the quotient is adjusted based on the remainder.

🔑 Example

Myles is helping his father with the supply order for his pizza shop. For next week, the shop will need 1,450 ounces of mozzarella cheese. Each package of cheese weighs 32 ounces. Complete Myles's work to find how many packages of mozzarella cheese he needs to order.

```
32)1,450
   - 320      ____ × 32          [    ]
   1,130
   - 320      ____ × 32          [    ]
     810
   - 320      ____ × 32          [    ]
     490
   - 320      ____ × 32          [    ]
     170
   - 160      ____ × 32        + [    ]
      10
```

1,450 ÷ 32 is _____ r _____.

So, he needs to order _____ packages of mozzarella cheese.

Math Talk MATHEMATICAL PRACTICES

What does the remainder represent? **Explain** how the remainder will affect your answer.

Try This! Use different partial quotients to solve the problem above.

```
32)1,450
```

Math Idea

Using different multiples of the divisor to find partial quotients provides many ways to solve a division problem. Some ways are quicker, but all result in the same answer.

Name _____

Share and Show

Divide. Use partial quotients.

1. $18\overline{)648}$

2. $62\overline{)3{,}186}$

3. $858 \div 57$

Math Talk MATHEMATICAL PRACTICES
Explain what the greatest possible whole-number remainder is if you divide any number by 23.

On Your Own

Divide. Use partial quotients.

4. $73\overline{)584}$

5. $51\overline{)1{,}831}$

6. $82\overline{)2{,}964}$

7. $892 \div 26$

8. $1{,}056 \div 48$

9. $2{,}950 \div 67$

Practice: Copy and Solve Divide. Use partial quotients.

10. $653 \div 42$ 11. $946 \div 78$ 12. $412 \div 18$ 13. $871 \div 87$

14. $1{,}544 \div 34$ 15. $2{,}548 \div 52$ 16. $2{,}740 \div 83$ 17. $4{,}135 \div 66$

Problem Solving REAL WORLD

Use the table to solve 18–20 and 22.

Each year each person in the U.S. eats about...
- 68 quarts of popcorn
- 53 pounds of bread
- 19 pounds of apples
- 14 pounds of turkey

18. How many years would it take for a person in the United States to eat 855 pounds of apples?

_____

19. How many years would it take for a person in the United States to eat 1,120 pounds of turkey?

_____

20. If 6 people in the United States each eat the average amount of popcorn for 5 years, how many quarts of popcorn will they eat?

_____

21. **H.O.T.** In a study, 9 people ate a total of 1,566 pounds of potatoes in 2 years. If each person ate the same amount each year, how many pounds of potatoes did each person eat in 1 year?

_____

22. **Write Math** ▸ **Sense or Nonsense?** In the United States, a person eats more than 40,000 pounds of bread in a lifetime if he or she lives to be 80 years old. Does this statement make sense, or is it nonsense? **Explain.**

_____

_____

_____

23. **Test Prep** The school auditorium has 448 seats arranged in 32 equal rows. How many seats are in each row?

Ⓐ 14,336   Ⓒ 416

Ⓑ 480   Ⓓ 14

SHOW YOUR WORK

FOR MORE PRACTICE:
Standards Practice Book, pp. P37–P38

Name _____

 Mid-Chapter Checkpoint

▶ **Concepts and Skills**

1. **Explain** how estimating the quotient helps you place the first digit in the quotient of a division problem.

_____

_____

2. **Explain** how to use multiplication to check the answer to a division problem.

_____

_____

_____

Divide.

3. $633 \div 3$

4. $487 \div 8$

5. $1,641 \div 4$

6. $2,765 \div 9$

Divide. Use partial quotients.

7. $156 \div 13$

8. $318 \div 53$

9. $1,562 \div 34$

10. $4,024 \div 68$

Fill in the bubble completely to show your answer.

11. Emma is planning a party for 128 guests. If 8 guests can be seated at each table, how many tables will be needed for seating at the party?

    Ⓐ 8

    Ⓑ 14

    Ⓒ 16

    Ⓓ 17

12. Tickets for the basketball game cost $14 each. If the sale of the tickets brought in $2,212, how many tickets were sold?

    Ⓐ 150

    Ⓑ 158

    Ⓒ 168

    Ⓓ 172

13. Margo used 864 beads to make necklaces for the art club. She made 24 necklaces with the beads. If each necklace has the same number of beads, how many beads did Margo use for each necklace?

    Ⓐ 24

    Ⓑ 36

    Ⓒ 37

    Ⓓ 60

14. Angie needs to buy 156 candles for a party. Each package has 8 candles. How many packages should Angie buy?

    Ⓐ 17

    Ⓑ 18

    Ⓒ 19

    Ⓓ 20

Name _____

Estimate with 2-Digit Divisors

Essential Question How can you use compatible numbers
to estimate quotients?

CONNECT You can estimate quotients using compatible
numbers that are found by using basic facts and patterns.

$$35 \div 5 = 7 \quad \leftarrow \text{basic fact}$$
$$350 \div 50 = 7$$
$$3,500 \div 50 = 70$$
$$35,000 \div 50 = 700$$

🔑 UNLOCK the Problem › REAL WORLD

The observation deck of the Willis Tower in Chicago,
Illinois, is 1,353 feet above the ground. Elevators lift
visitors to that level in 60 seconds. About how many
feet do the elevators travel per second?

◄ **Willis Tower,**
formerly known as
the Sears Tower, is
the tallest building
in the United States.

 Estimate. 1,353 ÷ 60

STEP 1

Use two sets of compatible
numbers to find two
different estimates.

$1,353 \div 60$	$1,353 \div 60$
↓	↓
$1,200 \div 60$	$1,800 \div 60$

STEP 2

Use patterns and basic
facts to help estimate.

$12 \div 6 = $ _____ $18 \div 6 = $ _____

$120 \div 60 = $ _____   _____ ÷ _____ = _____

$1,200 \div 60 = $ _____   _____ ÷ _____ = _____

The elevators travel about _____ to _____ feet per second.

The more reasonable estimate is _____ because

_____ is closer to 1,353 than _____ is.

So, the observation deck elevators in the Willis Tower travel

about _____ feet per second.

🔑 Example Estimate money.

Miriam has saved $650 to spend during her 18-day trip to Chicago. She doesn't want to run out of money before the trip is over, so she plans to spend about the same amount each day. Estimate how much she can spend each day.

Estimate. $18\overline{)\$650}$

$600 \div _____ = \$30$ or $_____ \div 20 = \$40$

So, Miriam can spend about $_____$ to $_____$ each day.

Math Talk MATHEMATICAL PRACTICES
Would it be more reasonable to have an estimate or an exact answer for this example? **Explain** your reasoning.

- Given Miriam's situation, which estimate do you think is the better one for her to use? **Explain** your reasoning. _____

_____

_____

Try This! Use compatible numbers.

Find two estimates.	Estimate the quotient.
$52\overline{)415}$	$38\overline{)\$2,764}$

Share and Show   ·

Use compatible numbers to find two estimates.

1. $22\overline{)154}$

$140 \div 20 = \_\_\_\_\_$

$160 \div 20 = \_\_\_\_\_$

2. $68\overline{)503}$

3. $81\overline{)7,052}$

✓ **4.** $33\overline{)291}$

✓ **5.** $58\overline{)2,365}$

6. $19\overline{)5,312}$

Name _____

On Your Own..

Use compatible numbers to find two estimates.

7. $42\overline{)396}$

8. $59\overline{)413}$

9. $28\overline{)232}$

10. $22\overline{)368}$

11. $78\overline{)375}$

12. $88\overline{)6,080}$

13. $5,821 \div 71$

14. $1,565 \div 67$

15. $7,973 \div 91$

Use compatible numbers to estimate the quotient.

16. $19\overline{)228}$

17. $25\overline{)\$595}$

18. $86\overline{)7,130}$

19. $83\overline{)462}$

20. $27\overline{)9,144}$

21. $68\overline{)710}$

22. $707 \div 36$

23. $1,198 \div 41$

24. $5,581 \div 72$

Problem Solving REAL WORLD

Use the picture to solve 25–26.

25. About how many meters tall is each floor of the Williams Tower?

_____

26. About how many meters tall is each floor of the Chrysler Building?

_____

27. H.O.T. Eli needs to save $235. To earn money, he plans to mow lawns and charge $21 for each. Write two estimates Eli could use to determine the number of lawns he needs to mow. Decide which estimate you think is the better one for Eli to use. **Explain** your reasoning.

_____

_____

_____

_____

_____

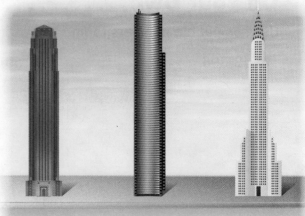

275 meters, 64 floors, Williams Tower, Texas

295 meters, 76 floors, Columbia Center, Washington

319 meters, 77 floors, Chrysler Building, New York

SHOW YOUR WORK

28. Write Math ► **Explain** how you know whether the quotient of 298 ÷ 31 is closer to 9 or to 10.

_____

_____

_____

29. Test Prep Anik built a tower of cubes. It was 594 millimeters tall. The height of each cube was 17 millimeters. About how many cubes did Anik use?

(A) 10

(C) 30

(B) 16

(D) 300

FOR MORE PRACTICE:
Standards Practice Book, pp. P39–P40

Name _____

Divide by 2-Digit Divisors

Essential Question How can you divide by 2-digit divisors?

UNLOCK the Problem REAL WORLD

Mr. Yates owns a smoothie shop. To mix a batch of his famous orange-punch smoothies, he uses 18 ounces of freshly squeezed orange juice. Each day he squeezes 560 ounces of fresh orange juice. How many batches of orange-punch smoothies can Mr. Yates make in a day?

- Underline the sentence that tells you what you are trying to find.
- Circle the numbers you need to use.

Divide. 560 ÷ 18   **Estimate.** _____

STEP 1   Use the estimate to place the first digit in the quotient.

18)‾560

The first digit of the quotient will be in the

_____ place.

STEP 2   Divide the tens.

$$\begin{array}{r} 3 \\ 18\overline{)560} \\ -54 \\ \hline 2 \end{array}$$

Divide.   _56 tens ÷ 18_

Multiply. _____

Subtract. _____

Check.   2 tens cannot be shared among 18 groups without regrouping.

STEP 3   Divide the ones.

$$\begin{array}{r} 31\ r2 \\ 18\overline{)560} \\ -54\downarrow \\ \hline 20 \\ -18 \\ \hline 2 \end{array}$$

Divide.   _____

Multiply. _____

Subtract. _____

Check.   _____

_____

Math Talk MATHEMATICAL PRACTICES
Explain what the remainder 2 represents.

Since 31 is close to the estimate of 30, the answer is reasonable.

So, Mr. Yates can make 31 batches of orange-punch smoothies each day.

🔒 Example

Every Wednesday, Mr. Yates orders fruit. He has set aside $1,250 to purchase Valencia oranges. Each box of Valencia oranges costs $41. How many boxes of Valencia oranges can Mr. Yates purchase?

You can use multiplication to check your answer.

Divide. 1,250 ÷ 41

DIVIDE	CHECK YOUR WORK

Estimate. _____

```
      30 r20
41)1,250
   -
   _____

   -
   _____
```

```
    30
  × 41
  ____
    30
+1,200
_____
```

```
       
+      
_____
 1,250 ✓
```

So, Mr. Yates can buy _____ boxes of Valencia oranges.

Try This! Divide. Check your answer.

A

$$63\overline{)756}$$

B

$$22\overline{)4,692}$$

Name _____

Share and Show

Divide. Check your answer.

1. 28)620

2. 64)842

3. 53)2,340

4. 723 ÷ 31

5. 1,359 ÷ 45

6. 7,925 ÷ 72

Math Talk MATHEMATICAL PRACTICES
Explain why you can use multiplication to check division.

On Your Own

Divide. Check your answer.

7. 16)346

8. 34)421

9. 77)851

10. 21)1,098

11. 32)6,466

12. 45)9,500

13. 483 ÷ 21

14. 2,292 ÷ 19

15. 4,255 ÷ 30

Practice: Copy and Solve **Divide. Check your answer.**

16. 775 ÷ 35

17. 820 ÷ 41

18. 805 ÷ 24

19. 1,166 ÷ 53

20. 1,989 ÷ 15

21. 3,927 ÷ 35

Problem Solving REAL WORLD

Use the list at the right to solve 22–24.

22. A smoothie shop receives a delivery of 980 ounces of grape juice. How many Royal Purple smoothies can be made with the grape juice?

_____

23. The shop has 1,260 ounces of cranberry juice and 650 ounces of passion fruit juice. If the juices are used to make Crazy Cranberry smoothies, which juice will run out first? How much of the other juice will be left over?

_____

_____

24. **H.O.T.** In the refrigerator, there are 680 ounces of orange juice and 410 ounces of mango juice. How many Orange Tango smoothies can be made? **Explain** your reasoning.

_____

_____

_____

_____

_____

25. **Test Prep** James has 870 action figures. He decides to divide them equally among 23 boxes. How many action figures will James have left over?

Ⓐ 19   Ⓒ 31

Ⓑ 23   Ⓓ 37

Smoothie Main Ingredients

Orange Tango Smoothie
18 ounces orange juice
12 ounces mango juice

Royal Purple Smoothie
22 ounces grape juice
8 ounces apple juice

Crazy Cranberry Smoothie
20 ounces cranberry juice
10 ounces passion fruit juice

SHOW YOUR WORK

Name _____

Interpret the Remainder

Essential Question When solving a division problem, when do you write the remainder as a fraction?

🔑 UNLOCK the Problem ⟩REAL WORLD

Scott and his family want to hike a trail that is 1,365 miles long. They will hike equal parts of the trail on 12 different hiking trips. How many miles will Scott's family hike on each trip?

- Circle the dividend you will use to solve the division problem.
- Underline the divisor you will use to solve the division problem.

When you solve a division problem with a remainder, the way you interpret the remainder depends on the situation and the question. Sometimes you need to use both the quotient and the remainder. You can do that by writing the remainder as a fraction.

🔓 One Way Write the remainder as a fraction.

First, divide to find the quotient and remainder.

Then, decide how to use the quotient and remainder to answer the question.

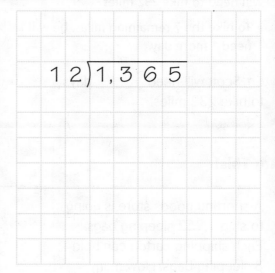

- The _____ represents the number of trips Scott and his family plan to take.

- The _____ represents the whole-number part of the number of miles Scott and his family will hike on each trip.

- The _____ represents the number of miles left over.

- The remainder represents 9 miles, which can also be divided into 12 parts and written as a fraction.

$$\frac{remainder}{divisor} \rightarrow \text{_____}$$

- Write the quotient with the remainder written as a fraction in simplest form.

_____

So, Scott and his family will hike _____ miles on each trip.

🔑 Another Way Use only the quotient.

The segment of the Appalachian Trail that runs through Pennsylvania is 232 miles long. Scott and his family want to hike 9 miles each day on the trail. How many days will they hike exactly 9 miles?

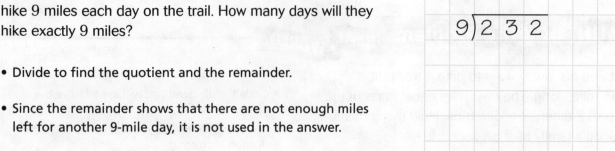

- Divide to find the quotient and the remainder.

- Since the remainder shows that there are not enough miles left for another 9-mile day, it is not used in the answer.

So, they will hike exactly 9 miles on each of _____ days.

🔑 Other Ways

Ⓐ Add 1 to the quotient.

What is the total number of days that Scott will need to hike 232 miles?

- To hike the 7 remaining miles, he will need 1 more day.

So, Scott will need _____ days to hike 232 miles.

Ⓑ Use the remainder as the answer.

If Scott hikes 9 miles each day except the last day, how many miles will he hike on the last day?

- The remainder is 7.

So, Scott will hike _____ miles on the last day.

Try This!

A sporting goods store is going to ship 1,252 sleeping bags. Each shipping carton can hold 8 sleeping bags. How many cartons are needed to ship all of the sleeping bags?

```
      1
8)1,252
 − 8
   45
 −
      2
 −
```

Since there are _____ sleeping bags left over,

_____ cartons will be needed for all of the sleeping bags.

Math Talk MATHEMATICAL PRACTICES
Explain why you would not write the remainder as a fraction when you find the number of cartons needed in the Try This.

Share and Show

Interpret the remainder to solve.

1. Erika and Bradley want to hike the Big Cypress Trail. They will hike a total of 75 miles. If Erika and Bradley plan to hike for 12 days, how many miles will they hike each day?

   a. Divide to find the quotient and remainder.

   b. Decide how to use the quotient and remainder to answer the question.

   _____

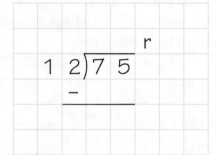

2. **What if** Erika and Bradley want to hike 14 miles each day? How many days will they hike exactly 14 miles?

   _____

3. Dylan's hiking club is planning to stay overnight at a camping lodge. Each large room can hold 15 hikers. There are 154 hikers. How many rooms will they need?

   _____

On Your Own

Interpret the remainder to solve.

4. The students in a class of 24 share 84 cookies equally among them. How many cookies did each student eat?

   _____

5. A campground has cabins that can each hold 28 campers. There are 148 campers visiting the campground. How many cabins are full if 28 campers are in each cabin?

   _____

6. A total of 123 fifth-grade students are going to Fort Verde State Historic Park. Each bus holds 38 students. All of the buses are full except one. How many students will be in the bus that is not full?

   _____

7. **H.O.T. What's the Error?** Sheila is going to divide a 36-inch piece of ribbon into 5 equal pieces. She says each piece will be 7 inches long.

   _____

   _____

UNLOCK the Problem REAL WORLD

8. Maureen has 243 ounces of trail mix. She puts an equal number of ounces in each of 15 bags. How many ounces of trail mix does Maureen have left over?

(A) 3 ounces (B) 15 ounces (C) 16 ounces (D) 17 ounces

a. What do you need to find? _____

_____

b. How will you use division to find how many ounces of trail mix are left over?

_____

_____

c. Show the steps you use to solve the problem.

d. Complete the sentences.

Maureen has _____ ounces of trail mix.

She puts an equal number in each of

_____ bags.

Each bag has _____ ounces.

Maureen has _____ ounces of trail mix left over.

e. Fill in the bubble completely to show your answer.

9. Mr. Field wants to give each of his 72 campers a certificate for completing an obstacle course. If there are 16 certificates in one package, how many packages will Mr. Field need?

(A) 4 (C) 16
(B) 5 (D) 17

10. James has 884 feet of rope. There are 12 teams of hikers. If James gives an equal amount of rope to each team, how much rope will each team receive?

(A) 12 feet (C) $73\frac{2}{3}$ feet
(B) 73 feet (D) 74 feet

Name _____

Adjust Quotients

Essential Question How can you adjust the quotient if your estimate
is too high or too low?

CONNECT When you estimate to decide where to place the first digit,
you can also try using the first digit of your estimate to find the first
digit of your quotient. Sometimes an estimate is too low or too high.

Divide. 3,382 ÷ 48

Estimate. 3,000 ÷ 50 = 60

Try 6 tens.

If an estimate is too low, the difference will be greater
than the divisor.

$$
\begin{array}{r}
6 \\
48\overline{)3,382} \\
-2\,88 \\
\hline
50
\end{array}
$$

Since the estimate is too
low, adjust by increasing the
number in the quotient.

Divide. 453 ÷ 65

Estimate. 490 ÷ 70 = 7

Try 7 ones.

If an estimate is too high, the product with the first
digit will be too large and cannot be subtracted.

$$
\begin{array}{r}
7 \\
65\overline{)453} \\
-455 \\
\hline
\end{array}
$$

Since the estimate is too
high, adjust by decreasing the
number in the quotient.

 UNLOCK the Problem REAL WORLD

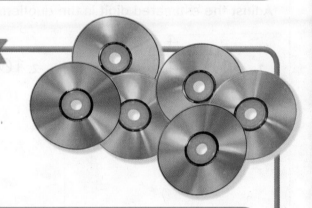

A new music group makes 6,127 copies of its first CD. The
group sells 75 copies of the CD at each of its shows. How
many shows does it take the group to sell all of the CDs?

Divide. 6,127 ÷ 75   **Estimate.** 6,300 ÷ 70 = 90

STEP 1 Use the estimate, 90. Try 9 tens.

• Is the estimate too high, too low, or correct?

_____

• Adjust the number in the quotient if needed.

_____

$$
75\overline{)6,127}
$$

STEP 2 Estimate the next digit in the quotient.
Divide the ones.
Estimate: 140 ÷ 70 = 2. Try 2 ones.

• Is the estimate too high, too low, or correct?

_____

• Adjust the number in the quotient if needed.

So, it takes the group _____ shows to sell all of the CDs.

Try This! When the difference is equal to or greater than the divisor, the estimate is too low.

Divide. 336 ÷ 48   **Estimate.** 300 ÷ 50 = 6

Use the estimate.	Adjust the estimated digit in the quotient if needed. Then divide.
Try 6 ones.	Try _____.
$\dfrac{6}{48\overline{)336}}$	
Since _____, the estimate is _____.	
336 ÷ 48 = _____	

Math Talk MATHEMATICAL PRACTICES

Explain why using the closest estimate could be useful in solving a division problem.

Share and Show MATH BOARD

Adjust the estimated digit in the quotient, if needed. Then divide.

1. $\dfrac{4}{41\overline{)1,546}}$

2. $\dfrac{2}{16\overline{)416}}$

3. $\dfrac{9}{34\overline{)2,831}}$

Divide.

4. $19\overline{)915}$

5. $28\overline{)1,825}$

6. $45\overline{)3,518}$

Math Talk MATHEMATICAL PRACTICES

Explain how you know whether an estimated quotient is too low or too high.

92

Name _____

On Your Own..

Adjust the estimated digit in the quotient, if needed. Then divide.

7. $\overset{2}{26\overline{)541}}$

8. $\overset{1}{43\overline{)688}}$

9. $\overset{6}{67\overline{)4,873}}$

Divide.

10. $15\overline{)975}$

11. $37\overline{)264}$

12. $22\overline{)6,837}$

Practice: Copy and Solve   Divide.

13. $452 \div 31$ 14. $592 \div 74$ 15. $785 \div 14$

16. $601 \div 66$ 17. $1,067 \div 97$ 18. $2,693 \div 56$

19. $1,488 \div 78$ 20. $2,230 \div 42$ 21. $4,295 \div 66$

 Algebra Write the unknown number for each �.

22. ▨ $\div 33 = 11$ 23. $1,092 \div 52 =$ ▨   24. $429 \div$ ▨ $= 33$

▨ = _____   ▨ = _____   ▨ = _____

🔑 UNLOCK the Problem · REAL WORLD

25. A banquet hall serves 2,394 pounds of turkey during a 3-week period. If the same amount is served each day, how many pounds of turkey does the banquet hall serve each day?

- **(A)** 50,274 pounds
- **(C)** 342 pounds
- **(B)** 798 pounds
- **(D)** 114 pounds

a. What do you need to find? _____

_____

b. What information are you given? _____

_____

c. What other information will you use?

_____

_____

_____

e. Divide to solve the problem.

d. Find how many days there are in 3 weeks.

There are _____ days in 3 weeks.

f. Fill in the bubble for the correct answer choice.

---

26. Marcos mixes 624 ounces of lemonade. He wants to fill the 52 cups he has with equal amounts of lemonade. How much lemonade should he put in each cup?

- **(A)** 8 ounces
- **(B)** 12 ounces
- **(C)** 18 ounces
- **(D)** 20 ounces

27. The Box of Sox company packs 18 pairs of socks in a box. How many boxes will the company need to pack 810 pairs of socks?

- **(A)** 40
- **(B)** 45
- **(C)** 55
- **(D)** 56

FOR MORE PRACTICE:
Standards Practice Book, pp. P45–P46

Name _____

Problem Solving • Division

Essential Question How can the strategy *draw a diagram* help you solve a division problem?

🔑 UNLOCK the Problem REAL WORLD

Sean and his family chartered a fishing boat for the day. Sean caught a blue marlin and an amberjack. The weight of the blue marlin was 12 times as great as the weight of the amberjack. The combined weight of both fish was 273 pounds. How much did each fish weigh?

Read the Problem

What do I need to find?	**What information do I need to use?**	**How will I use the information?**
I need to find _____ _____.	I need to know that Sean caught a total of _____ pounds of fish and the weight of the blue marlin was _____ times as great as the weight of the amberjack.	I can use the strategy _____ and then divide. I can draw and use a bar model to write the division problem that helps me find the weight of each fish.

Solve the Problem

I will draw one box to show the weight of the amberjack. Then I will draw a bar of 12 boxes of the same size to show the weight of the blue marlin. I can divide the total weight of the two fish by the total number of boxes.

amberjack [ ]

blue marlin [ ][ ][ ][ ][ ][ ][ ][ ][ ][ ][ ][ ] } 273 pounds

$$\begin{array}{r} 2 \\ 13\overline{)273} \\ -26 \\ \hline \\ - \\ \hline \end{array}$$

Write the quotient in each box. Multiply it by 12 to find the weight of the blue marlin.

So, the amberjack weighed _____ pounds and the

blue marlin weighed _____ pounds.

🔑 Try Another Problem

Jason, Murray, and Dana went fishing. Dana caught a red snapper. Jason caught a tuna with a weight 3 times as great as the weight of the red snapper. Murray caught a sailfish with a weight 12 times as great as the weight of the red snapper. If the combined weight of the three fish was 208 pounds, how much did the tuna weigh?

Read the Problem

What do I need to find?	What information do I need to use?	How will I use the information?

Solve the Problem

So, the tuna weighed _____ pounds.

- How can you check if your answer is correct? _____

_____

Math Talk Explain how you could use another strategy to solve this problem.

Name _____

Share and Show MATH BOARD

Choose a STRATEGY

Act It Out
Draw a Diagram
Make a Table
Solve a Simpler Problem
Work Backward
Guess, Check, and Revise

1. Paula caught a tarpon with a weight that was 10 times as great as the weight of a permit fish she caught. The total weight of the two fish was 132 pounds. How much did each fish weigh?

   First, draw one box to represent the weight of the permit fish and ten boxes to represent the weight of the tarpon.

   Next, divide the total weight of the two fish by the total number of boxes you drew. Place the quotient in each box.

   Last, find the weight of each fish.

   The permit fish weighed _____ pounds.

   The tarpon weighed _____ pounds.

SHOW YOUR WORK

2. **What if** the weight of the tarpon was 11 times the weight of the permit fish, and the total weight of the two fish was 132 pounds? How much would each fish weigh?

   permit fish: _____ pounds

   tarpon: _____ pounds

3. Jon caught four fish that weighed a total of 252 pounds. The kingfish weighed twice as much as the amberjack and the white marlin weighed twice as much as the kingfish. The weight of the tarpon was 5 times the weight of the amberjack. How much did each fish weigh?

   amberjack: _____ pounds

   kingfish: _____ pounds

   marlin: _____ pounds

   tarpon: _____ pounds

On Your Own....................

Use the table to solve 4–7.

4. Kevin is starting a saltwater aquarium with 36 fish. He wants to start with 11 times as many damselfish as clown fish. How many of each fish will Kevin buy? How much will he pay for the fish?

_____

5. Kevin used a store coupon to buy a 40-gallon tank, an aquarium light, and a filtration system. He paid a total of $240. How much money did Kevin save by using the coupon?

_____

6. **H.O.T.** Kevin bought 3 bags of gravel to cover the bottom of his fish tank. He has 8 pounds of gravel left over. How much gravel did Kevin use to cover the bottom of the tank?

_____

7. **Write Math** ▸ **Pose a Problem** Look back at Problem 6. Write a similar problem by changing the number of bags of gravel and the amount of gravel left.

_____

_____

_____

_____

_____

Kevin's Supply List for a Saltwater Aquarium	
40-gal tank	$170
Aquarium light	$30
Filtration system	$65
Thermometer	$2
15-lb bag of gravel	$13
Large rocks	$3 per lb
Clown fish	$20 each
Damselfish	$7 each

8. **Test Prep** Captain James offers a deep-sea fishing tour. He charges $2,940 for a 14-hour trip. How much does each hour of the tour cost?

Ⓐ $138   Ⓒ $210

Ⓑ $201   Ⓓ $294

FOR MORE PRACTICE:
Standards Practice Book, pp. P47–P48

Name _____

Chapter Review/Test

▶ Vocabulary

Choose the best term from the box.

1. You can use _____ to estimate quotients because they are easy to compute with mentally. **(p. 79)**

2. To decide where to place the first digit in the quotient, you can estimate or use _____. **(p. 61)**

▶ Concepts and Skills

Use compatible numbers to estimate the quotient.

3. $522 \div 6$

4. $1,285 \div 32$

5. $6,285 \div 89$

_____

_____

_____

Divide. Check your answer.

6. $2\overline{)554}$

7. $8\overline{)680}$

8. $5\overline{)462}$

9. $522 \div 18$

10. $529 \div 37$

11. $987 \div 15$

12. $1,248 \div 24$

13. $5,210 \div 17$

14. $8,808 \div 42$

© Houghton Mifflin Harcourt Publishing Company

GO Online Assessment Options **Chapter Test**

Fill in the bubble completely to show your answer.

15. Samira bought 156 ounces of trail mix. She wants to divide the amount equally into 24 portions. How many ounces of trail mix will be in each portion?

Ⓐ 6 ounces

Ⓑ $6\frac{1}{2}$ ounces

Ⓒ 7 ounces

Ⓓ 12 ounces

16. A school band performed 6 concerts. Every seat for each performance was sold. If a total of 1,248 seats were sold for all 6 concerts, how many seats were sold for each performance?

Ⓐ 28

Ⓑ 200

Ⓒ 206

Ⓓ 208

17. Dylan's dog weighs 12 times as much as his pet rabbit. The dog and rabbit weigh 104 pounds altogether. How much does Dylan's dog weigh?

Ⓐ 104 pounds

Ⓑ 96 pounds

Ⓒ 88 pounds

Ⓓ 8 pounds

18. Jamie is sewing 14 identical costumes for the school play. She needs 210 buttons to complete all of the costumes. How many buttons will she sew onto each costume?

Ⓐ 15

Ⓑ 14

Ⓒ 11

Ⓓ 9

Name __A-J__

Fill in the bubble completely to show your answer.

19. A book publishing company is shipping an order of 300 books.
The books are packaged in boxes that each can hold 24 books.
How many boxes are needed to ship the order of books?

Ⓐ 10

Ⓑ 11

Ⓒ 12

Ⓓ 13

$$300$$
$$\times\ 24$$

20. Richard is planning a trip to Italy. He thinks he will need $2,750 for
the trip. If the trip is 40 weeks away, which is the best estimate of
how much money Richard needs to save each week?

Ⓐ $60

Ⓑ $70

Ⓒ $600

Ⓓ $700

21. A school club raises $506 to spend on a field trip. There are
23 people going on the trip. How much money did the club raise
for each person going?

Ⓐ $27

Ⓑ $22

Ⓒ $18

Ⓓ $12

22. A local orange grower processes 2,330 oranges from his grove this
year. The oranges are packaged in crates that each hold 96 oranges.
All but one crate is full. How many oranges are in this last crate?

Ⓐ 24

Ⓑ 25

Ⓒ 26

Ⓓ 27

23. On Monday, 1,900 bottles of perfume are delivered to a warehouse.
The bottles are packed in boxes. Each box can hold 32 bottles.
How many boxes were delivered? **Explain** how you found your
answer.

_____

_____

_____

► Performance Task

24. Quincy needs 322 yards of ribbon to decorate quilts for a craft fair.
The ribbon comes in rolls of 15 yards.

Ⓐ How many rolls of ribbon should Quincy buy? **Explain** your answer.

_____

_____

Ⓑ Alice needs twice as many yards of ribbon as Quincy. How many
rolls of ribbon does Alice need? **Explain** your answer.

_____

_____

_____

Ⓒ Elena needs yellow, red, and blue ribbon. She needs 285 yards of
the three colors combined. Suggest numbers of rolls of each color
that would give her enough ribbon. (HINT: Break apart the 285 yards
into any combination of 3 groups that total this amount.)

_____

_____

_____

_____

© Houghton Mifflin Harcourt Publishing Company

Add and Subtract Decimals

Show What You Know

Check your understanding of important skills.

Name _____

▶ **2-Digit Addition and Subtraction** Find the sum or difference.

1.

Hundreds	Tens	Ones
☐	☐	
	5	8
+	7	6

2.

Hundreds	Tens	Ones
	☐	☐
	8	2
−	4	7

▶ **Decimals Greater Than One** Write the word form and the expanded form for each.

3. 3.4

_____

_____

4. 2.51

_____

_____

▶ **Relate Fractions and Decimals** Write as a decimal or a fraction.

5. 0.8 _____

6. $\frac{5}{100}$ _____

7. 0.46 _____

8. $\frac{6}{10}$ _____

9. 0.90 _____

10. $\frac{35}{100}$ _____

MATH DETECTIVE

WITH

CARMEN SANDIEGO™

Jason has 4 tiles. Each tile has a number printed on it. The numbers are 2, 3, 6, and 8. A decimal number is formed using the tiles and the clues. Be a Math Detective and find the number.

Clues

- The digit in the tens place is the greatest number.
- The digit in the tenths place is less than the digit in the hundredths place.
- The digit in the ones place is greater than the digit in the hundredths place.

Vocabulary Builder

▶ **Visualize It** •

Use the ✓ words to complete the tree map.

Review Words

✓ benchmark

✓ hundredth

✓ place value

✓ round

✓ tenth

Preview Words

sequence

term

✓ thousandth

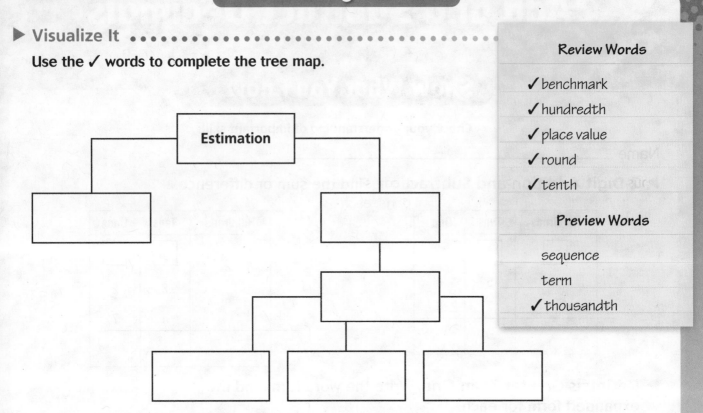

▶ **Understand Vocabulary** •

Read the description. Which word do you think is described?

1. One of one hundred equal parts _____

2. The value of each digit in a number based on the location of the digit

   _____

3. To replace a number with one that is simpler and is approximately

   the same size as the original number _____

4. An ordered set of numbers _____

5. One of ten equal parts _____

6. A familiar number used as a point of reference _____

7. One of one thousand equal parts _____

8. Each of the numbers in a sequence _____

GO Online • eStudent Edition • Multimedia eGlossary

Name _____

Thousandths

Essential Question How can you describe the relationship between two decimal place-value positions?

Investigate

Materials ■ color pencils ■ straightedge

Thousandths are smaller parts than hundredths. If one hundredth is divided into ten equal parts, each part is one **thousandth**.

Use the model at the right to show tenths, hundredths, and thousandths.

A. Divide the larger square into 10 equal columns or rectangles. Shade one rectangle. What part of the whole is the shaded rectangle? Write that part as a decimal and a fraction.

_____

B. Divide each rectangle into 10 equal squares. Use a second color to shade in one of the squares. What part of the whole is the shaded square? Write that part as a decimal and a fraction.

_____

C. Divide the enlarged hundredths square into 10 equal columns or rectangles. If each hundredths square is divided into ten equal rectangles, how many parts will the model have?

_____

Use a third color to shade one rectangle of the enlarged hundredths square. What part of the whole is the shaded rectangle? Write that part as a decimal and a fraction.

_____

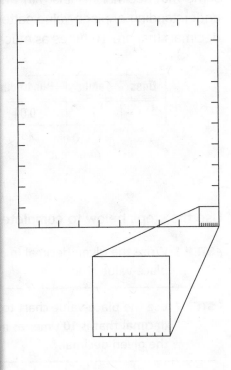

Math Talk MATHEMATICAL PRACTICES
There are 10 times as many hundredths as there are tenths. **Explain** how the model shows this.

© Houghton Mifflin Harcourt Publishing Company

Draw Conclusions

1. **Explain** what each shaded part of your model in the Investigate section shows. What fraction can you write that relates each shaded part to the next greater shaded part? _____

_____

2. **Identify** and describe a part of your model that shows one thousandth. **Explain** how you know.

_____

_____

_____

Make Connections

The relationship of a digit in different place-value positions is the same with decimals as it is with whole numbers. You can use your understanding of place-value patterns and a place-value chart to write decimals that are 10 times as much as or $\frac{1}{10}$ of any given decimal.

Ones •	Tenths	Hundredths	Thousandths
	?	0.04	?

10 times as much $\frac{1}{10}$ of

_____ is 10 times as much as 0.04.

_____ is $\frac{1}{10}$ of 0.04.

Use the steps below to complete the table.

STEP 1 Write the given decimal in a place-value chart.

STEP 2 Use the place-value chart to write a decimal that is 10 times as much as the given decimal.

STEP 3 Use the place-value chart to write a decimal that is $\frac{1}{10}$ of the given decimal.

Decimal	10 times as much as	$\frac{1}{10}$ of
0.03		
0.1		
0.07		

Math Talk   MATHEMATICAL PRACTICES

Describe the pattern you see when you move one decimal place value to the right and one decimal place value to the left.

Share and Show

Write the decimal shown by the shaded parts of each model.

1.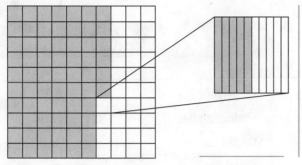

2.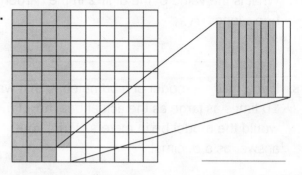

3.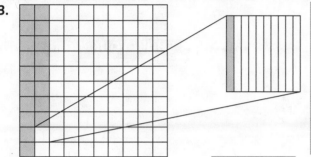

4.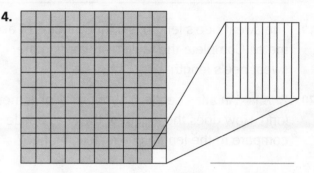

Complete the sentence.

5. 0.6 is 10 times as much as _____.

6. 0.007 is $\frac{1}{10}$ of _____.

7. 0.008 is $\frac{1}{10}$ of _____.

8. 0.5 is 10 times as much as _____.

Use place-value patterns to complete the table.

	Decimal	10 times as much as	$\frac{1}{10}$ of
9.	0.2		
10.	0.07		
11.	0.05		
12.	0.4		

	Decimal	10 times as much as	$\frac{1}{10}$ of
13.	0.06		
14.	0.9		
15.	0.3		
16.	0.08		

Problem Solving REAL WORLD

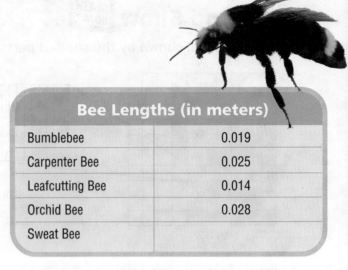

Use the table for 17–20.

17. What is the value of the digit 2 in the carpenter bee's length?

    _____

18. If you made a model of a bumblebee that was 10 times as large as the actual bee, how long would the model be in meters? Write your answer as a decimal.

    _____

Bee Lengths (in meters)	
Bumblebee	0.019
Carpenter Bee	0.025
Leafcutting Bee	0.014
Orchid Bee	0.028
Sweat Bee	

19. The sweat bee's length is 6 thousandths of a meter. Complete the table by recording the sweat bee's length.

20. **H.O.T.** An atlas beetle is about 0.14 of a meter long. How does the length of the atlas beetle compare to the length of a leafcutting bee?

    _____

    _____

    _____

················· SHOW YOUR WORK ·················

21. **Write Math** ▸ **Explain** how you can use place value to describe how 0.05 and 0.005 compare.

    _____

    _____

    _____

    _____

22. **Test Prep** What is the relationship between 1.0 and 0.1?

    (A) 0.1 is 10 times as much as 1.0

    (B) 1.0 is $\frac{1}{10}$ of 0.1

    (C) 0.1 is $\frac{1}{10}$ of 1.0

    (D) 1.0 is equal to 0.1

FOR MORE PRACTICE:
Standards Practice Book, pp. P53–P54

Name _____

Place Value of Decimals

Essential Question How do you read, write, and represent decimals through thousandths?

🔑 UNLOCK the Problem REAL WORLD

The Brooklyn Battery Tunnel in New York City is 1.726 miles long. It is the longest underwater tunnel for vehicles in the United States. To understand this distance, you need to understand the place value of each digit in 1.726.

You can use a place-value chart to understand decimals. Whole numbers are to the left of the decimal point. Decimals are to the right of the decimal point. The thousandths place is to the right of the hundredths place.

▲ The Brooklyn Battery Tunnel passes under the East River.

Tens	Ones	Tenths	Hundredths	Thousandths	
	1 ●	7	2	6	
1	1×1	$7 \times \frac{1}{10}$	$2 \times \frac{1}{100}$	$6 \times \frac{1}{1,000}$	} Value
	1.0	0.7	0.02	0.006	

The place value of the digit 6 in 1.726 is thousandths. The value of 6 in 1.726 is $6 \times \frac{1}{1,000}$, or 0.006.

Standard Form: 1.726
Word Form: one and seven hundred twenty-six thousandths

Expanded Form: $1 \times 1 + 7 \times \left(\frac{1}{10}\right) + 2 \times \left(\frac{1}{100}\right) + 6 \times \left(\frac{1}{1,000}\right)$

Math Talk MATHEMATICAL PRACTICES
Explain how the value of the last digit in a decimal can help you read a decimal.

Try This! Use place value to read and write decimals.

A **Standard Form:** 2.35
  Word Form: two and _____

  Expanded Form: $2 \times 1 +$ _____

B **Standard Form:** _____
  Word Form: three and six hundred fourteen thousandths

  Expanded Form: _____ $+ 6 \times \left(\frac{1}{10}\right) +$ _____ $+$ _____

Example Use a place-value chart.

The silk spun by a common garden spider is about 0.003 millimeter thick. A commonly used sewing thread is about 0.3 millimeter thick. How does the thickness of the spider silk and the thread compare?

STEP 1 Write the numbers in a place-value chart.

Ones	Tenths	Hundredths	Thousandths
•			
•			

STEP 2

Count the number of decimal place-value positions to the digit 3 in 0.3 and 0.003.

0.3 has _____ fewer decimal places than 0.003

2 fewer decimal places: $10 \times 10 =$ _____

0.3 is _____ times as much as 0.003

0.003 is _____ of 0.3

So, the thread is _____ times as thick as the garden spider's silk. The thickness of the garden spider's silk is _____ that of the thread.

You can use place-value patterns to rename a decimal.

Try This! Use place-value patterns.

Rename 0.3 using other place values.

0.300	3 tenths	$3 \times \frac{1}{10}$
0.300	_____ hundredths	_____ $\times \frac{1}{100}$
0.300	_____	_____

Name _____

Share and Show

1. Complete the place-value chart to find the value of each digit.

Ones	Tenths	Hundredths	Thousandths	
3 •	5	2	4	
3 × 1		2 × $\frac{1}{100}$		} Value
	0.5			

Write the value of the underlined digit.

2. 0.5<u>4</u>3

_____

3. 6.<u>2</u>34

_____

4. 3.95<u>4</u>

_____

Write the number in two other forms.

5. 0.253

_____

_____

_____

6. 7.632

_____

_____

_____

On Your Own

Write the value of the underlined digit.

7. 0.4<u>9</u>6

_____

8. 2.<u>7</u>26

_____

9. 1.06<u>6</u>

_____

10. 6.<u>3</u>99

_____

11. 0.00<u>2</u>

_____

12. 14.37<u>1</u>

_____

Write the number in two other forms.

13. 0.489

_____

_____

_____

14. 5.916

_____

_____

_____

Problem Solving REAL WORLD

Use the table for 15–17.

Average Annual Rainfall (in meters)	
California	0.564
New Mexico	0.372
New York	1.041
Wisconsin	0.820
Maine	

15. What is the value of the digit 7 in New Mexico's average annual rainfall?

_____

16. The average annual rainfall in Maine is one and seventy-four thousandths of a meter per year. Complete the table by writing that amount in standard form.

17. Which of the states has an average annual rainfall with the least number in the thousandths place?

_____

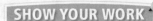

SHOW YOUR WORK

18. H.O.T. **What's the Error?** Damian wrote the number four and twenty-three thousandths as 4.23. **Describe** and correct his error.

_____

_____

_____

19. Write Math ► **Explain** how you know that the digit 6 in the numbers 3.675 and 3.756 does not have the same value.

_____

_____

_____

_____

20. Test Prep In 24.736, which digit is in the thousandths place?

  Ⓐ 3   Ⓒ 6

  Ⓑ 4   Ⓓ 7

FOR MORE PRACTICE:
Standards Practice Book, pp. P55–P56

Compare and Order Decimals

Essential Question How can you use place value to compare and order decimals?

🔑 UNLOCK the Problem REAL WORLD

The table lists some of the mountains in the United States that are over two miles high. How does the height of Cloud Mountain in New York compare to the height of Boundary Mountain in Nevada?

Mountain Heights	
Mountain and State	**Height (in miles)**
Boundary, Nevada	2.488
Cloud, New York	2.495
Grand Teton, Wyoming	2.607
Wheeler, New Mexico	2.493

▲ The Tetons are located in Grand Teton National Park.

🔑 One Way Use place value.

Line up the decimal points. Start at the left. Compare the digits in each place-value position until the digits are different.

STEP 1 Compare the ones.

2.495
↓   2 = 2
2.488

STEP 2 Compare the tenths.

2.495
↓   4 ◯ 4
2.488

STEP 3 Compare the hundredths.

2.495
↓   9 ◯ 8
2.488

Since 9 ◯ 8, then 2.495 ◯ 2.488, and 2.488 ◯ 2.495.

So, the height of Cloud Mountain is _____ the height of Boundary Mountain.

🔑 Another Way Use a place-value chart to compare.

Compare the height of Cloud Mountain to Wheeler Mountain.

Ones	•	Tenths	Hundredths	Thousandths
2	•	4	9	5
2	•	4	9	3

Math Talk MATHEMATICAL PRACTICES
Explain why it is important to line up the decimal points when comparing decimals.

2 = 2   4 = _____   9 = _____   5 > _____

Since 5 ◯ 3, then 2.495 ◯ 2.493, and 2.493 ◯ 2.495.

So, the height of Cloud Mountain is _____ the height of Wheeler Mountain.

Order Decimals You can use place value to order decimal numbers.

🔑 Example

Mount Whitney in California is 2.745 miles high, Mount Rainier in Washington is 2.729 miles high, and Mount Harvard in Colorado is 2.731 miles high. Order the heights of these mountains from least to greatest. Which mountain has the least height? Which mountain has the greatest height?

STEP 1

Line up the decimal points. There are the same number of ones. Circle the tenths and compare.

2.745   **Whitney**

2.729   **Rainier**

2.731   **Harvard**

There are the same number of tenths.

So, _____ has the least height and

_____ has the greatest height.

STEP 2

Underline the hundredths and compare. Order from least to greatest.

2.745   **Whitney**

2.729   **Rainier**

2.731   **Harvard**

Since ◯ < ◯ < ◯, the heights in order from least to

greatest are _____, _____, _____.

Math Talk MATHEMATICAL PRACTICES

Explain why you do not have to compare the digits in the thousandths place to order the heights of the 3 mountains.

Try This! Use a place-value chart.

What is the order of 1.383, 1.321, 1.456, and 1.32 from greatest to least?

- Write each number in the place-value chart. Compare the digits, beginning with the greatest place value.

- Compare the ones. The ones are the same.

- Compare the tenths. 4 > 3.

The greatest number is _____.
Circle the greatest number in the place-value chart.

- Compare the remaining hundredths. 8 > 2.

The next greatest number is _____.
Draw a rectangle around the number.

- Compare the remaining thousandths. 1 > 0.

Ones •	Tenths	Hundredths	Thousandths
1 •	3	8	3
1 •			
1 •			
1 •			

So, the order of the numbers from greatest to least is: _____.

Name _____

Share and Show . . . .

1. Use the place-value chart to compare the two numbers. What is the greatest place-value position where the digits differ?

Ones	Tenths	Hundredths	Thousandths
3 •	4	7	2
3 •	4	4	5

_____

Compare. Write <, >, or =.

2. 4.563 ◯ 4.536

3. 5.640 ◯ 5.64

✓ **4.** 8.673 ◯ 8.637

Name the greatest place-value position where the digits differ.
Name the greater number.

5. 3.579; 3.564

6. 9.572; 9.637

✓ **7.** 4.159; 4.152

_____ _____ _____

_____ _____ _____

Order from least to greatest.

8. 4.08; 4.3; 4.803; 4.038

9. 1.703; 1.037; 1.37; 1.073

_____ _____

On Your Own . . . .

Compare. Write <, >, or =.

10. 8.72 ◯ 8.720

11. 5.4 ◯ 5.243

12. 1.036 ◯ 1.306

13. 2.573 ◯ 2.753

14. 9.300 ◯ 9.3

15. 6.76 ◯ 6.759

Order from greatest to least.

16. 2.007; 2.714; 2.09; 2.97

17. 0.386; 0.3; 0.683; 0.836

_____ _____

18. 5.249; 5.43; 5.340; 5.209

19. 0.678; 1.678; 0.587; 0.687

_____ _____

 Algebra Find the unknown digit to make each statement true.

20. 3.59 > 3.5 ▨ 1 > 3.572

21. 6.837 > 6.83 ▨ > 6.835

22. 2.45 < 2. ▨ 6 < 2.461

Problem Solving REAL WORLD

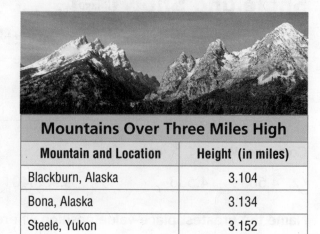

Use the Table for 23–26.

23. In comparing the height of the mountains, which is the greatest place value where the digits differ?

   _____

24. How does the height of Steele Mountain compare to the height of Blackburn Mountain? Compare the heights using words.

   _____

   _____

Mountains Over Three Miles High

Mountain and Location	Height (in miles)
Blackburn, Alaska	3.104
Bona, Alaska	3.134
Steele, Yukon	3.152

25. **Write Math** ▶ **Explain** how to order the height of the mountains from greatest to least.

   _____

   _____

   _____

   _____

26. **H.O.T.** **What if** the height of Blackburn Mountain were 0.05 mile greater. Would it then be the mountain with the greatest height? **Explain**.

   _____

   _____

   _____

27. **Test Prep** Mount Logan in the Yukon is 3.702 miles high. Mount McKinley in Alaska is 3.848 miles high and Pico de Orizaba in Mexico is 3.571 miles high. Order these mountains by height from greatest to least.

   (A) Logan, McKinley, Pico de Orizaba

   (B) McKinley, Logan, Pico de Orizaba

   (C) Pico de Orizaba, Logan, McKinley

   (D) Logan, Pico de Orizaba, McKinley

FOR MORE PRACTICE:
Standards Practice Book, pp. P57–P58

Name _____

Round Decimals

Essential Question How can you use place value to round decimals to a given place?

🔑 UNLOCK the Problem REAL WORLD

The Gold Frog of South America is one of the smallest frogs in the world. It is 0.386 of an inch long. What is this length rounded to the nearest hundredth of an inch?

- Underline the length of the Gold Frog.
- Is the frog's length about the same as the length or the width of a large paper clip?

_____

🔒 One Way Use a place-value chart.

- Write the number in a place-value chart and circle the digit in the place value to which you want to round.

- In the place-value chart, underline the digit to the right of the place to which you are rounding.

- If the digit to the right is less than 5, the digit in the place value to which you are rounding stays the same. If the digit to the right is 5 or greater, the digit in the rounding place increases by 1.

- Drop the digits after the place to which you are rounding.

Ones	Tenths	Hundredths	Thousandths
0 •	3	8	6

Think: Does the digit in the rounding place stay the same or increase by 1?

So, to the nearest hundredth of an inch, a Gold Frog is

about _____ of an inch long.

🔒 Another Way Use place value.

The Little Grass Frog is the smallest frog in North America. It is 0.437 of an inch long.

A What is the length of the frog to the nearest hundredth of an inch?

0.437   7 > 5
↓
0.44

So, to the nearest hundredth of an inch, the frog

is about _____ of an inch long.

B What is the length of the frog to the nearest tenth of an inch?

0.437   3 < 5
↓
0.4

So, to the nearest tenth of an inch, the frog is

about _____ of an inch long.

🔒 Example

The Goliath Frog is the largest frog in the world. It is found in the country of Cameroon in West Africa. The Goliath Frog can grow to be 11.815 inches long. How long is the Goliath Frog to the nearest inch?

STEP 1 Write 11.815 in the place-value chart.

Tens	Ones	Tenths	Hundredths	Thousandths
		•		

STEP 2 Find the place to which you want to round. Circle the digit.

STEP 3 Underline the digit to the right of the place value to which you are rounding. Then round.

> **Think:** Does the digit in the rounding place stay the same or increase by 1?

So, to the nearest inch, the Goliath Frog is about _____ inches long.

- **Explain** why any number less than 12.5 and greater than or equal to 11.5 would round to 12 when rounded to the nearest whole number.

_____

_____

_____

Try This! Round. 14.603

A To the nearest hundredth:

Tens	Ones	Tenths	Hundredths	Thousandths
		•		

Circle and underline the digits as you did above to help you round to the nearest hundredth.

So, 14.603 rounded to the nearest hundredth is _____.

B To the nearest whole number:

Tens	Ones	Tenths	Hundredths	Thousandths
		•		

Circle and underline the digits as you did above to help you round to the nearest whole number.

So, 14.603 rounded to the nearest whole number is _____.

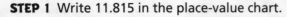

Name _____

Share and Show

Write the place value of the underlined digit. Round each
number to the place of the underlined digit.

1. 0.6<u>7</u>3

_____

_____

✓ 2. 4.<u>2</u>82

_____

_____

3. 12.<u>9</u>17

_____

_____

Name the place value to which each number was rounded.

4. 0.982 to 0.98

_____

5. 3.695 to 4

_____

✓ 6. 7.486 to 7.5

_____

On Your Own

Write the place value of the underlined digit. Round each
number to the place of the underlined digit.

7. 0.<u>5</u>92

_____

_____

8. <u>6</u>.518

_____

_____

9. 0.8<u>0</u>9

_____

_____

10. 3.<u>3</u>34

_____

_____

11. 12.<u>0</u>74

_____

_____

12. 4.4<u>9</u>4

_____

_____

Name the place value to which each number was rounded.

13. 0.328 to 0.33

_____

14. 2.607 to 2.61

_____

15. 12.583 to 13

_____

Round **16.748** to the place named.

16. tenths _____

17. hundredths _____

18. ones _____

19. **Write Math** ▶ **Explain** what happens when you round 4.999 to

the nearest tenth. _____

_____

Problem Solving REAL WORLD

Use the table for 20–22.

20. The speeds of two insects when rounded to the nearest whole number are the same. Which two insects are they?

_____

21. What is the speed of the housefly rounded to the nearest hundredth?

_____

22. **H.O.T.** **What's the Error?** Mark said that the speed of a dragonfly rounded to the nearest tenth was 6.9 meters per second. Is he correct? If not, what is his error?

_____

_____

_____

Insect Speeds (meters per second)

Insect	Speed
Dragonfly	6.974
Horsefly	3.934
Bumblebee	2.861
Honeybee	2.548
Housefly	1.967

SHOW YOUR WORK

23. **H.O.T.** **Write Math** ▸ A rounded number for the speed of an insect is 5.67 meters per second. What are the fastest and slowest speeds to the thousandths that could round to 5.67? **Explain.**

_____

_____

_____

_____

_____

24. **Test Prep** To which place value is the number rounded?

6.706 to 6.71

(A) ones   (C) hundredths

(B) tenths   (D) thousandths

Name _____

Decimal Addition

Essential Question How can you use base-ten blocks to model decimal addition?

CONNECT You can use base-ten blocks to help you find decimal sums.

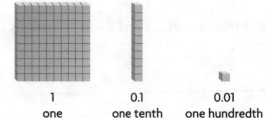

1	0.1	0.01
one	one tenth	one hundredth

Investigate

Materials ■ base-ten blocks

A. Use base-ten blocks to model the sum of 0.34 and 0.27.

B. Add the hundredths first by combining them.
- Do you need to regroup the hundredths? **Explain**.

_____

_____

_____

C. Add the tenths by combining them.
- Do you need to regroup the tenths? **Explain**.

_____

_____

_____

D. Record the sum. 0.34 + 0.27 = _____

Draw Conclusions . . . .

1. **What if** you combine the tenths first and then the hundredths?
Explain how you would regroup.

_____

_____

2. **H.O.T.** **Synthesize** If you add two decimals that are each greater than 0.5, will the sum be less than or greater than 1.0? **Explain**.

_____

_____

_____

Make Connections

You can use a quick picture to add decimals greater than 1.

STEP 1

Model the sum of 2.5 and 2.8 with a quick picture.

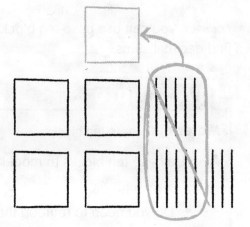

STEP 2

Add the tenths.

- Are there more than 9 tenths? _____
  If there are more than 9 tenths, regroup.

Add the ones.

STEP 3

Draw a quick picture of your answer. Then record.

2.5 + 2.8 = _____

Share and Show [MATH BOARD]

Complete the quick picture to find the sum.

1. 1.37 + 1.85 = _____

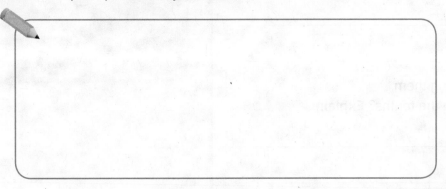

Math Talk MATHEMATICAL PRACTICES
Explain how you know where to write the decimal point in the sum.

Name _____

Add. Draw a quick picture.

2. $0.9 + 0.7 =$ _____

3. $0.65 + 0.73 =$ _____

4. $3.71 + 0.54 =$ _____

5. $1.05 + 0.78 =$ _____

6. $1.3 + 0.7 =$ _____

7. $2.72 + 0.51 =$ _____

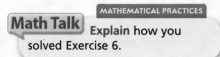

Math Talk MATHEMATICAL PRACTICES
Explain how you solved Exercise 6.

Problem Solving

H.O.T. Sense or Nonsense?

8. Robyn and Jim used quick pictures to model 1.85 + 2.73.

Robyn's Work	**Jim's Work**

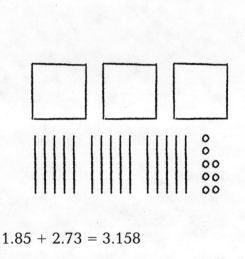

1.85 + 2.73 = 3.158

Does Robyn's work make sense?
Explain your reasoning.

_____

_____

_____

_____

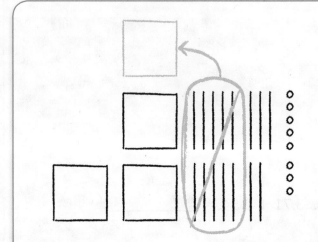

1.85 + 2.73 = 4.58

Does Jim's work make sense?
Explain your reasoning.

_____

_____

_____

_____

- **Explain** how you would help Robyn understand that regrouping is important when adding decimals.

_____

_____

_____

_____

Name _____

Decimal Subtraction

Essential Question How can you use base-ten blocks to model decimal subtraction?

CONNECT You can use base-ten blocks to help you find the difference between two decimals.

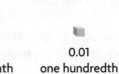

1	0.1	0.01
one	one tenth	one hundredth

Investigate

Materials ■ base-ten blocks

A. Use base-ten blocks to find $0.84 - 0.56$.
Model 0.84.

B. Subtract 0.56. Start by removing 6 hundredths.

- Do you need to regroup to subtract? **Explain**.

_____

_____

C. Subtract the tenths. Remove 5 tenths.

D. Record the difference. $0.84 - 0.56 =$ _____

Draw Conclusions . . . .

1. **What if** you remove the tenths first and then the hundredths?
   Explain how you would regroup.

_____

_____

2. **H.O.T.** **Synthesize** If two decimals are both less than 1.0, what do you know about the difference between them? **Explain**.

_____

_____

_____

Make Connections

You can use quick pictures to subtract decimals that need to be regrouped.

STEP 1

- Use a quick picture to model 2.82 − 1.47.

- Subtract the hundredths.

- Are there enough hundredths to remove? _____
  If there are not enough hundredths, regroup.

STEP 2

- Subtract the tenths.

- Are there enough tenths to remove? _____
  If there are not enough tenths, regroup.

- Subtract the ones.

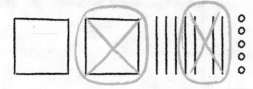

STEP 3

Draw a quick picture of your answer. Then record.

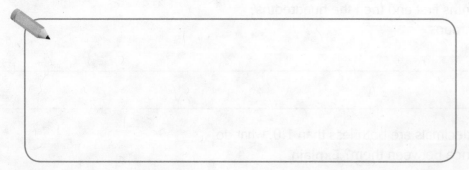

2.82 − 1.47 = _____

© Houghton Mifflin Harcourt Publishing Company

Math Talk MATHEMATICAL PRACTICES
Explain why you have to regroup in Step 1.

Name _____

Share and Show ....

Complete the quick picture to find the difference.

1. $0.62 - 0.18 =$ _____

Subtract. Draw a quick picture.

2. $3.41 - 1.74 =$ _____

3. $0.84 - 0.57 =$ _____

4. $0.93 - 0.38 =$ _____

5. $2.71 - 1.34 =$ _____

6. $4.05 - 1.61 =$ _____

7. $1.37 - 0.52 =$ _____

Math Talk MATHEMATICAL PRACTICES
Explain how you can use a quick picture to find $0.81 - 0.46$.

Problem Solving

Pose a Problem

8. Antonio left his MathBoard on his desk during lunch. The quick picture below shows the problem he was working on when he left.

Write a problem that can be solved using the quick picture above.

Pose a problem.   **Solve your problem.**

_____
_____
_____
_____
_____
_____
_____
_____

• **Describe** how you can change the problem by changing the quick picture.

_____

_____

_____

_____

Name _____

▶ **Concepts and Skills**

1. **Explain** how you can use base-ten blocks to find 1.54 + 2.37.

_____

_____

_____

Complete the sentence.

2. 0.04 is $\frac{1}{10}$ of _____.

3. 0.06 is 10 times as much as _____.

Write the value of the underlined digit.

4. 6.5<u>4</u>

5. 0.8<u>3</u>7

6. 8.70<u>2</u>

7. <u>9</u>.173

_____

_____

_____

_____

_____

_____

_____

_____

Compare. Write <, >, or =.

8. 6.52 ◯ 6.520

9. 3.589 ◯ 3.598

10. 8.463 ◯ 8.483

Write the place value of the underlined digit. Round each number to the place of the underlined digit.

11. 0.<u>7</u>24

12. <u>2</u>.576

13. 4.7<u>6</u>9

_____

_____

_____

_____

_____

_____

Draw a quick picture to find the sum or difference.

14. 2.46 + 0.78 = _____

15. 3.27 − 1.84 = _____

Fill in the bubble completely to show your answer.

16. Marco read that a honeybee can fly up to 2.548 meters per second. He rounded the number to 2.55. To which place value did Marco round the speed of a honeybee?

    (A) ones
    (C) hundredths
    (B) tenths
    (D) thousandths

17. What is the relationship between 0.04 and 0.004?

    (A) 0.04 is 10 times as much as 0.004

    (B) 0.04 is $\frac{1}{10}$ of 0.004

    (C) 0.004 is 10 times as much as 0.04

    (D) 0.04 is equal to 0.004

18. Jodi drew a quick picture to model the answer for 3.14 − 1.75. Which picture did she draw?

    (A)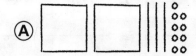
    (C)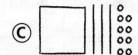

    (B)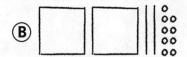
    (D)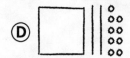

19. The average annual rainfall in California is 0.564 of a meter per year. What is the value of the digit 4 in that number?

    (A) 4×1
    (C) $4 \times \frac{1}{100}$
    (B) $4 \times \frac{1}{10}$
    (D) $4 \times \frac{1}{1,000}$

20. Jan ran 1.256 miles on Monday, 1.265 miles on Wednesday, and 1.268 miles on Friday. What were her distances from greatest to least?

    (A) 1.268 miles, 1.256 miles, 1.265 miles

    (B) 1.268 miles, 1.265 miles, 1.256 miles

    (C) 1.265 miles, 1.256 miles, 1.268 miles

    (D) 1.256 miles, 1.265 miles, 1.268 miles

130

Name _____

Estimate Decimal Sums and Differences

Essential Question How can you estimate decimal sums and differences?

 UNLOCK the Problem REAL WORLD

A singer is recording a CD. The lengths of the three songs are 3.4 minutes, 2.78 minutes, and 4.19 minutes. About how much recording time will be on the CD?

🔑 **Use rounding to estimate.**

Round to the nearest whole number. Then add.

$$
\begin{array}{r}
3.4 \\
2.78 \\
+4.19 \\
\hline
\end{array}
\qquad
\begin{array}{r}
3 \\
 \\
+ \\
\hline
\end{array}
$$

> **Remember**
>
> To round a number, determine the place to which you want to round.
> - If the digit to the right is less than 5, the digit in the rounding place stays the same.
> - If the digit to the right is 5 or greater, the digit in the rounding place increases by 1.

So, there will be about _____ minutes of recording time on the CD.

Try This! Use rounding to estimate.

A Round to the nearest whole dollar. Then subtract.

$$
\begin{array}{r}
\$27.95 \\
-\$11.72 \\
\hline
\end{array}
\qquad
\begin{array}{r}
 \\
- \\
\hline

\end{array}
$$

To the nearest dollar, $27.95 − $11.72 is about _____.

B Round to the nearest ten dollars. Then subtract.

$$
\begin{array}{r}
\$27.95 \\
-\$11.72 \\
\hline
\end{array}
\qquad
\begin{array}{r}
 \\
- \\
\hline

\end{array}
$$

To the nearest ten dollars, $27.95 − $11.72 is about _____.

- Do you want an overestimate or an underestimate when you estimate the total cost of items you want to buy? **Explain.**

_____

_____

_____

Use Benchmarks Benchmarks are familiar numbers used as points of reference. You can use the benchmarks 0, 0.25, 0.50, 0.75, and 1 to estimate decimal sums and differences.

🔑 Example 1 Use benchmarks to estimate. 0.18 + 0.43

Locate and graph a point on the number line for each decimal. Identify which benchmark each decimal is closer to.

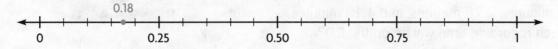

Think: 0.18 is between 0 and 0.25.

It is closer to _____.

Think: 0.43 is between _____ and

_____. It is closer to _____.

$$0.18 + 0.43$$
↓   ↓
_____ + _____ = _____

So, 0.18 + 0.43 is about _____.

🔑 Example 2 Use benchmarks to estimate. 0.76 − 0.22

Locate and graph a point on the number line for each decimal. Identify which benchmark each decimal is closer to.

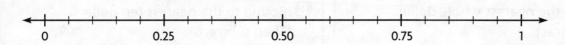

Think: 0.76 is between _____ and

_____. It is closer to _____.

Think: 0.22 is between 0 and 0.25. It is

closer to _____.

$$0.76 - 0.22$$
↓   ↓
_____ − _____ = _____

So, 0.76 − 0.22 is about _____.

© Houghton Mifflin Harcourt Publishing Company

MATHEMATICAL PRACTICES

Math Talk Use Example 2 to explain how using rounding or benchmarks to estimate a decimal difference can give you different answers.

Name _____

Share and Show

Use rounding to estimate.

1. 2.34
    1.9
    + 5.23

2. 10.39
    − 4.28

✓ 3. $19.75
    + $ 3.98

Use benchmarks to estimate.

4. 0.34
    0.1
    + 0.25

✓ 5. 10.39
    − 4.28

Math Talk MATHEMATICAL PRACTICES
Describe the difference between an estimate and an exact answer.

On Your Own

Use rounding to estimate.

6. 0.93
    + 0.18

7. 7.41
    − 3.88

8. 14.68
    − 9.93

Use benchmarks to estimate.

9. 12.41
    − 6.47

10. 8.12
    + 5.52

11. 9.75
    − 3.47

Practice: Copy and Solve Use rounding or benchmarks to estimate.

12. 12.83 + 16.24

13. $26.92 − $11.13

14. 9.41 + 3.82

 Estimate to compare. Write < or >.

15. 2.74 + 4.22 ◯ 3.13 + 1.87

16. 6.25 − 2.39 ◯ 9.79 − 3.84

_____ _____
estimate   estimate

_____ _____
estimate   estimate

Problem Solving REAL WORLD

Model • Reason • Make Sense

Use the table to solve 17–18. Show your work.

17. For the week of April 4, 1964, the Beatles had the top four songs. About how long would it take to listen to these four songs?

Top Songs

Number	Song Title	Song Length (in minutes)
1	"Can't Buy Me Love"	2.30
2	"She Loves You"	2.50
3	"I Want to Hold Your Hand"	2.75
4	"Please Please Me"	2.00

18. **What's the Error?** Isabelle says she can listen to the first three songs in the table in 6 minutes.

19. **Test Prep** Fran bought sneakers for $54.26 and a shirt for $34.34. If Fran started with $100, about how much money does she have left?

Ⓐ $5

Ⓑ $20

Ⓒ $35

Ⓓ $80

Connect to Science

Nutrition

Your body needs protein to build and repair cells. You should get a new supply of protein each day. The average 10-year-old needs 35 grams of protein daily. You can find protein in foods like meat, vegetables, and dairy products.

Grams of Protein per Serving

Type of Food	Protein (in grams)
1 scrambled egg	6.75
1 cup shredded wheat cereal	5.56
1 oat bran muffin	3.99
1 cup low-fat milk	8.22

Use estimation to solve.

20. Gina had a scrambled egg, an oat bran muffin, and a cup of low-fat milk for breakfast. About how many grams of protein did Gina have at breakfast?

21. Pablo had a cup of shredded wheat cereal, a cup of low-fat milk, and one other item for breakfast. He had about 21 grams of protein. What was the third item Pablo had for breakfast?

© Houghton Mifflin Harcourt Publishing Company

134 FOR MORE PRACTICE: Standards Practice Book, pp. P65–P66

Name _____

Add Decimals

Essential Question How can place value help you add decimals?

🔑 UNLOCK the Problem REAL WORLD

Henry recorded the amount of rain that fell over 2 hours.
In the first hour, Henry measured 2.35 centimeters of rain.
In the second hour, he measured 1.82 centimeters of rain.

Henry estimated that about 4 centimeters of rain fell in 2 hours.
What is the total amount of rain that fell? How can you use
this estimate to decide if your answer is reasonable?

Add. 2.35 + 1.82

- Add the hundredths first.

  5 hundredths + 2 hundredths = _____ hundredths.

- Then add the tenths and ones. Regroup as needed.

  3 tenths + 8 tenths = _____ tenths. Regroup.

  2 ones + 1 one + 1 regrouped one = _____ ones.

- Record the sum for each place value.

$$\begin{array}{r} 2.35 \\ +\ 1.82 \\ \hline \end{array}$$

_____

Draw a quick picture to check your work.

> **Math Talk** MATHEMATICAL PRACTICES
> Explain how you know
> when you need to regroup in a
> decimal addition problem.

So, _____ centimeters of rain fell.

Since _____ is close to the estimate, 4, the answer is reasonable.

Equivalent Decimals When adding decimals, you can use equivalent decimals to help keep the numbers aligned in each place. Add zeros to the right of the last digit as needed, so that the addends have the same number of decimal places.

Try This! Estimate. Then find the sum.

STEP 1

Estimate the sum.

$$20.4 + 13.76$$

Estimate: $20 + 14 = \underline{\hspace{1.5cm}}$

$20.40 + 13.76 = \underline{\hspace{2cm}}$

STEP 2

Find the sum.

Add the hundredths first.
Then, add the tenths, ones, and tens.
Regroup as needed.

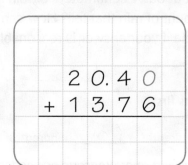

Think: 20.4 = 20.40

- Is your answer reasonable? **Explain.**

Share and Show

Estimate. Then find the sum.

1. Estimate: _____

$$\begin{array}{r} 2.5 \\ + 4.6 \\ \hline \end{array}$$

2. Estimate: _____

$$\begin{array}{r} 8.75 \\ + 6.43 \\ \hline \end{array}$$

 3. Estimate: _____

$$\begin{array}{r} 2.03 \\ + 7.89 \\ \hline \end{array}$$

4. Estimate: _____

$6.34 + 3.8 = \underline{\hspace{2cm}}$

 5. Estimate: _____

$5.63 + 2.6 = \underline{\hspace{2cm}}$

MATHEMATICAL PRACTICES

Math Talk **Explain** why it is important to remember to line up the place values in each number when adding or subtracting decimals.

Name _____

On Your Own ....

Estimate. Then find the sum.

6. Estimate: _____

$$12.3$$
$$+\ 4.9$$

7. Estimate: _____

$$19.2$$
$$+12.68$$

8. Estimate: _____

$$6.8$$
$$+7.4$$

9. Estimate: _____

$7.86 + 2.9 =$ _____

10. Estimate: _____

$4.3 + 2.49 =$ _____

11. Estimate: _____

$9.95 + 0.47 =$ _____

 Find the sum.

12. seven and twenty-five hundredths added to nine and four tenths

13. twelve and eight hundredths added to four and thirty-five hundredths

14. nineteen and seven tenths added to four and ninety-two hundredths

15. one and eighty-two hundredths added to fifteen and eight tenths

Practice: Copy and Solve Find the sum.

16. $7.99 + 8.34$

17. $15.76 + 8.2$

18. $9.6 + 5.49$

19. $33.5 + 16.4$

20. $9.84 + 21.52$

21. $3.89 + 4.6$

22. $42.19 + 8.8$

23. $16.74 + 5.34$

24. $27.58 + 83.9$

UNLOCK the Problem REAL WORLD

25. A city receives an average rainfall of 16.99 centimeters in August. One year, during the month of August, it had rained 8.33 centimeters by August 15th. Then it rained another 4.65 centimeters through the end of the month. What was the total rainfall in centimeters for the month?

(A) 3.68 centimeters

(B) 4.68 centimeters

(C) 12.98 centimeters

(D) 13.98 centimeters

a. What do you need to find? _____

b. What information are you given? _____

_____

c. How will you use addition to find the total number of centimeters of rain that fell?

_____

d. Show how you solved the problem.

e. Fill in the bubble for the correct answer choice above.

26. Tania measured the growth of her plant each week. The first week, the plant's height measured 2.65 decimeters. During the second week, Tania's plant grew 0.38 decimeter. How tall was Tania's plant at the end of the second week?

(A) 2.27 decimeters

(B) 3.03 decimeters

(C) 3.23 decimeters

(D) 3.93 decimeters

27. Maggie had $35.13. Then her mom gave her $7.50 for watching her younger brother. How much money does Maggie have now?

(A) $31.63

(B) $32.63

(C) $41.63

(D) $42.63

Name _____

Subtract Decimals

Essential Question How can place value help you subtract decimals?

🔑 UNLOCK the Problem REAL WORLD

Hannah has 3.36 kilograms of apples and 2.28 kilograms of oranges. Hannah estimates she has about 1 more kilogram of apples than oranges. How many more kilograms of apples than oranges does Hannah have? How can you use this estimate to decide if your answer is reasonable?

- What operation will you use to solve the problem?

  _____

- Circle Hannah's estimate to check that your answer is reasonable.

Subtract. 3.36 − 2.28

- Subtract the hundredths first. If there are not enough hundredths, regroup 1 tenth as 10 hundredths.

  _____ hundredths − 8 hundredths = 8 hundredths

- Then subtract the tenths and ones. Regroup as needed.

  _____ tenths − 2 tenths = 0 tenths

  _____ ones − 2 ones = 1 one

- Record the difference for each place value.

$$\begin{array}{r} 3.36 \\ -\ 2.28 \\ \hline \end{array}$$

Draw a quick picture to check your work.

So, Hannah has _____ more kilograms of apples than oranges.

Since _____ is close to 1, the answer is reasonable.

Math Talk MATHEMATICAL PRACTICES
Explain how you know when to regroup in a decimal subtraction problem.

Try This! Use addition to check.

Since subtraction and addition are inverse operations, you can check subtraction by adding.

STEP 1

Find the difference.

Subtract the hundredths first.

Then, subtract the tenths, ones, and tens. Regroup as needed.

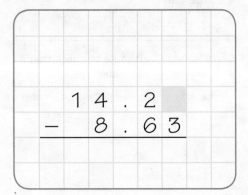

```
  1 4 . 2
−    8 . 6 3
```

STEP 2

Check your answer.

Add the difference to the number you subtracted. If the sum matches the number you subtracted from, your answer is correct.

```
            ← difference
+ 8.63      ← number subtracted
            ← number subtracted from
```

- Is your answer correct? **Explain.**

_____

_____

Share and Show . . . .

Estimate. Then find the difference.

1. Estimate: _____

   5.83
   −2.18

2. Estimate: _____

   4.45
   −1.86

☑ 3. Estimate: _____

   4.03
   −2.25

Find the difference. Check your answer.

4. 0.70
   − 0.43

5. 13.2
   − 8.04

☑ 6. 15.8
   − 9.67

Name _____

On Your Own...

Estimate. Then find the difference.

7. Estimate: _____

$$
\begin{array}{r}
4.08 \\
-1.74 \\
\hline
\end{array}
$$

8. Estimate: _____

$$
\begin{array}{r}
13.54 \\
-6.7 \\
\hline
\end{array}
$$

9. Estimate: _____

$$
\begin{array}{r}
19.64 \\
-8.12 \\
\hline
\end{array}
$$

Find the difference. Check your answer.

10.
$$
\begin{array}{r}
16.05 \\
-1.5 \\
\hline
\end{array}
$$

11.
$$
\begin{array}{r}
7.3 \\
-5.4 \\
\hline
\end{array}
$$

12.
$$
\begin{array}{r}
21.4 \\
-16.97 \\
\hline
\end{array}
$$

 Find the difference.

13. three and seventy-two hundredths subtracted from five and eighty-one hundredths

14. one and six hundredths subtracted from eight and thirty-two hundredths

 Algebra Write the unknown number for *n*.

15. $5.28 - 3.4 = n$

16. $n - 6.47 = 4.32$

17. $11.57 - n = 7.51$

$n =$ _____

$n =$ _____

$n =$ _____

Practice: Copy and Solve Find the difference.

18. $8.42 - 5.14$

19. $16.46 - 13.87$

20. $34.27 - 17.51$

21. $15.83 - 11.45$

22. $12.74 - 10.54$

23. $48.21 - 13.65$

UNLOCK the Problem REAL WORLD

24. In peanut butter, how many more grams of protein are there than grams of carbohydrates? Use the label at the right.

PEANUT BUTTER **Nutrition Facts**		
Serving Size 2 Tbsp (32.0 g)		
Amount Per Serving		
Calories		190
Calories from Fat		190
		% Daily Value*
Total Fat 16g		25%
Saturated Fat 3g		18%
Polyunsaturated Fat 4.4g		
Monounsaturated Fat 7.8g		
Cholesterol 0mg		0%
Sodium 5mg		0%
Total Carbohydrates 6.2g		2%
Dietary Fiber 1.9g		8%
Sugars 2.5g		8%
Protein 8.1g		
*Based on a 2,000 calorie diet		

a. What do you need to know? _____

_____

_____

b. How will you use subtraction to find how many more grams of protein there are than grams of carbohydrates?

_____

_____

c. Show how you solved the problem.

d. Complete each sentence.

The peanut butter has _____ grams of protein.

The peanut butter has _____ grams of carbohydrates.

There are _____ more grams of protein than grams of carbohydrates in the peanut butter.

25. Kyle is building a block tower. Right now the tower stands 0.89 meter tall. How much higher does the tower need to be to reach a height of 1.74 meters?

_____

26. Test Prep Allie is 158.7 centimeters tall. Her younger brother is 9.53 centimeters shorter than she is. How tall is Allie's younger brother?

(A) 159.27 centimeters

(B) 159.23 centimeters

(C) 149.27 centimeters

(D) 149.17 centimeters

FOR MORE PRACTICE:
Standards Practice Book, pp. P69–P70

Name _____

Patterns with Decimals

Essential Question How can you use addition or subtraction to describe a pattern or create a sequence with decimals?

UNLOCK the Problem · REAL WORLD

A state park rents canoes for guests to use at the lake. It costs $5.00 to rent a canoe for 1 hour, $6.75 for 2 hours, $8.50 for 3 hours, and $10.25 for 4 hours. If this pattern continues, how much should it cost Jason to rent a canoe for 7 hours?

A **sequence** is an ordered list of numbers. A **term** is each number in a sequence. You can find the pattern in a sequence by comparing one term with the next term.

STEP 1

Write the terms you know in a sequence. Then look for a pattern by finding the difference from one term in the sequence to the next.

+ $1.75   difference between terms

$5.00 $6.75 $8.50 $10.25

↑   ↑   ↑   ↑

1 hour   2 hours   3 hours   4 hours

STEP 2

Write a rule that describes the pattern in the sequence.

Rule: _____

_____

STEP 3

Extend the sequence to solve the problem.

$5.00, $6.75, $8.50, $10.25, _____, _____, _____

So, it should cost _____ to rent a canoe for 7 hours.

- What observation can you make about the pattern in the sequence that will help you write a rule?

_____

_____

🔑 Example Write a rule for the pattern in the sequence.
Then find the unknown terms in the sequence.

29.6, 28.3, 27, 25.7, _____, _____, _____, 20.5, 19.2

STEP 1 Look at the first few terms in the sequence.

Think: Is the sequence increasing or decreasing from one term to the next?

_____

STEP 2 Write a rule that describes the pattern in the sequence.

What operation can be used to describe a sequence that increases?

_____

What operation can be used to describe a sequence that decreases?

_____

Rule: _____

STEP 3 Use your rule to find the unknown terms.
Then complete the sequence above.

• **Explain** how you know whether your rule for a sequence

would involve addition or subtraction. _____

_____

_____

Try This!

Ⓐ Write a rule for the sequence. Then find the
unknown term.

65.9, 65.3, _____, 64.1, 63.5, 62.9

Rule: _____

Ⓑ Write the first four terms of the sequence.

Rule: start at 0.35, add 0.15

_____, _____, _____, _____

144

Name _____

Share and Show

Write a rule for the sequence.

1. 0.5, 1.8, 3.1, 4.4, …

   Think: Is the sequence increasing or decreasing?

   Rule: _____

✓ 2. 23.2, 22.1, 21, 19.9, …

   Rule: _____

Write a rule for the sequence. Then find the unknown term.

3. 31.5, 25.2, 18.9, _____, 6.3

   Rule: _____

4. 0.25, 0.75, _____, 1.75, 2.25

   Rule: _____

5. 0.3, 1.5, _____, 3.9, 5.1

   Rule: _____

✓ 6. 19.5, 18.8, 18.1, 17.4, _____

   Rule: _____

Math Talk MATHEMATICAL PRACTICES
What operation, other than addition, suggests an increase from one term to the next?

On Your Own

Write a rule for the sequence. Then find the unknown term.

7. 1.8, 4.1, _____, 8.7, 11

   Rule: _____

8. 6.85, 5.73, 4.61, _____, 2.37

   Rule: _____

9. 33.4, _____, 28.8, 26.5, 24.2

   Rule: _____

10. 15.9, 16.1, 16.3, _____, 16.7

   Rule: _____

Write the first four terms of the sequence.

11. **Rule:** start at 10.64, subtract 1.45

   _____, _____, _____, _____

12. **Rule:** start at 0.87, add 2.15

   _____, _____, _____, _____

13. **Rule:** start at 19.3, add 1.8

   _____, _____, _____, _____

14. **Rule:** start at 29.7, subtract 0.4

   _____, _____, _____, _____

MATHEMATICAL PRACTICES

Problem Solving  REAL WORLD ....

H.O.T. **Pose a Problem**

15. Bren has a deck of cards. As shown below, each card is labeled with a rule describing a pattern in a sequence. Select a card and decide on a starting number. Use the rule to write the first five terms in your sequence.

| Add 1.6 | Add 0.33 | Add 6.5 | Add 0.25 | Add 1.15 |

Sequence: _____, _____, _____, _____, _____

Write a problem that relates to your sequence and requires the sequence be extended to solve.

Pose a Problem

Solve your problem.

• **Explain** how you solved your problem. _____

_____

© Houghton Mifflin Harcourt Publishing Company

FOR MORE PRACTICE:
Standards Practice Book, pp. P71–P72

Problem Solving • Add and Subtract Money

Essential Question How can the strategy *make a table* help you
organize and keep track of your bank account balance?

UNLOCK the Problem ⟩ REAL WORLD

At the end of May, Mrs. Freeman had an account balance of
$442.37. Since then, she has written a check for $63.92 and made
a deposit of $350.00. Mrs. Freeman says she has $729.45 in her
account. Make a table to determine if Mrs. Freeman is correct.

Read the Problem	Solve the Problem

Read the Problem

What do I need to find?

I need to find _____

_____

What information do I need to use?

I need to use the _____

_____

_____

How will I use the information?

I need to make a table and use the information to

_____

_____

_____

Solve the Problem

Mrs. Freeman's Checkbook			
May balance			$442.37
Check	$63.92		−$63.92
Deposit		$350.00	

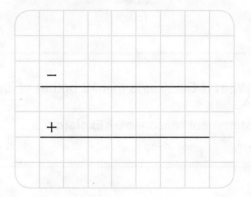

Mrs. Freeman's correct balance is _____.

1. How can you tell if your answer is reasonable? _____

_____

Try Another Problem

Nick is buying juice for himself and 5 friends. Each bottle of juice costs $1.25. How much does 6 bottles of juice cost? Make a table to find the cost of 6 bottles of juice.

Use the graphic below to solve the problem.

Read the Problem	Solve the Problem
What do I need to find?	
What information do I need to use?	
How will I use the information?	So, the total cost of 6 bottles of juice is _____.

2. **What if** Ginny says that 12 bottles of juice cost $25.00? Is Ginny's

   statement reasonable? **Explain.** _____

   _____

3. If Nick had $10, how many bottles of juice could he buy? _____

Math Talk

Explain how you could use another strategy to solve this problem.

Share and Show

1. Sara wants to buy a bottle of apple juice from a vending machine. She needs exactly $2.30. She has the following bills and coins:

Make and complete a table to find all the ways Sara could pay for the juice.

First, draw a table with a column for each type of bill or coin.

Next, fill in your table with each row showing a different way Sara can make exactly $2.30.

2. **What if** Sara decides to buy a bottle of water that costs $1.85? What are all the different ways she can make exactly $1.85 with the bills and coins she has? Which coin must Sara use?

3. At the end of August, Mr. Diaz had a balance of $441.62. Since then, he has written two checks for $157.34 and $19.74 and made a deposit of $575.00. Mr. Diaz says his balance is $739.54. Find Mr. Diaz's correct balance.

On Your Own......

Use the following information to solve 4–7.

At Open Skate Night, admission is $3.75 with a membership card and $5.00 without a membership card. Skate rentals are $3.00.

4. Aidan paid the admission for himself and two friends at Open Skate Night. Aidan had a membership card, but his friends did not. Aidan paid with a $20 bill. How much change should Aidan receive?

_____

5. The Moores and Cotters were at Open Skate Night. The Moores paid $6 more for skate rentals than the Cotters did. Together, the two families paid $30 for skate rentals. How many pairs of skates did the Moores rent?

_____

6. **H.O.T.** Jennie and 5 of her friends are going to Open Skate Night. Jennie does not have a membership card. Only some of her friends have membership cards. What is the total amount that Jennie and her friends might pay for admission?

_____

_____

_____

7. **Test Prep** Sean and Hope each have a membership card for Open Skate Night. Sean has his own skates, but Hope will have to rent skates. Sean gives the clerk $15 for their admission and skate rental. How much change should he receive?

(A) $3.50   (C) $5.00

(B) $4.50   (D) $6.50

Choose a **STRATEGY**

Act It Out

Draw a Diagram

Make a Table

Solve a Simpler Problem

Work Backward

Guess, Check, and Revise

SHOW YOUR WORK

© Houghton Mifflin Harcourt Publishing Company

FOR MORE PRACTICE:
Standards Practice Book, pp. P73–P74

Name _____

Choose a Method

Essential Question Which method could you choose to find decimal sums and differences?

🔑 UNLOCK the Problem REAL WORLD

At a track meet, Steven entered the long jump. His jumps were 2.25 meters, 1.81 meters, and 3.75 meters. What was the total distance Steven jumped?

To find decimal sums, you can use properties and mental math or you can use paper and pencil.

- Underline the sentence that tells you what you are trying to find.
- Circle the numbers you need to use.
- What operation will you use?

_____

🔑 One Way Use properties and mental math.

Add. 2.25 + 1.81 + 3.75

$$2.25 + 1.81 + 3.75$$

$$= 2.25 + 3.75 + 1.81 \qquad \text{Commutative Property}$$

$$= (\underline{\hspace{1cm}} + \underline{\hspace{1cm}}) + 1.81 \qquad \text{Associative Property}$$

$$= \underline{\hspace{1cm}} + 1.81$$

$$= \underline{\hspace{1cm}}$$

🔑 Another Way Use place-value.

Add. 2.25 + 1.81 + 3.75

$$\begin{array}{r} 2.25 \\ 1.81 \\ + 3.75 \\ \hline \end{array}$$

So, the total distance Steven jumped was _____ meters.

Math Talk MATHEMATICAL PRACTICES
Explain why you might choose to use the properties to solve this problem.

© Houghton Mifflin Harcourt Publishing Company

Try This!

In 1924, William DeHart Hubbard won a gold medal with a long jump of 7.44 meters. In 2000, Roman Schurenko won the bronze medal with a jump of 8.31 meters. How much longer was Schurenko's jump than Hubbard's?

A Use place-value.

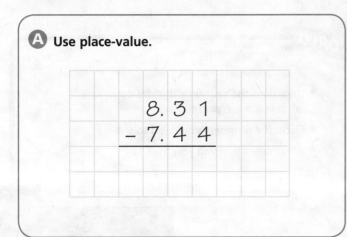

$$\begin{array}{r} 8.\,3\;1 \\ -\;7.\,4\;4 \\ \hline \end{array}$$

B Use a calculator.

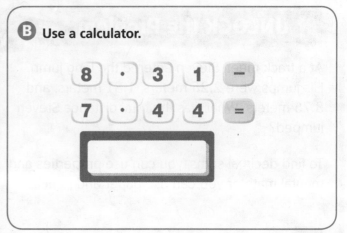

So, Schurenko's jump was _____ meter longer than Hubbard's.

- **Explain** why you cannot use the Commutative Property or the Associative Property to find the difference between two decimals.

_____

_____

Share and Show

Find the sum or difference.

1. $4.19 + 0.58$

2. $9.99 - 4.1$

3. $5.7 + 2.25 + 1.3$

4. $28.6 - 9.84$

5. $\$15.79 + \$32.81$

6. $38.44 - 25.86$

Name _____

On Your Own ·····································

Find the sum or difference.

7. $18.39
+$ 7.56

8. 8.22 − 4.39

9. 93.6 − 79.84

10. 1.82
2.28
+2.18

11. 2.35
− 0.16

12. 5.16
+4.54

13. 15.3
− 6.53

14. 2.64
+8.41

Practice: Copy and Solve Find the sum or difference.

15. 6.3 + 2.98 + 7.7

16. 27.96 − 16.2

17. 12.63 + 15.04

18. 9.24 − 2.68

19. $18 − $3.55

20. 9.73 − 2.52

21. $54.78 + $43.62

22. 7.25 + 0.25 + 1.5

23. 14.56 − 7.8

24. 3.35 + 1.4 + 3.65

25. $22.50 − $8.99

26. 9.77 + 5.54

 Algebra Find the missing number.

27. $n − 9.02 = 3.85$

28. $n + 31.53 = 62.4$

29. $9.2 + n + 8.4 = 20.8$

$n =$ _____

$n =$ _____

$n =$ _____

Problem Solving REAL WORLD

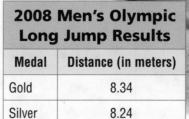

Use the table to solve 30–32.

2008 Men's Olympic Long Jump Results	
Medal	Distance (in meters)
Gold	8.34
Silver	8.24
Bronze	8.20

30. How much farther did the gold medal winner jump than the silver medal winner?

_____

31. **Write Math** ➤ The fourth-place competitor's jump measured 8.19 meters. If his jump had been 0.10 meter greater, what medal would he have received? **Explain** how you solved the problem.

SHOW YOUR WORK

_____

_____

_____

_____

32. In the 2004 Olympics, the gold medalist for the men's long jump had a jump of 8.59 meters. How much farther did the 2004 gold medalist jump compared to the 2008 gold medalist?

_____

33. Jake cuts a length of 1.12 meters from a 3-meter board. How long is the board now?

_____

34. **Test Prep** In the long jump, Danny's first attempt was 5.47 meters. His second attempt was 5.63 meters. How much farther did Danny jump on his second attempt than on his first?

(A) 11.1 meters   (C) 5.16 meters

(B) 10.1 meters   (D) 0.16 meter

Name _____

Chapter Review/Test

▶ Vocabulary

Choose the best term from the box.

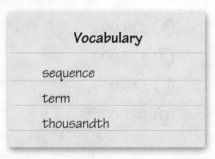

Vocabulary
sequence
term
thousandth

1. If one hundredth is divided into ten equal parts, each part is

   one _____. (p. 105)

2. An ordered list of numbers is called a _____. (p. 143)

▶ Concepts and Skills

3. **Explain** how the value of a decimal changes as you move to the left
   or the right in a place-value chart.

   _____

   _____

**Write the place value of the underlined digit. Round each number to the
place of the underlined digit.**

4. 0.7<u>3</u>5

5. <u>9</u>.283

6. 4.0<u>7</u>9

Find the sum or difference.

7. $12.87 − $5.75

8. $32.64 + $18.78

9. 9.28 − 0.54

10. 14.36 + 7.87

11. 10.05 − 6.38

12. 3.25 + 6.75 + 8.75

GO Online Assessment Options **Chapter Test**

Fill in the bubble completely to show your answer.

13. Doug bought a pair of sneakers for $47.82 and a shirt for $13.36. If Doug had $100 before his purchase, about how much money does Doug have left now?

Ⓐ $29.00

Ⓑ $39.00

Ⓒ $48.00

Ⓓ $61.00

14. Since September, Mrs. Bishop has written a check for $178.23 and made a deposit of $363.82. Her balance was $660.00. Which amount should Mrs. Bishop's checkbook balance show now?

Ⓐ $481.77

Ⓑ $483.77

Ⓒ $845.59

Ⓓ $847.59

15. Helen earns $12 each weekend babysitting her brother. After the third weekend, Helen buys a new CD for $12.48. How much money does Helen have left after buying the CD?

Ⓐ $36.00

Ⓑ $24.00

Ⓒ $23.52

Ⓓ $11.52

16. Morgan jogged 51.2 kilometers one week. Karen jogged 53.52 kilometers the same week. How many more kilometers did Karen jog that week than Morgan?

Ⓐ 48.4 kilometers

Ⓑ 12.3 kilometers

Ⓒ 2.32 kilometers

Ⓓ 2.3 kilometers

17. Angelo measured the amount of rain that fell on July 14th. His rain gauge recorded 1.54 centimeters. If 1.73 centimeters fell between July 1st and July 13th, which model shows the total amount of rain that fell from July 1st through July 14th?

Ⓐ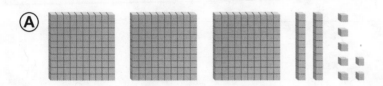

Ⓑ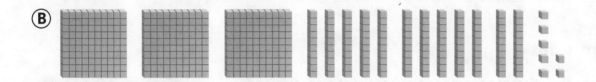

Ⓒ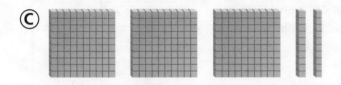

Ⓓ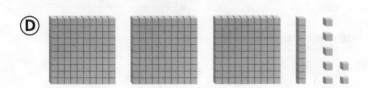

18. The Ruby Throated Hummingbird has an average weight of just 4.253 grams. What is its average weight rounded to the nearest tenth?

Ⓐ 4.3 grams

Ⓑ 4.253 grams

Ⓒ 4.25 grams

Ⓓ 4.2 grams

▶ Constructed Response

19. The Smiths are on a summer road trip. They travel 10.9 hours the first day, 8.6 hours the second day, and 12.4 hours the final day. About how may hours does the Smith family travel over the 3-day trip?

_____

Explain how you found your answer.

_____

_____

▶ Performance Task

20. The prices for different beverages and snacks at a snack stand in a park are shown in the table.

Ⓐ Blair buys a pretzel and fruit juice. Jen buys popcorn and iced tea. Find the difference between the cost of the snacks Blair buys and the cost of the snacks Jen buys.

Park Snacks	
Item	**Price**
Fruit Juice	$0.89
Iced Tea	$1.29
Lemonade	$1.49
Pretzel	$2.50
Popcorn	$1.25

_____

Ⓑ For which two beverages is the difference between the prices the greatest? What is the difference?

_____

Ⓒ **What if** a frosty beverage was being added to the menu that would cost $0.20 more than the fruit juice? How much would the frosty beverage cost? **Explain** how you can determine the cost by using mental math.

_____

_____

_____

© Houghton Mifflin Harcourt Publishing Company

Show What You Know

Check your understanding of important skills.

Name _____

▶ **Meaning of Multiplication** Complete.

1.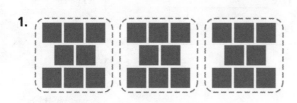

_____ groups of _____ = _____

2.

_____ groups of _____ = _____

▶ **Decimals Greater Than One** Write the word form and the expanded form for each.

3. 1.7

_____

_____

4. 5.62

_____

_____

▶ **Multiply by 3-Digit Numbers** Multiply.

5. 321
 × 4

6. 387
 × 5

7. 126
 × 13

8. 457
 × 35

Staghorn Coral is a type of branching coral. It can add as much as 0.67 feet to its branches each year. Be a Math Detective and find how much a staghorn coral can grow in 5 years.

Vocabulary Builder

▶ **Visualize It** •

Complete the Flow Map using the words with a ✓.

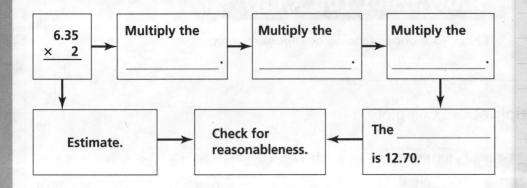

Review Words
decimal
expanded form
✓hundredths
multiplication
✓ones
pattern
place value
✓product
✓tenths
thousandths

6.35
×   2

Multiply the _____.

Multiply the _____.

Multiply the _____.

Estimate.

Check for reasonableness.

The _____ is 12.70.

▶ **Understand Vocabulary** •

Read the description. What term do you think it describes?

1. It is the process used to find the total number of items in a

   given number of groups. _____

2. It is a way to write a number that shows the value of

   each digit. _____

3. It is one of one hundred equal parts. _____

4. This is the result when you multiply two numbers.

   _____

5. It is the value of a digit in a number based on the location

   of the digit. _____

GO Online • eStudent Edition • Multimedia eGlossary

Multiplication Patterns with Decimals

Essential Question How can patterns help you place the decimal point in a product?

UNLOCK the Problem REAL WORLD

Cindy is combining equal-sized rectangles from different fabric patterns to make a postage-stamp quilt. Each rectangle has an area of 0.75 of a square inch. If she uses 1,000 rectangles to make the quilt, what will be the area of the quilt?

 Use the pattern to find the product.

$1 \times 0.75 = 0.75$

$10 \times 0.75 = 7.5$

$100 \times 0.75 = 75.$

$1,000 \times 0.75 = 750.$

The quilt will have an area of _____ square inches.

1. As you multiply by increasing powers of 10, how does the position of the decimal point change in the product? _____

_____

Place value patterns can be used to find the product of a number and the decimals 0.1 and 0.01.

 Example 1

Jorge is making a scale model of the Willis Tower in Chicago for a theater set. The height of the tower is 1,353 feet. If the model is $\frac{1}{100}$ of the actual size of the building, how tall is the model?

$1 \times 1,353 = 1,353$

$0.1 \times 1,353 = 135.3$

$0.01 \times 1,353 = \boxed{}$ ← $\frac{1}{100}$ of 1,353

- What fraction of the actual size of the building is the model?

_____

- Write the fraction a decimal.

_____

Jorge's model of the Willis Tower is _____ feet tall.

2. As you multiply by decreasing powers of 10, how does the position of the decimal point change in the product?

_____

_____

🔑 Example 2

Three friends are selling items at an arts and crafts fair. Josey makes $45.75 selling jewelry. Mark makes 100 times as much as Josey makes by selling his custom furniture. Chance makes a tenth of the money Mark makes by selling paintings. How much money does each friend make?

Josey: $45.75

Mark: _____ × $45.75   **Chance:** _____ × _____

Think: $1 \times \$45.75 =$ _____   Think: $1 \times$ _____ = _____

$10 \times \$45.75 =$ _____   _____ × _____ = _____

$100 \times \$45.75 =$ _____

So, Josey makes $45.75, Mark makes _____,

and Chance makes _____.

Try This! Complete the pattern.

A $10^0 \times 4.78 =$ _____

$10^1 \times 4.78 =$ _____

$10^2 \times 4.78 =$ _____

$10^3 \times 4.78 =$ _____

B $38 \times 1 =$ _____

$38 \times 0.1 =$ _____

$38 \times 0.01 =$ _____

Share and Show ·······················

Complete the pattern.

1. $10^0 \times 17.04 = 17.04$

$10^1 \times 17.04 = 170.4$

$10^2 \times 17.04 = 1{,}704$

$10^3 \times 17.04 =$ _____

Think: The decimal point moves one place to the _____ for each increasing power of 10.

© Houghton Mifflin Harcourt Publishing Company

Name _____

Complete the pattern.

2. $1 \times 3.19 = $ _____

   $10 \times 3.19 = $ _____

   $100 \times 3.19 = $ _____

   $1{,}000 \times 3.19 = $ _____

3. $45.6 \times 10^0 = $ _____

   $45.6 \times 10^1 = $ _____

   $45.6 \times 10^2 = $ _____

   $45.6 \times 10^3 = $ _____

4. $1 \times 6{,}391 = $ _____

   $0.1 \times 6{,}391 = $ _____

   $0.01 \times 6{,}391 = $ _____

Math Talk MATHEMATICAL PRACTICES
Explain how you know that when you multiply the product of 10 × 34.1 by 0.1, the result will be 34.1.

On Your Own

Complete the pattern.

5. $1.06 \times 1 = $ _____

   $1.06 \times 10 = $ _____

   $1.06 \times 100 = $ _____

   $1.06 \times 1{,}000 = $ _____

6. $1 \times 90 = $ _____

   $0.1 \times 90 = $ _____

   $0.01 \times 90 = $ _____

7. $10^0 \times \$0.19 = $ _____

   $10^1 \times \$0.19 = $ _____

   $10^2 \times \$0.19 = $ _____

   $10^3 \times \$0.19 = $ _____

8. $580 \times 1 = $ _____

   $580 \times 0.1 = $ _____

   $580 \times 0.01 = $ _____

9. $10^0 \times 80.72 = $ _____

   $10^1 \times 80.72 = $ _____

   $10^2 \times 80.72 = $ _____

   $10^3 \times 80.72 = $ _____

10. $1 \times 7{,}230 = $ _____

    $0.1 \times 7{,}230 = $ _____

    $0.01 \times 7{,}230 = $ _____

H.O.T. **Algebra** Find the value of n.

11. $n \times \$3.25 = \$325.00$

12. $0.1 \times n = 89.5$

13. $10^3 \times n = 630$

$n = $ _____

$n = $ _____

$n = $ _____

Problem Solving REAL WORLD

 What's the Error?

14. Kirsten is making lanyards for a convention. She needs to make 1,000 lanyards and knows that 1 lanyard uses 1.75 feet of cord. How much cord will Kirsten need?

Kirsten's work is shown below.

$1 \times 1.75 = 1.75$

$10 \times 1.75 = 10.75$

$100 \times 1.75 = 100.75$

$1,000 \times 1.75 = 1,000.75$

Find and describe Kirsten's error.

_____

_____

_____

_____

_____

_____

Solve the problem using the correct pattern.

So, Kirsten needs _____ feet of cord to make 1,000 lanyards.

- **Describe** how Kirsten could have solved the problem without writing out the pattern needed.

_____

_____

FOR MORE PRACTICE:
Standards Practice Book, pp. P81–P82

Name _____

Multiply Decimals and Whole Numbers

Essential Question How can you use a model to multiply a whole number and a decimal?

Investigate

Materials ■ decimal models ■ color pencils

Giant tortoises move very slowly. They can travel a distance of about 0.17 mile in 1 hour. If it travels at the same speed, how far could a giant tortoise move in 4 hours?

A. Complete the statement to describe the problem.

I need to find how many total miles are in _____ groups

of _____.

- Write an expression to represent the problem. _____

B. Use the decimal model to find the answer.

- What does each small square in the decimal model represent?

_____

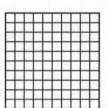

C. Shade a group of _____ squares to represent the distance a giant tortoise can move in 1 hour.

D. Use a different color to shade each additional

group of _____ squares until you

have _____ groups of _____ squares.

E. Record the total number of squares shaded. _____ squares

So, the giant tortoise can move _____ mile in 4 hours.

Math Talk MATHEMATICAL PRACTICES
Explain how the model helps you determine if your answer is reasonable.

Draw Conclusions

1. Explain why you used only one decimal model to show the product.

_____

_____

2. Explain how the product of 4 groups of 0.17 is similar to the product of 4 groups of 17. How is it different?

_____

_____

3. H.O.T. **Compare** the product of 0.17 and 4 with each of the factors. Which number has the greatest value? **Explain** how this is different than multiplying two whole numbers.

_____

_____

Make Connections

You can draw a quick picture to solve decimal multiplication problems.

Find the product. 3 × 0.46

STEP 1 Draw 3 groups of 4 tenths and 6 hundredths. Remember that a square is equal to 1.

STEP 2 Combine the hundredths and rename.

There are _____ hundredths. I will rename

_____ hundredths as _____.

Cross out the hundredths you renamed.

STEP 3 Combine the tenths and rename.

There are _____ tenths. I will rename

_____ tenths as _____.

Cross out the tenths you renamed.

STEP 4 Record the value shown by your completed quick picture.

So, 3 × 0.46 = _____.

© Houghton Mifflin Harcourt Publishing Company

Math Talk MATHEMATICAL PRACTICES
Explain how renaming decimals is like renaming whole numbers.

166

Name _____

Share and Show

Use the decimal model to find the product.

1. $5 \times 0.06 =$ _____

2. $2 \times 0.38 =$ _____

3. $4 \times 0.24 =$ _____

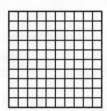

Find the product. Draw a quick picture.

4. $4 \times 0.6 =$ _____

5. $2 \times 0.67 =$ _____

6. $3 \times 0.62 =$ _____

7. $4 \times 0.32 =$ _____

8. **Write Math** ▶ **Describe** how you solved Exercise 7 using place

value and renaming. _____

_____

_____

Problem Solving REAL WORLD

Use the table for 9–11.

9. Each day a bobcat drinks about 3 times as much water as a Canada goose drinks. How much water can a bobcat drink in one day?

_____

Water Consumption	
Animal	**Average Amount (liters per day)**
Canada Goose	0.24
Cat	0.15
Mink	0.10
Opossum	0.30
Bald Eagle	0.16

10. River otters drink about 5 times as much water as a bald eagle drinks in a day. How much water can a river otter drink in one day?

_____

11. [Write Math] ▶ **Explain** how you could use a quick picture to find the amount of water that a cat drinks in 5 days.

_____

_____

_____

12. **Test Prep** Jared has a parakeet that weighs 1.44 ounces. Susie has a Senegal parrot that weighs 3 times as much as Jared's parakeet. How many ounces does Susie's parrot weigh?

(A) 0.32 ounce   (C) 4.32 ounces

(B) 0.43 ounce   (D) 43.2 ounces

FOR MORE PRACTICE:
Standards Practice Book, pp. P83–P84

Name _____

Multiplication with Decimals and Whole Numbers

Essential Question How can you use drawings and place value to multiply a decimal and a whole number?

🔑 UNLOCK the Problem REAL WORLD

In 2010, the United States Mint released a newly designed Lincoln penny. A Lincoln penny has a mass of 2.5 grams. If there are 5 Lincoln pennies on a tray, what is the total mass of the pennies?

- How much mass does one penny have?

  _____

- How many pennies are on the tray?

  _____

- Use grouping language to describe what you are asked to find.

  _____

🔓 One Way Use place value.

Multiply. 5×2.5

MODEL

THINK AND RECORD

STEP 1 Estimate the product by rounding the decimal to the nearest whole number.

$$5 \times \underline{\hspace{1cm}} = \underline{\hspace{1cm}}$$

STEP 2 Multiply the tenths by 5.

$$\begin{array}{r} 2.5 \\ \times\ \ 5 \\ \hline \end{array}$$
← 5×5 tenths = 25 tenths, or 2 ones and 5 tenths

STEP 3 Multiply the ones by 5.

$$\begin{array}{r} 2.5 \\ \times\ \ 5 \\ \hline 2.5 \\ \end{array}$$
← 5×2 ones = 10 ones, or 1 ten

STEP 4 Add the partial products.

$$\begin{array}{r} 2.5 \\ \times\ \ 5 \\ \hline 2.5 \\ +\ 10 \\ \hline \end{array}$$

So, 5 Lincoln pennies have a mass of _____ grams.

Math Talk MATHEMATICAL PRACTICES
Explain how the estimate helps you determine if the answer is reasonable.

Chapter 4   169

🔒 Another Way Use place value patterns.

Having a thickness of 1.35 millimeters, the dime is the thinnest coin produced by the United States Mint. If you stacked 8 dimes, what would be the total thickness of the stack?

Multiply. 8 × 1.35

STEP 1	**STEP 2**	**STEP 3**
Write the decimal factor as a whole number.	Multiply as with whole numbers.	Place the decimal point.
Think: 1.35 × 100 = 135		**Think:** 0.01 of 135 is 1.35. Find 0.01 of 1,080 and record the product.

$$
\begin{array}{c} 1.35 \\ \times\ \ 8 \\ \hline ? \end{array} \xrightarrow{\times\,100}
\begin{array}{c} 135 \\ \times\ \ 8 \\ \hline 1{,}080 \end{array} \xrightarrow{\times\,0.01}
\begin{array}{c} 1.35 \\ \times\ \ 8 \\ \hline \end{array}
$$

$$\xrightarrow{\times\,100} \qquad\qquad \xrightarrow{\times\,0.01}$$

A stack of 15 dimes would have a thickness of _____ millimeters.

1. **Explain** how you know the product of 8 × 1.35 is greater than 8.

_____

_____

2. **What if** you multiplied 0.35 by 8? Would the product be less than or greater than 8? **Explain.**

_____

_____

Share and Show

Place the decimal point in the product.

1. $\begin{array}{r} 6.81 \\ \times\ \ 7 \\ \hline 4\ 7\ 6\ 7 \end{array}$ **Think:** The place value of the decimal factor is hundredths.

2. $\begin{array}{r} 3.7 \\ \times\ \ 2 \\ \hline 7\ 4 \end{array}$

3. $\begin{array}{r} 19.34 \\ \times\ \ \ 5 \\ \hline 9\ 6\ 7\ 0 \end{array}$

Find the product.

4. 6.32
    × 3

✓ 5. 4.5
     × 8

✓ 6. 40.7
     × 5

Math Talk MATHEMATICAL PRACTICES
Explain how you can determine if your answer to Exercise 6 is reasonable.

On Your Own...

Find the product.

7. 4.93
    × 7

8. 8.2
    × 6

9. 0.49
    × 4

10. 9.08
     × 9

11. 7.55
     × 8

12. 15.37
     × 5

Practice: Copy and Solve Find the product.

13. 8×7.2

14. 3×1.45

15. 9×8.6

16. 6×0.79

17. 4×9.3

18. 7×0.81

19. 6×2.08

20. 5×23.66

Problem Solving REAL WORLD

Use the table for 21–23.

21. Sari has a bag containing 6 half dollars. What is the weight of the half dollars in Sari's bag?

_____

Coin	Weight (in grams)
Nickel	5.00
Dime	2.27
Quarter	5.67
Half Dollar	11.34
Dollar	8.1

22. Felicia is running a game booth at a carnival. One of the games requires participants to guess the weight, in grams, of a bag of 9 dimes. What is the actual weight of the dimes in the bag?

_____

23. **H.O.T.** Chance has $2 in quarters. Blake has $5 in dollar coins. Whose coins have the greatest weight? **Explain**.

_____

_____

_____

_____

SHOW YOUR WORK

24. **Write Math** ► Julie multiplies 6.27 by 7 and claims the product is 438.9. **Explain** without multiplying how you know Julie's answer is not correct. Find the correct answer.

_____

_____

_____

25. **Test Prep** Every day on his way to and from school, Milo walks a total of 3.65 miles. If he walks to school 5 days, how many miles will Milo have walked?

(A) 1.825 miles   (C) 182.5 miles

(B) 18.25 miles   (D) 1,825 miles

FOR MORE PRACTICE:
Standards Practice Book, pp. P85–P86

Name _____

Multiply Using Expanded Form

Essential Question How can you use expanded form and place value to multiply a decimal and a whole number?

🔑 UNLOCK the Problem REAL WORLD

The length of a day is the amount of time it takes a planet to make a complete rotation on its axis. On Jupiter, there are 9.8 Earth hours in a day. How many Earth hours are there in 46 days on Jupiter?

You can use a model and partial products to solve the problem.

▲ A day on Jupiter is called a Jovian day.

🔒 One Way Use a model.

Multiply. 46 × 9.8

THINK	MODEL	RECORD

STEP 1

Rewrite the factors in expanded form, and label the model.

46 = _____ + _____

9.8 = _____ + _____
_____

STEP 2

Multiply to find the area of each section. The area of each section represents a partial product.
_____

STEP 3

Add the partial products.

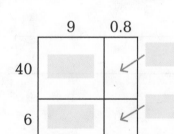

$$9.8$$
$$\times 46$$

← 40 × 9
← 40 × 0.8
← 6 × 9
← 6 × 0.8
+

So, there are _____ Earth hours in 46 days on Jupiter.

---

1. What if you wanted to find the number of Earth hours in 125 days on Jupiter? How would your model change?

_____

_____

Another Way Use place value patterns.

A day on the planet Mercury lasts about 58.6 Earth days. How many Earth days are there in 14 days on Mercury?

Multiply. 14 × 58.6

STEP 1

Write the decimal factor as a whole number.

_____

STEP 2

Multiply as with whole numbers.

_____

STEP 3

Place the decimal point.

The decimal product is _____ of the whole number product.

▲ It takes Mercury 88 Earth days to complete an orbit of the Sun.

$$58.6 \xrightarrow{\times 10} 586$$

$$\begin{array}{r} 586 \\ \times\ 14 \\ \hline 2{,}344 \\ +5{,}860 \\ \hline 8{,}204 \end{array}$$

$$\begin{array}{r} 58.6 \\ \times\ 14 \\ \hline \end{array}$$

586 ×0.1 → 58.6

8,204 ×0.1

58.6 ×10 ?

So, there are _____ Earth days in 14 days on Mercury.

2. **What if** you rewrite the problem as (10 + 4) × 58.6 and used the Distributive Property to solve? **Explain** how this is similar to your model using place value. _____

_____

Try This! Find the product.

Ⓐ Use a model.

52 × 0.35 = _____

Ⓑ Use place value patterns.

16 × 9.18 = _____

174

Name _____

Share and Show

Draw a model to find the product.

1. $19 \times 0.75 =$ _____

	0.7	0.05
10		
9		

2. $27 \times 8.3 =$ _____

Find the product.

3. $18 \times 8.7 =$ _____

4. $23 \times 56.1 =$ _____

5. $47 \times 5.92 =$ _____

Math Talk MATHEMATICAL PRACTICES
Describe how you could use an estimate to determine if your answer to Exercise 3 is reasonable.

On Your Own

Draw a model to find the product.

6. $71 \times 8.3 =$ _____

7. $28 \times 0.19 =$ _____

Find the product.

8. $19 \times 0.65 =$ _____

9. $34 \times 98.3 =$ _____

10. $26 \times 16.28 =$ _____

© Houghton Mifflin Harcourt Publishing Company

UNLOCK the Problem REAL WORLD

11. While researching facts on the planet Earth, Kate learned that a true Earth day is about 23.93 hours long. How many hours are in 2 weeks on Earth?

a. What are you being asked to find?

_____

_____

b. What information do you need to know to solve the problem? _____

_____

_____

c. Write an expression to represent the problem to be solved. _____

d. Show the steps you used to solve the problem.

e. Complete the sentences.

On Earth, there are about _____

hours in a day, _____ days in 1 week,

and _____ days in two weeks.

Since _____ × _____ =

_____, there are about

_____ hours in 2 weeks on Earth.

---

12. Michael's favorite song is 3.19 minutes long. If he listens to the song 15 times on repeat, how long will he have listened to the same song?

_____

13. **Test Prep** A car travels 56.7 miles in an hour. If it continues at the same speed, how far will the car travel in 12 hours?

Ⓐ 68.004 miles

Ⓑ 680.04 miles

Ⓒ 680.4 miles

Ⓓ 6,804 miles

Problem Solving • Multiply Money

Essential Question How can the strategy *draw a diagram* help you solve
a decimal multiplication problem?

🔑 UNLOCK the Problem   REAL WORLD

A group of friends go to a local fair. Jayson spends
$3.75. Maya spends 3 times as much as Jayson.
Tia spends $5.25 more than Maya. How much does
Tia spend?

Use the graphic organizer below to help you solve
the problem.

Read the Problem	Solve the Problem
What do I need to find? I need to find _____ _____ _____.	The amount of money Maya and Tia spend depends on the amount Jayson spends. Draw a diagram to compare the amounts without calculating. Then, use the diagram to find the amount each person spends.

What information do I need to use?

I need to use the amount spent by _____

to find the amount spent by _____ and

_____ at the fair.

Jayson $3.75

Maya   _____ _____ _____

Tia   _____ _____ _____ $5.25

How will I use the information?

I can draw a diagram to show _____

_____

_____.

Jayson: $3.75

Maya: 3 × _____ = _____

Tia: _____ + $5.25 = _____

So, Tia spent _____ at the fair.

🔐 Try Another Problem

Julie's savings account has a balance of $57.85 in January. By March, her balance is 4 times as much as her January balance. Between March and November, Julie deposits a total of $78.45. If she does not withdraw any money from her account, what should Julie's balance be in November?

Read the Problem	Solve the Problem
What do I need to find?	
What information do I need to use?	
How will I use the information?	So, Julie's savings account balance will be _____ in November.

• How does the diagram help you determine if your answer is reasonable? _____

_____

_____

Math Talk MATHEMATICAL PRACTICES
Describe a different diagram you could use to solve the problem.

Share and Show

1. Manuel collects $45.18 for a fundraiser. Gerome collects $18.07 more than Manuel. Cindy collects 2 times as much as Gerome. How much money does Cindy collect for the fundraiser?

   First, draw a diagram to show the amount Manuel collects.

   Then, draw a diagram to show the amount Gerome collects.

   Next, draw a diagram to show the amount Cindy collects.

   Finally, find the amount each person collects.

   Cindy collects _____ for the fundraiser.

SHOW YOUR WORK

2. **What if** Gerome collects $9.23 more than Manuel? If Cindy still collects 2 times as much as Gerome, how much money would Cindy collect?

   _____

3. It costs $5.15 to rent a kayak for 1 hour at a local state park. The price per hour stays the same for up to 5 hours of rental. After 5 hours, the cost is decreased to $3.75 per hour. How much would it cost to rent a kayak for 6 hours?

   _____

4. Jenn buys a pair of jeans for $24.99. Her friend Karen spends $3.50 more for the same pair of jeans. Vicki paid the same price as Karen for the jeans but bought 2 pairs. How much did Vicki spend?

   _____

On Your Own.....................

Use the sign for 5–8.

5. Austin shops at Surfer Joe's Surf Shop before going to the beach. He buys 2 T-shirts, a pair of board shorts, and a towel. If he gives the cashier $60, how much change will Austin get back?

_____

6. Maria buys 3 T-shirts and 2 pairs of sandals at Surfer Joe's Surf Shop. How much does Maria spend?

_____

7. Nathan receives a coupon in the mail for $10 off of a purchase of $100 or more. If he buys 3 pairs of board shorts, 2 towels, and a pair of sunglasses, will he spend enough to use the coupon? How much will his purchase cost?

_____

8. **H.O.T.** Moya spends $33.90 on 3 different items. If she did not buy board shorts, which three items did Moya buy?

_____

9. **Test Prep** At a donut shop in town, each donut costs $0.79. If Mr. Thomas buys a box of 8 donuts, how much will he pay for the donuts?

Ⓐ $6.32

Ⓑ $8.79

Ⓒ $63.20

Ⓓ $87.90

Choose a STRATEGY

Act It Out

Draw a Diagram

Make a Table

Solve a Simpler Problem

Work Backward

Guess, Check, and Revise

Surfer Joe's Surf Shop

T-shirt $12.75
Board Shorts $25.99
Sandals $8.95
Towel $5.65
Sunglasses $15.50

SHOW YOUR WORK

FOR MORE PRACTICE:
Standards Practice Book, pp. P89–P90

Name _____

 Mid-Chapter Checkpoint

▶ **Concepts and Skills**

1. **Explain** how you can use a quick picture to find 3×2.7. _____

_____

_____

Complete the pattern.

2. $1 \times 3.6 =$ _____

   $10 \times 3.6 =$ _____

   $100 \times 3.6 =$ _____

   $1,000 \times 3.6 =$ _____

3. $10^0 \times 17.55 =$ _____

   $10^1 \times 17.55 =$ _____

   $10^2 \times 17.55 =$ _____

   $10^3 \times 17.55 =$ _____

4. $1 \times 29 =$ _____

   $0.1 \times 29 =$ _____

   $0.01 \times 29 =$ _____

Find the product.

5. $\begin{array}{r} 3.14 \\ \times \quad 8 \\ \hline \end{array}$

6. 17×0.67

7. 29×7.3

Draw a diagram to solve.

8. Julie spends $5.62 at the store. Micah spends 5 times as much as Julie. Jeremy spends $6.72 more than Micah. How much money does each person spend?

   Julie: $5.62

   Micah: _____

   Jeremy: _____

Fill in the bubble completely to show your answer.

9. Sarah is cutting ribbons for a pep rally. The length of each ribbon needs to be 3.68 inches. If she needs 1,000 ribbons, what is the length of ribbon Sarah needs?

   (A) 3.68 inches

   (B) 36.8 inches

   (C) 368 inches

   (D) 3,680 inches

10. Adam is carrying books to the classroom for his teacher. Each books weighs 3.85 pounds. If he carries 4 books, how many pounds is Adam carrying?

    (A) 12.2 pounds

    (B) 13.2 pounds

    (C) 15.2 pounds

    (D) 15.4 pounds

11. A car travels 54.9 miles in an hour. If the car continues at the same speed for 12 hours, how many miles will it travel?

    (A) 54.9 miles

    (B) 549 miles

    (C) 658.8 miles

    (D) 6,588 miles

12. Charlie saves $21.45 each month for 6 months. In the seventh month, he only saves $10.60. How much money will Charlie have saved after 7 months?

    (A) $150.15

    (B) $139.30

    (C) $128.70

    (D) $118.10

Name _____

Decimal Multiplication

Essential Question How can you use a model to multiply decimals?

Investigate

Materials ■ color pencils

The distance from Charlene's house to her school is 0.8 mile. Charlene rides her bike 7 tenths of the distance and walks the rest of the way. How far does Charlene ride her bike to school?

You can use a decimal square to multiply decimals.

Multiply. 0.7 × 0.8

A. Draw a square with 10 equal columns.

- What decimal value does each column represent? _____

B. Using a color pencil, shade columns on the grid to represent the distance to Charlene's school.

- The distance to the school is 0.8 mile.

  How many columns did you shade? _____

C. Divide the square into 10 equal rows.

- What decimal value does each row represent? _____

D. Using a different color, shade rows that overlap the shaded columns to represent the distance to school that Charlene rides her bike.

- What part of the distance to school does Charlene ride

  her bike? _____

- How many rows of the shaded columns did you shade?

  _____

E. Count the number of squares that you shaded twice.

There are _____ squares. Each square represents _____.

Record the value of the squares as the product. 0.7 × 0.8 = _____

So, Charlene rides her bike for _____ mile.

Draw Conclusions

1. **Explain** how dividing the decimal square into 10 equal columns and rows shows that tenths multiplied by tenths is equal to hundredths.

   _____

   _____

   _____

2. **H.O.T. Comprehension** Why is the part of the model representing the product less than either factor?

   _____

   _____

   _____

Make Connections

You can use decimal squares to multiply decimals greater than 1.

Multiply. 0.3 × 1.4

STEP 1

Shade columns to represent 1.4.

How many tenths are in 1.4?

_____

STEP 2

Shade rows that overlap the shaded columns to represent 0.3.

How many rows of the shaded

columns did you shade? _____

_____

STEP 3

Count the number of squares that you shaded twice. Record the product at the right.

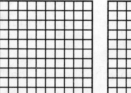

0.3 × 1.4 = _____

Math Talk **Explain** why the product is less than only one of the decimal factors.

Share and Show

Multiply. Use the decimal model.

1. $0.8 \times 0.4 =$ _____

2. $0.1 \times 0.7 =$ _____

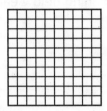

3. $0.4 \times 1.6 =$ _____

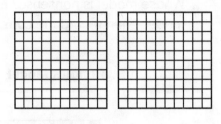

4. $0.3 \times 0.4 =$ _____

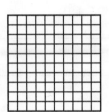

5. $0.9 \times 0.6 =$ _____

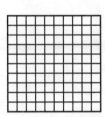

6. $0.5 \times 1.2 =$ _____

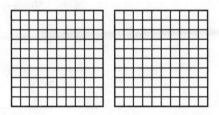

7. $0.8 \times 0.9 =$ _____

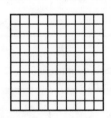

8. $0.5 \times 0.3 =$ _____

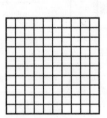

9. $0.5 \times 1.5 =$ _____

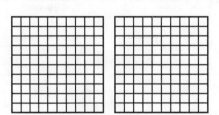

10. **Write Math** ▸ **Explain** why when you multiply and find one tenth of one tenth, it is equal to one hundredth.

_____

_____

_____

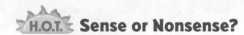
Problem Solving REAL WORLD

H.O.T. Sense or Nonsense?

11. Randy and Stacy used models to find 0.3 of 0.5. Both Randy's and
Stacy's models are shown below. Whose model makes sense?
Whose model is nonsense? **Explain** your reasoning below each
model. Then record the correct answer.

Randy's Model

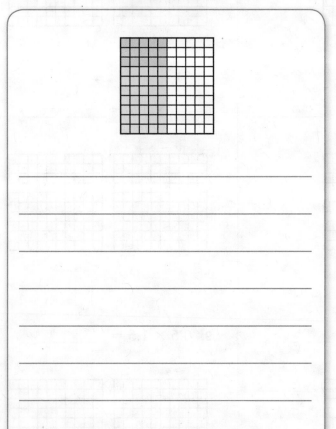

Stacy's Model

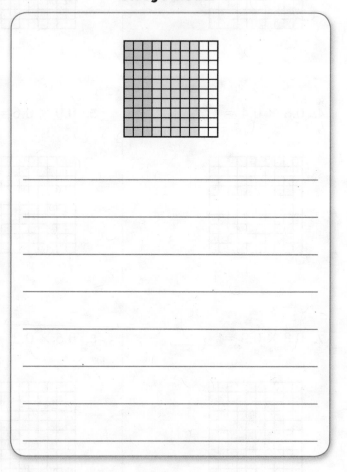

0.3 × 0.5 = _____

- For the answer that is nonsense, describe the error the student made.

_____

_____

Name _____

Multiply Decimals

Essential Question What strategies can you use to place a decimal point in a product?

CONNECT You can use what you have learned about patterns and place value to place the decimal point in the product when you multiply two decimals.

$1 \times 0.1 = 0.1$

$0.1 \times 0.1 = 0.01$

$0.01 \times 0.1 = 0.001$

Remember

When a number is multiplied by a decimal, the decimal point moves one place to the left in the product for each decreasing place value being multiplied.

🔑 UNLOCK the Problem REAL WORLD

A male leopard seal is measured and has a length of 2.8 meters. A male elephant seal is about 1.5 times as long. What length is the male elephant seal?

Multiply. 1.5×2.8

🔒 One Way Use place value.

STEP 1

Multiply as with whole numbers.

_____

STEP 2

Place the decimal point.

Think: Tenths are being multiplied by tenths. Use the pattern 0.1×0.1.

Place the decimal point so the value of the decimal is _____.

$$
\begin{array}{r}
28 \\
\times 15 \\
\hline
140 \\
+ 280 \\
\hline
420
\end{array}
$$

$28 \xrightarrow{\times 0.1} 2.8$ 1 place value

$\times 15 \xrightarrow{\times 0.1} \times 1.5$ 1 place value

$\boxed{}$ 1 + 1, or 2 place values

$420 \xleftarrow{\times 0.01}$

So, the length of a male elephant seal is about _____ meters.

- **What if** you multiplied 2.8 by 1.74? What would be the place value of the product? **Explain** your answer.

_____

_____

🔑 Another Way Use estimation.

You can use an estimate to place the decimal point in a product.

Multiply. 7.8 × 3.12

STEP 1

Esimate by rounding each factor to the nearest whole number.

7.8 × 3.12
↓ ↓
_____ × _____ = _____

$$\begin{array}{r} 312 \\ \times\ 78 \\ \hline \end{array} \qquad \begin{array}{r} 3.12 \\ \times\ 7.8 \\ \hline \end{array}$$

STEP 2

Multiply as with whole numbers.

_____

STEP 3

Use the estimate to place the decimal point.

Think: The product should be close to your estimate.

7.8 × 3.12 = _____

Share and Show 📝 MATH BOARD ・・・・・・・・・・・・・・・・・・・・・・・・・・

Place the decimal point in the product.

1.
$$\begin{array}{r} 3.62 \\ \times\ 1.4 \\ \hline 5\ 0\ 6\ 8 \end{array}$$
Think: A hundredth is being multiplied by a tenth. Use the pattern 0.01 × 0.1.

2.
$$\begin{array}{r} 6.8 \\ \times\ 1.2 \\ \hline 8\ 1\ 6 \end{array}$$
Estimate: 1 × 7 = _____

Find the product.

3.
$$\begin{array}{r} 0.9 \\ \times\ 0.8 \\ \hline \end{array}$$

✅ 4.
$$\begin{array}{r} 84.5 \\ \times\ 5.5 \\ \hline \end{array}$$

✅ 5.
$$\begin{array}{r} 2.39 \\ \times\ 2.7 \\ \hline \end{array}$$

Math Talk MATHEMATICAL PRACTICES

Explain how you might know the place value of the product for Exercise 5 before you solve.

Name _____

On Your Own ·

Find the product.

6. 7.9
    × 3.4

7. 9.2
    × 5.6

8. 3.45
    × 9.7

9. 45.3
    × 0.8

10. 6.98
    × 2.5

11. 7.02
    × 3.4

12. 14.9
    × 0.35

13. 50.99
    × 3.7

14. 18.43
    × 1.9

Practice: Copy and Solve **Find the product.**

15. 3.4×5.2

16. 0.9×2.46

17. 9.1×5.7

18. 4.8×6.01

19. 7.6×18.7

20. 1.5×9.34

21. 0.77×14.9

22. 3.3×58.14

Problem Solving REAL WORLD

23. Charlie has an adult Netherlands dwarf rabbit that weighs 1.2 kilograms. Cliff's adult Angora rabbit weighs 2.9 times as much as Charlie's rabbit. How much does Cliff's rabbit weigh?

_____

24. John has pet rabbits in an enclosure that has an area of 30.72 square feet. The enclosure Taylor is planning to build for his rabbits will be 2.2 times as large as John's. What will be the area of the enclosure Taylor is planning to build?

_____

SHOW YOUR WORK

25. H.O.T. A zoo is planning a new building for the penguin exhibit. First, they made a model that was 1.3 meters tall. Then, they made a more detailed model that was 1.5 times as tall as the first model. The building will be 2.5 times as tall as the height of the detailed model. What will be the height of the building?

_____

26. **Write Math** ▸ Leslie and Paul both solve the multiplication problem 5.5×4.6. Leslie says the answer is 25.30. Paul says the answer is 25.3. Whose answer is correct? **Explain** your reasoning.

_____

_____

_____

27. **Test Prep** A vine in Mr. Jackson's garden is 3.6 feet long. When it is measured again, it is 2.1 times as long. How long is the vine?

Ⓐ 5.7 feet   Ⓒ 7.5 feet

Ⓑ 6.6 feet   Ⓓ 7.56 feet

FOR MORE PRACTICE:
Standards Practice Book, pp. P93–P94

© Houghton Mifflin Harcourt Publishing Company

Zeros in the Product

Essential Question How do you know you have the correct number of decimal places in your product?

 UNLOCK the Problem REAL WORLD

CONNECT When decimals are multiplied, the product may not have enough digits to place the decimal point. In these cases, you may need to write additional zeros.

Students are racing typical garden snails and measuring the distance the snails travel in 1 minute. Chris's snail travels a distance of 0.2 foot. Jamie's snail travels 0.4 times as far as Chris's snail. How far does Jamie's snail travel?

• Using the given information, describe what you are being asked to find.

_____

_____

_____

Multiply. 0.4×0.2

STEP 1

Multiply as with whole numbers.

_____

STEP 2

Determine the position of the decimal point in the product.

Since tenths are being multiplied by tenths, the product will show _____.

STEP 3

Place the decimal point.

Are there enough digits in the product

to place the decimal point? _____

Write zeros, as needed, to the left of the whole number product to place the decimal point.

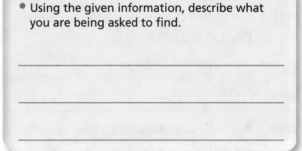

$$\begin{array}{ccc} 2 & \xrightarrow{\times 0.1} & 0.2 \quad \text{1 place value} \\ \times 4 & \xrightarrow{\times 0.1} & \times 0.4 \quad \text{1 place value} \\ \hline 8 & \xrightarrow{\times 0.01} & \boxed{}8 \quad \text{1 + 1, or 2 place values} \end{array}$$

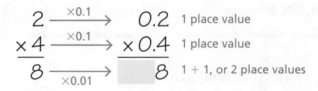

Math Talk MATHEMATICAL PRACTICES

Explain how you know when to write zeros in the product to place a decimal point.

So, Jamie's snail travels a distance of _____ foot.

🔑 Example   Multiply money.

Multiply. 0.2 × $0.30

STEP 1 Multiply as with whole numbers.

Think: The factors are 30 hundredths and 2 tenths.

What are the whole numbers you will multiply?

_____

STEP 2 Determine the position of the decimal
point in the product.

Since hundredths are being multiplied by tenths,

the product will show _____.

STEP 3 Place the decimal point. Write zeros to
the left of the whole number product
as needed.

Since the problem involves dollars and cents,
what place value should you use to show cents?

_____

So, 0.2 × $0.30 is _____.

$$\begin{array}{r} \$0.30 \\ \times\quad 0.2 \\ \hline \end{array}$$

Try This! Find the product.

0.2 × 0.05 = _____

What steps did you take to find the product?

_____

_____

_____

_____

_____

_____

Math Talk MATHEMATICAL PRACTICES
Explain why the
answer to the Try This! can have
a digit with a place value of
hundredths or thousandths and
still be correct.

Name _____

Share and Show

Write zeros in the product.

1.   0.05   **Think:** Hundredths
   × 0.7   are multiplied by
  ———— tenths. What should
     35   be the place value
            of the product?

2.   0.2
   × 0.3
  ————
     6

3.   0.02
   × 0.2
  ————
     4

Find the product.

4. $0.05
  × 0.8

5.   0.09
  × 0.7

6.   0.2
  × 0.1

Math Talk   MATHEMATICAL PRACTICES
Explain why 0.04 × 0.2
has the same product as 0.4 × 0.02.

On Your Own

Find the product.

7.   0.3
  × 0.3

8.   0.05
  × 0.3

9.   0.02
  × 0.4

10. $0.40
  × 0.1

11.   0.09
  × 0.2

12. $0.05
  × 0.6

13.   0.04
  × 0.5

14.   0.06
  × 0.8

 Algebra Find the value of *n*.

15. $0.03 \times 0.6 = n$

16. $n \times 0.2 = 0.08$

17. $0.09 \times n = 0.063$

$n =$ _____

$n =$ _____

$n =$ _____

🔑 UNLOCK the Problem   REAL WORLD

18. On an average day, a garden snail can travel about 0.05 mile. If a snail travels 0.2 times as far as the average distance in a day, how far can it travel?

Ⓐ 0.7 mile

Ⓑ 0.25 mile

Ⓒ 0.1 mile

Ⓓ 0.01 mile

a. What are you being asked to find? _____

b. What information will you use to solve the problem? _____

_____

_____

c. How will you use multiplication and place value to solve the problem? _____

_____

d. Show how you will solve the problem.

e. Fill in the bubble for the correct answer choice above.

---

19. In a science experiment, Tania uses 0.8 ounce of water to create a reaction. She wants the next reaction to be 0.1 times the size of the previous reaction. How much water should she use?

Ⓐ 0.08 ounce

Ⓑ 0.09 ounce

Ⓒ 0.8 ounce

Ⓓ 0.9 ounce

20. Michael multiplies 0.2 by a number. He records the product as 0.008. What number did Michael use?

Ⓐ 0.016

Ⓑ 0.04

Ⓒ 0.28

Ⓓ 0.4

FOR MORE PRACTICE:
Standards Practice Book, pp. P95–P96

Name _____

▶ Check Concepts

1. **Explain** how estimation helps you to place the decimal point when

   multiplying 3.9×5.3. _____

   _____

Complete the pattern.

2. $1 \times 7.45 =$ _____

   $10 \times 7.45 =$ _____

   $100 \times 7.45 =$ _____

   $1{,}000 \times 7.45 =$ _____

3. $10^0 \times 376.2 =$ _____

   $10^1 \times 376.2 =$ _____

   $10^2 \times 376.2 =$ _____

   $10^3 \times 376.2 =$ _____

4. $1 \times 191 =$ _____

   $0.1 \times 191 =$ _____

   $0.01 \times 191 =$ _____

Find the product.

5. $5 \times 0.89 =$ _____

6. $9 \times 2.35 =$ _____

7. $23 \times 8.6 =$ _____

8. $7.3 \times 0.6 =$ _____

9. $0.09 \times 0.7 =$ _____

10. $0.8 \times \$0.40 =$ _____

Draw a diagram to solve.

11. In January, Dawn earns $9.25 allowance. She earns
    3 times as much in February. If during March, she
    earns $5.75 more than she did in February, how
    much allowance does Dawn earn in March?

    _____

GO Online **Assessment Options Chapter Test**

Fill in the bubble completely to show your answer.

12. Janet hikes a trail at a local forest each day. The trail is 3.6 miles long, and she has hiked 5 days in the past week. How many miles has Janet hiked in the past week?

    Ⓐ 18 miles

    Ⓑ 15.3 miles

    Ⓒ 11 miles

    Ⓓ 8.6 miles

13. To earn money for his vacation, Grayson works at a local shop on weekends. His job is to cut bricks of fudge into 0.25 pound squares. If he cuts 36 equal-sized squares on Saturday, how many pounds of fudge has Grayson cut?

    Ⓐ 7.25 pounds

    Ⓑ 9 pounds

    Ⓒ 90 pounds

    Ⓓ 72.5 pounds

14. James is making a scale model of his bedroom. The model is 0.6 feet wide. If the actual room is 17.5 times as wide as the model, what is the width of James's room?

    Ⓐ 18.1 feet

    Ⓑ 17.11 feet

    Ⓒ 16.9 feet

    Ⓓ 10.5 feet

15. The cost of admission to the matinee showing at a movie theater is $6.75. If 7 friends want to see the matinee showing of their favorite movie, how much will it cost?

    Ⓐ $11.25

    Ⓑ $14.75

    Ⓒ $42.75

    Ⓓ $47.25

Fill in the bubble completely to show your answer.

16. On Friday, Gail talks for 38.4 minutes on her cell phone.
On Saturday, she uses 5.5 times as many minutes as she did on
Friday. How long does Gail talk on her cell phone on Saturday?

Ⓐ 2.112 minutes

Ⓑ 21.12 minutes

Ⓒ 211.2 minutes

Ⓓ 2,112 minutes

17. Harry walks to a produce market to buy bananas. If a pound of
bananas costs $0.49, how much will Harry pay for 3 pounds of
bananas?

Ⓐ $1.47

Ⓑ $3.49

Ⓒ $5.49

Ⓓ $10.47

18. At Anne's Fabric Emporium, a yard of chiffon fabric costs $7.85. Lee
plans to purchase 0.8 yard for a craft project. How much money will
Lee spend on chiffon fabric?

Ⓐ $0.63

Ⓑ $6.28

Ⓒ $7.05

Ⓓ $8.65

19. Mitchell has $18.79 in his savings account. Jeremy has 3 times as
much as Mitchell. Maritza has $4.57 more than Jeremy. How much
money does Maritza have in her savings account?

Ⓐ $13.71

Ⓑ $32.50

Ⓒ $56.37

Ⓓ $60.94

▶ Constructed Response

20. A river otter eats about 0.15 times its weight in food each day. At the Baytown Zoo, the male river otter weighs 5 pounds. About how much food will the otter at the zoo consume each day? **Explain** how you found your answer.

_____

_____

_____

▶ Performance Task

21. The cost of admission to the Baytown Zoo is shown below. Use the table to answer the questions.

Baytown Zoo Admission	
	(Cost per Person)
Senior Citizen	$10.50
Adult	$15.75
Child	$8.25

Ⓐ A family of 2 adults and 1 child plans to spend the day at the Baytown Zoo. How much does admission for the family cost? **Explain** how you found your answer.

_____

_____

Ⓑ **Describe** another way you could solve the problem.

_____

_____

Ⓒ **What if** 2 more tickets for admission are purchased? If the two additional tickets cost $16.50, determine what type of tickets the family purchases. **Explain** how you can determine the answer without calculating.

_____

_____

_____

198

5 Divide Decimals

Show What You Know ✓

Check your understanding of important skills.

Name _____

▶ **Division Facts** Find the quotient.

1. $6\overline{)24}$ = _____

2. $7\overline{)56}$ = _____

3. 18 ÷ 9 = _____

4. 35 ÷ 5 = _____

▶ **Estimate with 1-Digit Divisors** Estimate the quotient.

5. $6\overline{)253}$

6. $4\overline{)1,165}$

7. $7\overline{)1,504}$

_____

_____

_____

▶ **Division** Divide.

8. $34\overline{)785}$

9. $27\overline{)1,581}$

10. $41\overline{)4,592}$

Instead of telling Carmen her age, Sora gave her this clue. Be a Math Detective and find Sora's age.

Clue

My age is 10 more than one-tenth of one-tenth of one-tenth of 3,000.

Vocabulary Builder

▶ **Visualize It** ●

Complete the bubble map using review words.

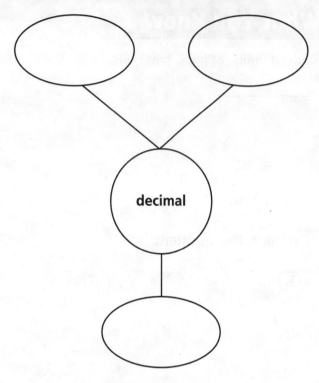

▶ **Understand Vocabulary** ●

Complete the sentences using the review words.

1. A _____ is a symbol used to separate the ones place from the tenths place in decimal numbers.

2. Numbers that are easy to compute with mentally are called

   _____.

3. A _____ is one of ten equal parts.

4. A number with one or more digits to the right of the decimal

   point is called a _____.

5. The _____ is the number that is to be divided in a division problem.

6. A _____ is one of one hundred equal parts.

7. You can _____ to find a number that is close to the exact amount.

GO Online ● eStudent Edition ● Multimedia eGlossary

Name _____

Division Patterns with Decimals

Essential Question How can patterns help you place the decimal point in a quotient?

🔑 UNLOCK the Problem — REAL WORLD

The Healthy Wheat Bakery uses 560 pounds of flour to make 1,000 loaves of bread. Each loaf contains the same amount of flour. How many pounds of flour are used in each loaf of bread?

You can use powers of ten to help you find quotients. Dividing by a power of 10 is the same as multiplying by 0.1, 0.01, or 0.001.

> • Underline the sentence that tells you what you are trying to find.
>
> • Circle the numbers you need to use.

🔒 One Way Use place-value patterns.

Divide. 560 ÷ 1,000

Look for a pattern in these products and quotients.

$560 \times 1 = 560$ $560 \div 1 = 560$

$560 \times 0.1 = 56.0$ $560 \div 10 = 56.0$

$560 \times 0.01 = 5.60$ $560 \div 100 = 5.60$

$560 \times 0.001 = 0.560$ $560 \div 1,000 = 0.560$

So, _____ pound of flour is used in each loaf of bread.

1. As you divide by increasing powers of 10, how does the position of the decimal point change in the quotients?

_____

🔒 Another Way Use exponents.

Divide. $560 \div 10^3$

Look for a pattern. $560 \div 10^0 = 560$

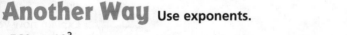

$560 \div 10^1 = 56.0$

$560 \div 10^2 = 5.60$

$560 \div 10^3 = $ _____

> **Remember**
> The zero power of 10 equals 1.
> $$10^0 = 1$$
> The first power of 10 equals 10.
> $$10^1 = 10$$

2. Each divisor, or power of 10, is 10 times the divisor before it. How do the quotients compare?

_____

CONNECT Dividing by 10 is the same as multiplying by 0.1 or finding $\frac{1}{10}$ of a number.

 **Example**

Liang used 25.5 pounds of tomatoes to make a large batch of salsa. He used one-tenth as many pounds of onions as pounds of tomatoes. He used one-hundredth as many pounds of green peppers as pounds of tomatoes. How many pounds of each ingredient did Liang use?

Tomatoes: 25.5 pounds

Onions: 25.5 pounds ÷ _____

Think: 25.5 ÷ 1 = _____

25.5 ÷ 10 = _____

Green Peppers: 25.5 pounds ÷ _____

Think: _____ ÷ 1 = _____

_____ ÷ 10 = _____

_____ ÷ 100 = _____ .

So, Liang used 25.5 pounds of tomatoes, _____ pounds of onions,

and _____ pound of green peppers.

Try This! Complete the pattern.

Ⓐ 32.6 ÷ 1 = _____

32.6 ÷ 10 = _____

32.6 ÷ 100 = _____

Ⓑ 50.2 ÷ 10^0 = _____

50.2 ÷ 10^1 = _____

50.2 ÷ 10^2 = _____

Math Talk MATHEMATICAL PRACTICES
Explain how you can determine where to place the decimal point in the quotient $47.3 ÷ 10^2$.

Share and Show

Complete the pattern.

1. $456 ÷ 10^0 = 456$

$456 ÷ 10^1 = 45.6$

$456 ÷ 10^2 = 4.56$

$456 ÷ 10^3 =$ _____

Think: The dividend is being divided by an increasing power of 10, so the decimal

point will move to the _____ one place for each increasing power of 10.

202

Name _____

Complete the pattern.

2. $225 \div 10^0 =$ _____

    $225 \div 10^1 =$ _____

    $225 \div 10^2 =$ _____

    $225 \div 10^3 =$ _____

3. $605 \div 10^0 =$ _____

    $605 \div 10^1 =$ _____

    $605 \div 10^2 =$ _____

    $605 \div 10^3 =$ _____

4. $74.3 \div 1 =$ _____

    $74.3 \div 10 =$ _____

    $74.3 \div 100 =$ _____

> **Math Talk**   MATHEMATICAL PRACTICES
>
> **Explain** what happens to the value of a number when you divide by 10, 100, or 1,000.

On Your Own ....

Complete the pattern.

5. $156 \div 1 =$ _____

    $156 \div 10 =$ _____

    $156 \div 100 =$ _____

    $156 \div 1{,}000 =$ _____

6. $32 \div 1 =$ _____

    $32 \div 10 =$ _____

    $32 \div 100 =$ _____

    $32 \div 1{,}000 =$ _____

7. $16 \div 10^0 =$ _____

    $16 \div 10^1 =$ _____

    $16 \div 10^2 =$ _____

    $16 \div 10^3 =$ _____

8. $12.7 \div 1 =$ _____

    $12.7 \div 10 =$ _____

    $12.7 \div 100 =$ _____

9. $92.5 \div 10^0 =$ _____

    $92.5 \div 10^1 =$ _____

    $92.5 \div 10^2 =$ _____

10. $86.3 \div 10^0 =$ _____

    $86.3 \div 10^1 =$ _____

    $86.3 \div 10^2 =$ _____

H.O.T.   **Algebra** Find the value of n.

11. $268 \div n = 0.268$

    $n =$ _____

12. $n \div 10^2 = 0.123$

    $n =$ _____

13. $n \div 10^1 = 4.6$

    $n =$ _____

Problem Solving REAL WORLD

Use the table to solve 14–16.

14. If each muffin contains the same amount of cornmeal, how many kilograms of cornmeal are in each corn muffin?

_____

15. **H.O.T.** If each muffin contains the same amount of sugar, how many kilograms of sugar, to the nearest thousandth, are in each corn muffin?

_____

16. **H.O.T.** The bakery decides to make only 100 corn muffins on Tuesday. How many kilograms of sugar will be needed?

_____

Dry Ingredients for 1,000 Corn Muffins

Ingredient	Number of Kilograms
Cornmeal	150
Flour	110
Sugar	66.7
Baking powder	10
Salt	4.17

17. **Write Math** ▶ **Explain** how you know that the quotient $47.3 \div 10^1$ is equal to the product 47.3×0.1.

_____

_____

_____

_____

_____

SHOW YOUR WORK

18. **Test Prep** Ella used 37.2 pounds of apples to make applesauce. She used one-tenth as many pounds of sugar as pounds of apples. How many pounds of sugar did Ella use?

Ⓐ 372 pounds

Ⓑ 3.72 pounds

Ⓒ 0.372 pound

Ⓓ 0.0372 pound

Name _____

Divide Decimals by Whole Numbers

Essential Question How can you use a model to divide a decimal by a whole number?

Investigate

Materials ■ decimal models ■ color pencils

Angela has enough wood to make a picture frame with a perimeter of 2.4 meters. She wants the frame to be a square. What will be the length of each side of the frame?

A. Shade decimal models to show 2.4.

B. You need to share your model among _____ equal groups.

C. Since 2 wholes cannot be shared among 4 groups without regrouping, cut your model apart to show the tenths.

There are _____ tenths in 2.4.

Share the tenths equally among the 4 groups.

There are _____ ones and _____ tenths in each group.

Write a decimal for the amount in each group. _____

D. Use your model to complete the number sentence.

2.4 ÷ 4 = _____

So, the length of each side of the frame will be _____ meter.

Draw Conclusions

1. **Explain** why you needed to cut apart the model in Step C.

   _____

   _____

2. **Explain** how your model would be different if the perimeter were 4.8 meters.

   _____

   _____

Make Connections

You can also use base-ten blocks to model division of a decimal by a whole number.

Materials ■ base-ten blocks

Kyle has a roll of ribbon 3.21 yards long. He cuts the ribbon into 3 equal lengths. How long is each piece of ribbon?

Divide. 3.21 ÷ 3

STEP 1

Use base-ten blocks to show 3.21.

Remember that a flat represents one, a long represents one tenth, and a small cube represents one hundredth.

There are _____ one(s), _____ tenth(s), and

_____ hundredth(s).

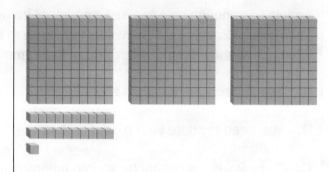

STEP 2 Share the ones.

Share an equal number of ones among 3 groups.

There is _____ one(s) shared in each group and _____ one(s) left over.

STEP 3 Share the tenths.

Two tenths cannot be shared among 3 groups without regrouping. Regroup the tenths by replacing them with hundredths.

There are _____ tenth(s) shared in each group and

_____ tenth(s) left over.

There are now _____ hundredth(s).

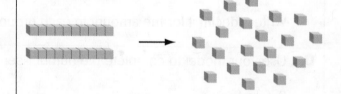

STEP 4 Share the hundredths.

Share the 21 hundredths equally among the 3 groups.

There are _____ hundredth(s) shared in each group

and _____ hundredth(s) left over.

So, each piece of ribbon is _____ yards long.

© Houghton Mifflin Harcourt Publishing Company

Math Talk MATHEMATICAL PRACTICES
Explain why your answer makes sense.

Name _____

Share and Show

Use the model to complete the number sentence.

1. $1.6 \div 4 =$ _____

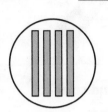 

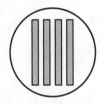

2. $3.42 \div 3 =$ _____

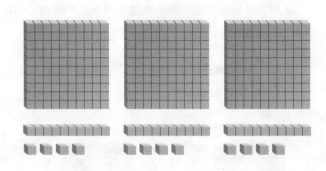

Divide. Use base-ten blocks.

3. $1.8 \div 3 =$ _____

4. $3.6 \div 4 =$ _____

5. $2.5 \div 5 =$ _____

6. $2.4 \div 8 =$ _____

7. $3.78 \div 3 =$ _____

8. $1.33 \div 7 =$ _____

9. $4.72 \div 4 =$ _____

10. $2.52 \div 9 =$ _____

11. $6.25 \div 5 =$ _____

Math Talk MATHEMATICAL PRACTICES
Explain how you can use inverse operations to find $1.8 \div 3$.

Problem Solving REAL WORLD

H.O.T. What's the Error?

12. Aida is making banners from a roll of paper that is 4.05 meters long. She will cut the paper into 3 equal lengths. How long will each banner be?

Look how Aida solved the problem. Find the error.

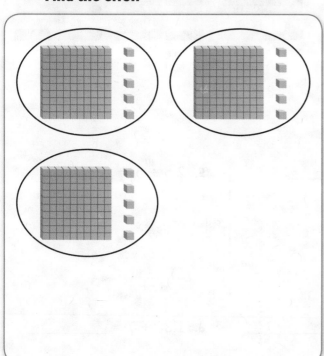

Solve the problem and correct the error.

So, Aida said that each banner would be _____ meters long,

but each banner should be _____ meters long.

• **Describe** Aida's error. _____

_____

_____

• **What if** the roll of paper were 4.35 meters long? How long would each banner be?

_____

Name _____

Estimate Quotients

Essential Question How can you estimate decimal quotients?

🔑 UNLOCK the Problem REAL WORLD

Carmen likes to ski. The ski resort where she goes to ski got 3.2 feet of snow during a 5-day period. The *average* daily snowfall for a given number of days is the quotient of the total amount of snow and the number of days. Estimate the average daily snowfall.

You can estimate decimal quotients by using compatible numbers. When choosing compatible numbers, you can look at the whole-number part of a decimal dividend or rename the decimal dividend as tenths or hundredths.

🔑 **Estimate.** 3.2 ÷ 5

Carly and her friend Marco each find an estimate. Since the divisor is greater than the dividend, they both first rename 3.2 as tenths.

3.2 is _____ tenths.

CARLY'S ESTIMATE	MARCO'S ESTIMATE
30 tenths is close to 32 tenths and divides easily by 5. Use a basic fact to find 30 tenths ÷ 5.	35 tenths is close to 32 tenths and divides easily by 5. Use a basic fact to find 35 tenths ÷ 5.
30 tenths ÷ 5 is _____ tenths or _____.	35 tenths ÷ 5 is _____ tenths or _____.
So, the average daily snowfall is about	So, the average daily snowfall is about
_____ foot.	_____ foot.

1. Whose estimate do you think is closer to the exact quotient?

   Explain your reasoning. _____

   _____

2. **Explain** how you would rename the dividend in 29.7 ÷ 40 to choose compatible numbers and estimate the quotient.

   _____

   _____

Estimate with 2-Digit Divisors

When you estimate quotients with compatible numbers, the number you use for the dividend can be greater than the dividend or less than the dividend.

Example

A group of 31 students is going to visit the museum. The total cost for the tickets is $144.15. About how much money will each student need to pay for a ticket?

Estimate. $144.15 ÷ 31

A **Use a whole number greater than the dividend.**

Use 30 for the divisor. Then find a number close to and greater than $144.15 that divides easily by 30.

$144.15 ÷ 31
   ↓   ↓
 $150   ÷ 30 = $ _____

So, each student will pay about $ _____ for a ticket.

B **Use a whole number less than the dividend.**

Use 30 for the divisor. Then find a number close to and less than $144.15 that divides easily by 30.

$144.15 ÷ 31
   ↓   ↓
 $120   ÷ 30 = $ _____

So, each student will pay about $ _____ for a ticket.

3. Which estimate do you think will be a better estimate of the cost

of a ticket? **Explain** your reasoning. _____

_____

_____

Share and Show . . . .

Use compatible numbers to estimate the quotient.

1. 28.8 ÷ 9

_____ ÷ _____ = _____

2. 393.5 ÷ 41

_____ ÷ _____ = _____

Estimate the quotient.

3. $161.7 \div 7$

4. $17.9 \div 9$

5. $145.4 \div 21$

Math Talk | MATHEMATICAL PRACTICES
Explain why you might want to find an estimate for a quotient.

On Your Own ·

Estimate the quotient.

6. $15.5 \div 4$

7. $394.8 \div 7$

8. $410.5 \div 18$

9. $72.1 \div 7$

10. $32.4 \div 52$

11. $\$134.42 \div 28$

12. $21.8 \div 4$

13. $3.4 \div 5$

14. $\$759.92 \div 42$

15. $157.5 \div 38$

16. $379.2 \div 6$

17. $108.4 \div 21$

Problem Solving REAL WORLD

Use the table to solve 18–20.

18. Estimate the average daily snowfall for Alaska's greatest 7-day snowfall.

_____

19. How does the estimate of the average daily snowfall for Wyoming's greatest 7-day snowfall compare to the estimate of the average daily snowfall for South Dakota's greatest 7-day snowfall?

_____

_____

Greatest 7-Day Snowfall	
State	**Amount (in inches)**
Alaska	186.9
Wyoming	84.5
South Dakota	112.7

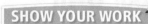
SHOW YOUR WORK

20. **H.O.T.** The greatest monthly snowfall total in Alaska is 297.9 inches. This happened in February, 1953. Compare the daily average snowfall for February, 1953, with the average daily snowfall for Alaska's greatest 7-day snowfall. Use estimation.

_____

_____

21. **Write Math** ▶ **What's the Error?** During a 3-hour storm, it snowed 2.5 inches. Jacob said that it snowed an average of about 8 inches per hour.

_____

_____

_____

22. **Test Prep** A plant grew 23.8 inches over 8 weeks. Which is the best estimate of the average number of inches the plant grew each week?

Ⓐ 0.2 inch   Ⓒ 2 inches

Ⓑ 0.3 inch   Ⓓ 3 inches

FOR MORE PRACTICE:
Standards Practice Book, pp. P105–P106

Name _____

Division of Decimals by Whole Numbers

Essential Question How can you divide decimals by whole numbers?

🔑 UNLOCK the Problem REAL WORLD

In a swimming relay, each swimmer swims an equal part of the total distance. Brianna and 3 other swimmers won a relay in 5.68 minutes. What is the average time each girl swam?

> • How many swimmers are part of the relay team?
>
> _____

🔓 One Way Use place value.

MODEL	THINK AND RECORD

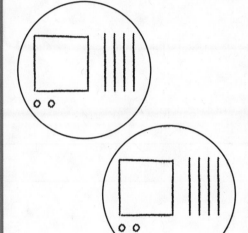

STEP 1 Share the ones.

$$4\overline{)5.68} \quad \begin{array}{l} 1 \\ \end{array}$$

-4

Divide. 5 ones ÷ 4

Multiply. 4 × 1 one(s)

Subtract. 5 ones − 4 ones

Check. _____ one(s) cannot be shared among 4 groups without regrouping.

STEP 2 Share the tenths.

$$4\overline{)5.68}$$
$$1$$
$$-4 \downarrow$$
$$-$$

Divide. _____ tenths ÷ 4

Multiply. 4 × _____ tenths

Subtract. _____ tenths − _____ tenths

Check. _____ tenth(s) cannot be shared among 4 groups.

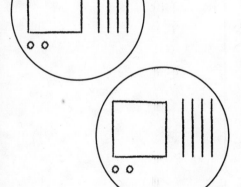

STEP 3 Share the hundredths.

$$4\overline{)5.68}$$
$$1$$
$$-4 \downarrow$$
$$16$$
$$-16 \downarrow$$
$$-$$

Divide. 8 hundredth(s) ÷ 4

Multiply. 4 × _____ hundredths

Subtract. _____ hundredths − _____ hundredths

Check. _____ hundredth(s) cannot be shared among 4 groups.

Place the decimal point in the quotient to separate the ones and the tenths.

So, each girl swam an average of _____ minutes.

🔒 Another Way Use an estimate.

Divide as you would with whole numbers.

Divide. $40.89 ÷ 47

- Estimate the quotient. 4,000 hundredths ÷ 50 = 80 hundredths, or $0.80

- Divide the tenths.

- Divide the hundredths. When the remainder is zero and there are no more digits in the dividend, the division is complete.

- Use your estimate to place the decimal point. Place a zero to show there are no ones.

$$47\overline{)40.89}$$

So, $40.89 ÷ 47 is _____.

- **Explain** how you used the estimate to place the decimal point in the quotient.

  _____

  _____

Try This! Divide. Use multiplication to check your work.

$$23\overline{)79.35}$$

Check.

$$\begin{array}{r} \\ \times\ \ 23 \\ \hline \\ + \\ \hline \\ \end{array}$$

Share and Show ··

Write the quotient with the decimal point placed correctly.

1. 4.92 ÷ 2 = 246 _____

2. 50.16 ÷ 38 = 132 _____

Name _____

Divide.

3. $5\overline{)8.65}$

 4. $3\overline{)2.52}$

 5. $27\overline{)97.2}$

Math Talk MATHEMATICAL PRACTICES
Explain how you can check that the decimal point is placed correctly in the quotient.

On Your Own . . . .

Divide.

6. $6\overline{)8.94}$

7. $5\overline{)3.75}$

8. $19\overline{)55.1}$

9. $23\overline{)52.9}$

10. $8\overline{)\$8.24}$

11. $5\overline{)44.5}$

Practice: Copy and Solve Divide.

12. $3\overline{)\$7.71}$

13. $14\overline{)79.8}$

14. $33\overline{)25.41}$

15. $7\overline{)15.61}$

16. $14\overline{)137.2}$

17. $34\overline{)523.6}$

H.O.T. **Algebra** Write the unknown number for each ▦.

18. ▦ ÷ 5 = 1.21

19. 46.8 ÷ 1.2 = ▦

20. 34.1 ÷ ▦ = 22

▦ = _____

▦ = _____

▦ = _____

UNLOCK the Problem   REAL WORLD

21. The standard width of 8 lanes in swimming pools used for competitions is 21.92 meters. The standard width of 9 lanes is 21.96 meters. How much wider is each lane when there are 8 lanes than when there are 9 lanes?

Ⓐ 0.30 meter   Ⓒ 2.74 meters

Ⓑ 2.44 meters   Ⓓ 22.28 meters

a. What are you asked to find? _____

_____

b. What operations will you use to solve the problem? _____

_____

c. Show the steps you used to solve the problem.

d. Complete the sentences.

Each lane is _____ meters wide when there are 8 lanes.

Each lane is _____ meters wide when there are 9 lanes.

Since _____ – _____ = _____, the

lanes are _____ meter(s) wider when there are 8 lanes than when there are 9 lanes.

e. Fill in the bubble for the correct answer choice.

---

22. Robert pays $32.04 for 6 student tickets to the basketball game. What is the cost of each student ticket?

Ⓐ $192.24   Ⓒ $26.04

Ⓑ $53.40   Ⓓ $5.34

23. Jasmine uses 14.24 pounds of fruit for 16 servings of fruit salad. If each serving contains the same amount of fruit, how much fruit is in each serving?

Ⓐ 0.089 pound   Ⓒ 1.76 pounds

Ⓑ 0.89 pound   Ⓓ 17.6 pounds

Name _____

▶ Concepts and Skills

1. **Explain** how the position of the decimal point changes in a quotient as you divide by increasing powers of 10.

_____

2. **Explain** how you can use base-ten blocks to find $2.16 ÷ 3$.

_____

_____

_____

_____

Complete the pattern.

3. $223 ÷ 1 =$ _____

   $223 ÷ 10 =$ _____

   $223 ÷ 100 =$ _____

   $223 ÷ 1,000 =$ _____

4. $61 ÷ 1 =$ _____

   $61 ÷ 10 =$ _____

   $61 ÷ 100 =$ _____

   $61 ÷ 1,000 =$ _____

5. $57.4 ÷ 10^0 =$ _____

   $57.4 ÷ 10^1 =$ _____

   $57.4 ÷ 10^2 =$ _____

Estimate the quotient.

6. $31.9 ÷ 4$

7. $6.1 ÷ 8$

8. $492.6 ÷ 48$

Divide.

9. $5\overline{)4.35}$

10. $8\overline{)9.92}$

11. $61\overline{)207.4}$

Fill in the bubble completely to show your answer.

12. The Westside Bakery uses 440 pounds of sugar to make 1,000 cakes. Each cake contains the same amount of sugar. How many pounds of sugar are used in each cake?

Ⓐ 0.044 pound

Ⓑ 0.44 pound

Ⓒ 4.4 pounds

Ⓓ 44 pounds

13. Elise pays $21.75 for 5 student tickets to the fair. What is the cost of each student ticket?

Ⓐ $4.35

Ⓑ $16.75

Ⓒ $43.40

Ⓓ $108.75

14. Jason has a piece of wire that is 62.4 inches long. He cuts the wire into 3 equal pieces. Which is the best estimate of the length of each piece of wire?

Ⓐ 2 inches

Ⓑ 3 inches

Ⓒ 20 inches

Ⓓ 30 inches

15. Elizabeth uses 33.75 ounces of granola for 15 servings of trail mix. If each serving contains the same amount of granola, how much granola is in each serving?

Ⓐ 0.225 ounce

Ⓑ 2.25 ounces

Ⓒ 18.75 ounces

Ⓓ 33.9 ounces

Name _____

Decimal Division

Essential Question How can you use a model to divide by a decimal?

Investigate

Materials ■ decimal models ■ color pencils

Leigh is making reusable shopping bags. She has 3.6 yards of fabric. She needs 0.3 yard of fabric for each bag. How many shopping bags can she make from the 3.6 yards of fabric?

A. Shade decimal models to show 3.6.

B. Cut apart your model to show the tenths. Separate the tenths into as many groups of 3 tenths as you can.

There are _____ groups of _____ tenths.

C. Use your model to complete the number sentence.

3.6 ÷ 0.3 = _____

So, Leigh can make _____ shopping bags.

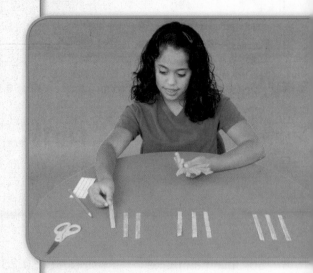

Draw Conclusions ·

1. **Explain** why you made each group equal to the divisor.

   _____

   _____

2. **Identify** the problem you would be modeling if each strip in the model represents 1.

   _____

Remember
The divisor can tell the number of same-sized groups, or it can tell the number in each group.

3. Dennis has 2.7 yards of fabric to make bags that require 0.9 yard of fabric each. **Describe** a decimal model you can use to find how many bags he can make.

   _____

   _____

Make Connections ....

You can also use a model to divide by hundredths.

Materials ■ decimal models ■ color pencils

Julie has $1.75 in nickels. How many stacks of $0.25 can she make from $1.75?

STEP 1

Shade decimal models to show 1.75.

There are _____ one(s) and _____ hundredth(s).

STEP 2

Cut apart your model to show groups of 0.25.

There are _____ groups of _____ hundredths.

STEP 3

Use your model to complete the number sentence.

1.75 ÷ 0.25 = _____

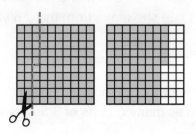

So, Julie can make _____ stacks of $0.25 from $1.75.

Math Talk MATHEMATICAL PRACTICES
Explain how to use decimal models to find 3 ÷ 0.75.

Share and Show MATH BOARD ●●●●●●●●●●●●●●●●●●●●●●●●●●

Use the model to complete the number sentence.

1. 1.2 ÷ 0.3 = _____

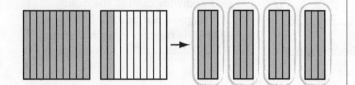

2. 0.45 ÷ 0.09 = _____

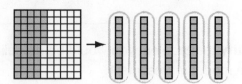

3. 0.96 ÷ 0.24 = _____

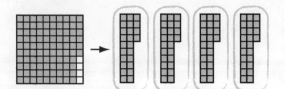

4. 1 ÷ 0.5 = _____

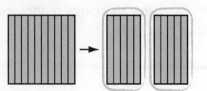

© Houghton Mifflin Harcourt Publishing Company

Name _____

Divide. Use decimal models.

5. $1.8 \div 0.6 =$ _____

6. $1.2 \div 0.3 =$ _____

7. $0.24 \div 0.04 =$ _____

8. $1.75 \div 0.35 =$ _____

9. $2 \div 0.4 =$ _____

10. $2.7 \div 0.9 =$ _____

11. $1.24 \div 0.62 =$ _____

12. $0.84 \div 0.14 =$ _____

✓ **13.** $1.6 \div 0.4 =$ _____

Use the model to find the unknown value.

14. $2.4 \div$ _____ $= 3$

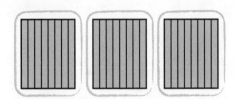

15. _____ $\div 0.32 = 4$

16. H.O.T. Make a model to find $0.6 \div 0.15$. **Describe** your model.

_____

_____

17. Write Math ➤ **Explain**, using the model, what the equation represents in Exercise 15.

_____

_____

Problem Solving REAL WORLD

Pose a Problem

18. Emilio buys 1.2 kilograms of grapes. He separates the grapes into packages that contain 0.3 kilogram of grapes each. How many packages of grapes does Emilio make?

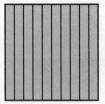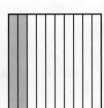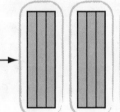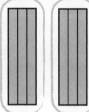

$1.2 \div 0.3 = 4$

Emilio made 4 packages of grapes.

Write a new problem using a different amount for the weight in each package. The amount should be a decimal with tenths. Use a total amount of 1.5 kilograms of grapes. Then use decimal models to solve your problem.

Pose a problem.

Solve your problem. Draw a picture of the model you used to solve your problem.

- **Explain** why you chose the amount you did for your problem.

Name _____

Divide Decimals

Essential Question How can you place the decimal point in the quotient?

When you multiply both the divisor and the dividend by the same power of 10, the quotient stays the same.

divisor		dividend			divisor		dividend		
6	÷	3	= 2		120	÷	30	= 4	
↓ × 10		↓ × 10			↓ × 0.1		↓ × 0.1		
60	÷	30	= 2		12	÷	3	= 4	
↓ × 10		↓ × 10			↓ × 0.1		↓ × 0.1		
600	÷	300	= 2		1.2	÷	0.3	= 4	

UNLOCK the Problem REAL WORLD

Matthew has $0.72. He wants to buy stickers that cost $0.08 each. How many stickers can he buy?

- Multiply both the dividend and the divisor by the power of 10 that makes the divisor a whole number. Then divide.

  0.72 ÷ 0.08 = ☐

  ↓ × 100 ↓ × 100

  72 ÷ 8 = ☐

So, Matthew can buy _____ stickers.

- What do you multiply hundredths by to get a whole number?

  _____

1. **Explain** how you know that the quotient 0.72 ÷ 0.08 is equal to the quotient 72 ÷ 8.

_____

_____

Try This! Divide. 0.56 ÷ 0.7

- Multiply the divisor by a power of 10 to make it a whole number. Then multiply the dividend by the same power of 10.

  0.7 × _____ = _____

  0.56 × _____ = _____

- Divide.

  ☐
  07.)5.6
  ↪ ↪

🔑 Example

Sherri hikes on the Pacific Coast trail. She plans to hike 3.72 miles. If she hikes at an average speed of 1.2 miles per hour, how long will she hike?

Divide. 3.72 ÷ 1.2

Estimate. _____

STEP 1	STEP 2	STEP 3
Multiply the divisor by a power of 10 to make it a whole number. Then, multiply the dividend by the same power of 10.	Write the decimal point in the quotient above the decimal point in the new dividend.	Divide.

STEP 1

1.2 × _____ = _____

3.72 × _____ = _____

STEP 2

$$12\overline{)37.2}$$

STEP 3

$$12\overline{)37.2}$$

So, Sherri will hike _____ hours.

2. **Describe** what happens to the decimal point in the divisor and in the dividend when you multiply by 10.

_____

3. **Explain** how you could have used the estimate to place the decimal point.

_____

_____

Try This!

Divide. Check your answer.

$$0.14\overline{)1.96}$$

Multiply the divisor and the

dividend by _____.

$$\begin{array}{r} 0.14 \\ \times\ \underline{} \\ \\ +\ \underline{} \\ \end{array}$$

Name _____

Share and Show

Copy and complete the pattern.

1. $45 \div 9 =$ _____

   $4.5 \div$ _____ $= 5$

   _____ $\div 0.09 = 5$

2. $175 \div 25 =$ _____

   $17.5 \div$ _____ $= 7$

   _____ $\div 0.25 = 7$

3. $164 \div 2 =$ _____

   $16.4 \div$ _____ $= 82$

   _____ $\div 0.02 = 82$

Divide.

✓ 4. $1.6\overline{)9.6}$

5. $0.3\overline{)0.24}$

✓ 6. $3.45 \div 1.5$

Math Talk MATHEMATICAL PRACTICES
Explain how you know that your quotient for Exercise 5 will be less than 1.

On Your Own

Divide.

7. $0.6\overline{)13.2}$

8. $0.3\overline{)0.9}$

9. $0.26\overline{)1.56}$

10. $0.45\overline{)5.85}$

11. $0.3\overline{)0.69}$

12. $3.6 \div 0.4$

13. $1.26 \div 2.1$

14. $7.84 \div 0.28$

15. $9.28 \div 2.9$

Problem Solving REAL WORLD

Use the table to solve 16–19.

16. Connie paid $1.08 for pencils. How many pencils did she buy?

_____

17. Albert has $2.16. How many more pencils can he buy than markers?

_____

18. How many erasers can Ayita buy for the same amount that she would pay for one notepad?

_____

19. **H.O.T.** Ramon paid $3.25 for notepads and $1.44 for markers. What is the total number of items he bought?

_____

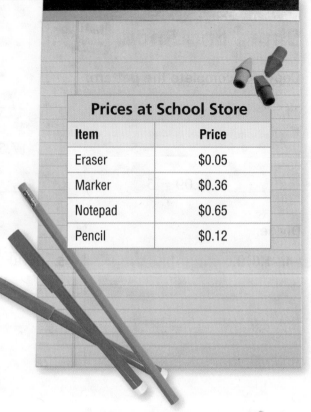

Prices at School Store

Item	Price
Eraser	$0.05
Marker	$0.36
Notepad	$0.65
Pencil	$0.12

20. **Write Math** ▸ **What's the Error?** Katie divided 4.25 by 0.25 and got a quotient of 0.17.

_____

_____

_____

_____

SHOW YOUR WORK

21. **Test Prep** Marcus bought apples that cost $0.45 per pound. He paid $1.35 for the apples. How many pounds of apples did he buy?

Ⓐ 0.3 pound

Ⓑ 2.8 pounds

Ⓒ 3 pounds

Ⓓ 30 pounds

Write Zeros in the Dividend

Essential Question When do you write a zero in the dividend to find a quotient?

CONNECT When decimals are divided, the dividend may not have enough digits for you to complete the division. In these cases, you can write zeros to the right of the last digit.

🔑 UNLOCK the Problem · REAL WORLD

The equivalent fractions show that writing zeros to the right of a decimal does not change the value.

$$90.8 = 90\frac{8 \times 10}{10 \times 10} = 90\frac{80}{100} = 90.80$$

During a fund-raising event, Adrian rode his bicycle 45.8 miles in 4 hours. Find his speed in miles per hour by dividing the distance by the time.

Divide. 45.8 ÷ 4   **Estimate.** 44 ÷ 4 = _____

STEP 1

Write the decimal point in the quotient above the decimal point in the dividend.

$$4\overline{)45.8}$$

STEP 2

Divide the tens, ones, and tenths.

$$4\overline{)45.8}$$

$$-\underline{}$$

$$-\underline{}$$

$$-\underline{}$$

STEP 3

Write a zero in the dividend and continue dividing.

$$4\overline{)45.80}$$
$$\underline{-4}$$
$$05$$
$$\underline{-4}$$
$$18$$
$$\underline{-16}\downarrow$$

$$-\underline{}$$

So, Adrian's speed was _____ miles per hour.

Math Talk MATHEMATICAL PRACTICES
Explain how you would model this problem using base-ten blocks.

Chapter 5 **227**

CONNECT When you divide whole numbers, you can show the amount that is left over by writing a remainder or a fraction. By writing zeros in the dividend, you can also show that amount as a decimal.

🔑 Example   Write zeros in the dividend.

Divide. 372 ÷ 15

- Divide until you have an amount less than the divisor left over.
- Insert a decimal point and a zero at the end of the dividend.
- Place a decimal point in the quotient above the decimal point in the dividend.
- Continue dividing.

$$
\begin{array}{r}
24. \\
15\overline{)372.0} \\
-30 \\
\hline
72 \\
-60 \\
\hline
\\
- \\
\hline
\end{array}
$$

So, 372 ÷ 15 = _____.

- Sarah has 78 ounces of rice. She puts an equal amount of rice in each of 12 bags. What amount of rice does she put in each bag? **Explain** how you would write the answer using a decimal.

_____

_____

Try This! Divide. Write a zero at the end of the dividend as needed.

Divide. 1.23 ÷ 0.06

$$
006.\overline{)123.}
$$

$$
\begin{array}{r}
20. \\
6\overline{)123.0} \\
-12 \\
\hline
03 \\
-0 \\
\hline
30 \\
- \\
\hline
\end{array}
$$

Divide. 10 ÷ 0.8

$$
08.\overline{)100.}
$$

$$
8.\overline{)100.}
$$

Name _____

Share and Show

Write the quotient with the decimal point placed correctly.

1. $5 \div 0.8 = 625$

2. $26.1 \div 6 = 435$

3. $0.42 \div 0.35 = 12$

4. $80 \div 50 = 16$

Divide.

5. $4 \overline{)32.6}$

6. $1.2 \overline{)9}$

☑ **7.** $15 \overline{)42}$

☑ **8.** $0.14 \overline{)0.91}$

Math Talk MATHEMATICAL PRACTICES

Explain why you would write a zero in the dividend when dividing decimals.

On Your Own

Divide.

9. $8 \overline{)84}$

10. $2.5 \overline{)4}$

11. $5 \overline{)16.2}$

12. $0.6 \overline{)2.7}$

13. $18 \div 7.5$

14. $34.8 \div 24$

15. $5.16 \div 0.24$

16. $81 \div 18$

Practice: Copy and Solve Divide.

17. $1.6 \overline{)20}$

18. $15 \overline{)4.8}$

19. $0.54 \overline{)2.43}$

20. $28 \overline{)98}$

21. $1.8 \div 12$

22. $3.5 \div 2.5$

23. $40 \div 16$

24. $2.24 \div 0.35$

Problem Solving REAL WORLD

Solve.

25. Jerry takes trail mix on hikes. A package of dried apricots weighs 25.5 ounces. Jerry divides the apricots equally among 6 bags of trail mix. How many ounces of apricots are in each bag?

_____

26. **H.O.T.** Amy has 3 pounds of raisins. She divides the raisins equally into 12 bags. How many pounds of raisins are in each bag? Tell how many zeros you had to write at the end of the dividend.

_____

27. **Write Math** ▶ Find $65 \div 4$. Write your answer using a remainder, a fraction, and a decimal. Then tell which form of the answer you prefer. **Explain** your choice.

_____

_____

_____

_____

28. **Test Prep** Todd has a piece of rope that is 1.6 meters long. He cuts the rope into 5 equal pieces. What is the length of each piece?

(A) 0.8 meter

(B) 0.32 meter

(C) 3.2 meters

(D) 8 meters

Connect to Science

Rate of Speed Formula

The formula for velocity, or rate of speed, is $r = d \div t$, where r represents rate of speed, d represents distance, and t represents time. For example, if an object travels 12 feet in 10 seconds, you can find its rate of speed by using the formula.

$r = d \div t$

$r = 12 \div 10$

$r = 1.2$ feet per second

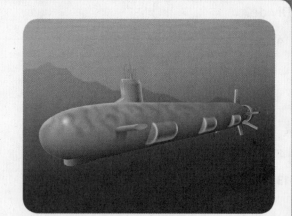

Use division and the formula for rate of speed to solve.

29. A car travels 168 miles in 3.2 hours. Find the car's rate of speed in miles per hour.

_____

30. A submarine travels 90 kilometers in 4 hours. Find the submarine's rate of speed in kilometers per hour.

_____

Problem Solving • Decimal Operations

Essential Question How do you use the strategy *work backward* to solve multistep decimal problems?

🔑 UNLOCK the Problem REAL WORLD

Carson spent $15.99 for 2 books and 3 pens. The books cost $4.95 each and sales tax was $1.22. Carson also used a coupon for $0.50 off his purchase. If each pen had the same cost, how much did each pen cost?

Read the Problem

What do I need to find?	What information do I need to use?	How will I use the information?

Solve the Problem

• Make a flowchart to show the information. Then using inverse operations, work backward to solve.

Cost of 3 pens	*plus* →	Cost of 2 books	*plus* →	Amount of tax	*minus* →	Amount of Coupon	*equals* →	Total Spent
3 × cost of each pen	+	2 × ☐	+	☐	−	☐	=	☐

Total Spent	*plus* →	Amount of Coupon	*minus* →	Amount of tax	*minus* →	Cost of 2 books	*equals* →	Cost of 3 pens
☐	+	☐	−	☐	−	☐	=	☐

• Divide the cost of 3 pens by 3 to find the cost of each pen.

_____ ÷ 3 = _____

> **Math Talk** MATHEMATICAL PRACTICES
> Explain why the amount of the coupon was added when you worked backward.

So, the cost of each pen was _____.

🔒 Try Another Problem

Last week, Vivian spent a total of $20.00. She spent $9.95 for tickets to the school fair, $5.95 for food, and the rest for 2 rings that were on sale at the school fair. If each ring had the same cost, how much did each ring cost?

Read the Problem

What do I need to find?	What information do I need to use?	How will I use the information?

Solve the Problem

So, the cost of each ring was _____.

Math Talk MATHEMATICAL PRACTICES
Explain how you can check your answer.

Name _____

Share and Show

1. Hector spent $36.75 for 2 DVDs with the same cost. The sales tax was $2.15. Hector also used a coupon for $1.00 off his purchase. How much did each DVD cost?

   First, make a flowchart to show the information and show how you would work backward.

   Then, work backward to find the cost of 2 DVDs.

   _____

   Finally, find the cost of one DVD.

   _____

   So, each DVD costs _____.

2. **What if** Hector spent $40.15 for the DVDs, the sales tax was $2.55, and he didn't have a coupon? How much would each DVD cost?

   _____

3. Sophia spent $7.30 for school supplies. She spent $3.00 for a notebook and $1.75 for a pen. She also bought 3 large erasers. If each eraser had the same cost, how much did she spend for each eraser?

   _____

SHOW YOUR WORK

On Your Own

Choose a
STRATEGY

Act It Out
Draw a Diagram
Make a Table
Solve a Simpler Problem
Work Backward
Guess, Check, and Revise

4. The change from a gift purchase was $3.90. Each of 6 students donated an equal amount for the gift. How much change should each student receive?

_____

5. If you divide this mystery number by 4, add 8, and multiply by 3, you get 42. What is the mystery number?

_____

SHOW YOUR WORK

6. H.O.T. A mail truck picks up two boxes of mail from the post office. The total weight of the boxes is 32 pounds. One box is 8 pounds heavier than the other box. How much does each box weigh?

_____

7. Stacy buys 3 CDs in a set for $29.98. She saved $6.44 by buying the set instead of buying the individual CDs. If each CD costs the same amount, how much does each of the 3 CDs cost when purchased individually?

_____

8. A school cafeteria sold 1,280 slices of pizza the first week, 640 the second week, and 320 the third week. If this pattern continues, in what week will the cafeteria sell 40 slices? **Explain** how you got your answer.

_____

_____

_____

9. **Test Prep** While working at the school store, John sold a jacket for $40.00 and notebooks for $1.50 each. If he collected $92.50, how many notebooks did he sell?

(A) 3.5   (C) 35

(B) 6.1   (D) 61

Name _____

▶ **Concepts and Skills**

Complete the pattern.

1. $341 \div 1 =$ _____

$341 \div 10 =$ _____

$341 \div 100 =$ _____

$341 \div 1,000 =$ _____

2. $15 \div 1 =$ _____

$15 \div 10 =$ _____

$15 \div 100 =$ _____

$15 \div 1,000 =$ _____

3. $68.2 \div 10^0 =$ _____

$68.2 \div 10^1 =$ _____

$68.2 \div 10^2 =$ _____

Estimate the quotient.

4. $49.3 \div 6$

5. $3.5 \div 4$

6. $396.5 \div 18$

Divide.

7. $6\overline{)3.24}$

8. $5\overline{)6.55}$

9. $26\overline{)96.2}$

10. $1.08 \div 0.4$

11. $8.84 \div 0.68$

12. $7.31 \div 1.7$

13. $9.18 \div 0.9$

14. $12.7 \div 5$

15. $8.33 \div 0.34$

GO Online **Assessment Options**
Chapter Test

Fill in the bubble completely to show your answer.

16. The Orchard Pie Company uses 95 pounds of apples to make 100 pies. Each pie contains the same amount of apples. How many pounds of apples are used in each pie?

    (A) 0.095 pound

    (B) 0.95 pound

    (C) 9.5 pounds

    (D) 95 pounds

17. During a special sale, all CDs have the same price. Mr. Ortiz pays $228.85 for 23 CDs. Which is the best estimate of the price of each CD?

    (A) $9

    (B) $10

    (C) $12

    (D) $13

18. Ryan earns $20.16 working for 3 hours. How much does he earn per hour?

    (A) $60.48

    (B) $6.82

    (C) $6.72

    (D) $6.71

19. Anna hikes 6.4 miles during a 4-day vacation. If she hikes the same distance each day, how many miles does she hike each day?

    (A) 1.06 miles

    (B) 1.1 miles

    (C) 1.4 miles

    (D) 1.6 miles

Name _____

Fill in the bubble completely to show your answer.

20. Karina pays $1.92 for pencil erasers. The erasers cost $0.08 each. How many erasers does she buy?

Ⓐ 2.4

Ⓑ 2.5

Ⓒ 24

Ⓓ 25

21. Wyatt has 25.4 ounces of fruit juice. He divides the juice equally into 4 glasses. How much juice is in each glass?

Ⓐ 6 ounces

Ⓑ 6.35 ounces

Ⓒ 6.4 ounces

Ⓓ 6.45 ounces

22. Jacob walks 70.4 feet in 0.2 hour. If he walks at the same rate the whole time, what is his speed in feet per hour?

Ⓐ 352 feet per hour

Ⓑ 140.8 feet per hour

Ⓒ 35.2 feet per hour

Ⓓ 14.08 feet per hour

23. Meghan earns $20.00 by walking dogs. She uses all of her earnings to buy a shirt for $12.85 and some stickers for $0.65 each. How many stickers does she buy?

Ⓐ 4.65

Ⓑ 11

Ⓒ 46

Ⓓ 110

Constructed Response

24. Percy buys tomatoes that cost $0.58 per pound. He pays $2.03 for the tomatoes. How many pounds of tomatoes does he buy? Show your work using words, pictures, or numbers. **Explain** how you know your answer is reasonable.

_____

_____

_____

Performance Task

25. Isabella is buying art supplies. The table at the right shows the prices of the items she wants to buy.

Art Supplies	
Item	**Price**
Glass beads	$0.28 per ounce
Paintbrush	$0.95
Poster board	$0.75
Jar of paint	$0.99

A Isabella spends $2.25 on poster boards. How many poster boards does she buy?

_____

B Isabella spends $4.87 on paintbrushes and paint. How many of each item does she buy? **Explain** how you found your answer.

_____

_____

_____

C Isabella spends less than $14.00 for glass beads, paintbrushes, poster board, and paint. She spends $1.68 on beads and $3.96 on paint. She buys more than 3 poster boards and more than 3 paintbrushes. Find how many ounces of glass beads and how many jars of paint she buys. Then, suggest the number of poster boards and paintbrushes she might buy for the total spent.

_____

_____

_____

Operations with Fractions

Developing fluency with addition and subtraction of fractions, and developing understanding of the multiplication of fractions and of division of fractions in limited cases (unit fractions divided by whole numbers and whole numbers divided by unit fractions)

Board operator at a recording studio ▶

The Rhythm Track

Math and music both involve numbers and patterns of change. In music, these patterns are called rhythm. We hear rhythm as a number of beats.

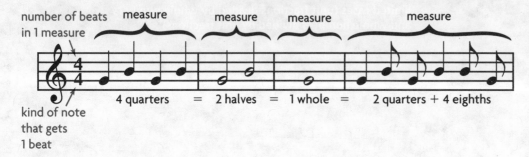

number of beats in 1 measure

kind of note that gets 1 beat

4 quarters = 2 halves = 1 whole = 2 quarters + 4 eighths

Get Started

The time signature at the beginning of a line of music looks like a fraction. It tells the number of beats in each measure and the kind of note that fills 1 beat. When the time signature is $\frac{4}{4}$, each $\frac{1}{4}$ note or quarter note, is 1 beat.

In the music below, different kinds of notes make up each measure. The measures are not marked. Check the time signature. Then draw lines to mark each measure.

Important Facts

$$\text{♩} = \frac{1}{2}$$

$$\text{♩} = \frac{1}{4}$$

$$\text{♪} = \frac{1}{8}$$

$$\text{♬} = \frac{1}{16}$$

Chapter 6

Add and Subtract Fractions with Unlike Denominators

Show What You Know

Check your understanding of important skills.

Name _____

▶ **Part of a Whole** Write a fraction to name the shaded part.

1.   number of shaded parts _____

   number of total parts _____

   fraction _____

2.   number of shaded parts _____

   number of total parts _____

   fraction _____

▶ **Add and Subtract Fractions** Write the sum or difference in simplest form.

3. $\dfrac{3}{6} + \dfrac{1}{6} =$ _____

4. $\dfrac{4}{10} + \dfrac{1}{10} =$ _____

5. $\dfrac{7}{8} - \dfrac{3}{8} =$ _____

6. $\dfrac{9}{12} - \dfrac{2}{12} =$ _____

▶ **Multiples** Write the first six nonzero multiples.

7. 5 _____

8. 3 _____

9. 7 _____

MATH DETECTIVE WITH CARMEN SANDIEGO™

There are 30 senators and 60 members of the House of Representatives in the Arizona Legislature. Suppose 20 senators and 25 representatives came to a committee meeting. Be a math detective to write a fraction that compares the number of legislators that attended to the total number of legislators.

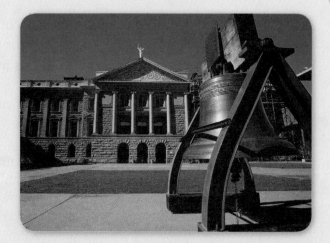

GO Online   Assessment Options: **Soar to Success Math**

Vocabulary Builder

▶ Visualize It

Use the ✓ words to complete the H-diagram.

Review Words

benchmark

✓common multiple

✓denominators

✓difference

✓equivalent fractions

mixed number

✓numerators

✓simplest form

✓sum

Preview Words

✓common denominator

```
┌──────────────────┐     ┌──────────────────┐
│ Add and Subtract │     │ Add and Subtract │
│ Fractions with Like    │ Fractions with Unlike │
│                  │     │                  │
│         ┌────────┘─────└────────┐         │
│         │                       │         │
│         └────────┐─────┌────────┘         │
│                  │     │                  │
└──────────────────┘     └──────────────────┘
```

▶ **Understand Vocabulary**

Draw a line to match the word with its definition.

1. common multiple

2. benchmark

3. simplest form

4. mixed number

5. common denominator

6. equivalent fractions

- a number that is made up of a whole number and a fraction

- a number that is a multiple of two or more numbers

- a common multiple of two or more denominators

- the form of a fraction in which the numerator and denominator have only 1 as their common factor

- a familiar number used as a point of reference

- fractions that name the same amount or part

GO Online • eStudent Edition • Multimedia eGlossary

Name _____

Addition with Unlike Denominators

Essential Question How can you use models to add fractions that have different denominators?

Investigate

Hilary is making a tote bag for her friend. She uses $\frac{1}{2}$ yard of blue fabric and $\frac{1}{4}$ yard of red fabric. How much fabric does Hilary use?

Materials ▪ fraction strips ▪ MathBoard

A. Find $\frac{1}{2} + \frac{1}{4}$. Place a $\frac{1}{2}$ strip and a $\frac{1}{4}$ strip under the 1-whole strip on your MathBoard.

B. Find fraction strips, all with the same denominator, that are equivalent to $\frac{1}{2}$ and $\frac{1}{4}$ and fit exactly under the sum $\frac{1}{2} + \frac{1}{4}$. Record the addends, using like denominators.

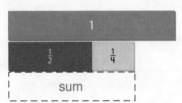

C. Record the sum in simplest form. $\frac{1}{2} + \frac{1}{4} =$ _____

So, Hilary uses _____ yard of fabric.

Math Talk MATHEMATICAL PRACTICES How can you tell if the sum of the fractions is less than 1?

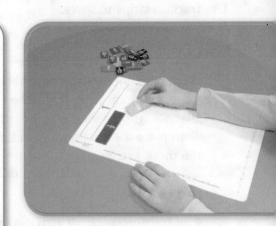

Draw Conclusions

1. **Describe** how you would determine what fraction strips, all with the same denominator, would fit exactly under $\frac{1}{2} + \frac{1}{3}$. What are they?

_____

_____

_____

2. **H.O.T.** **Explain** the difference between finding fraction strips with the same denominator for $\frac{1}{2} + \frac{1}{3}$ and $\frac{1}{2} + \frac{1}{4}$.

_____

_____

_____

Make Connections

Sometimes, the sum of two fractions is greater than 1. When adding fractions with unlike denominators, you can use the 1-whole strip to help determine if a sum is greater than 1 or less than 1.

Use fraction strips to solve. $\frac{3}{5} + \frac{1}{2}$

STEP 1

Work with another student. Place three $\frac{1}{5}$ fraction strips under the 1-whole strip on your MathBoard. Then place a $\frac{1}{2}$ fraction strip beside the three $\frac{1}{5}$ strips.

STEP 2

Find fraction strips, all with the same denominator, that are equivalent to $\frac{3}{5}$ and $\frac{1}{2}$. Place the fraction strips under the sum. At the right, draw a picture of the model and write the equivalent fractions.

$$\frac{3}{5} = \underline{\hspace{1cm}} \qquad \frac{1}{2} = \underline{\hspace{1cm}}$$

STEP 3

Add the fractions with like denominators. Use the 1-whole strip to rename the sum in simplest form.

Think: How many fraction strips with the same denominator are equal to 1 whole?

$$\frac{3}{5} + \frac{1}{2} = \underline{\hspace{1.5cm}} + \underline{\hspace{1.5cm}}$$

$$= \underline{\hspace{1.5cm}}, \text{ or } \underline{\hspace{1.5cm}}$$

 Math Talk

In what step did you find out that the answer is greater than 1? **Explain.**

Share and Show

Use fraction strips to find the sum. Write your answer in simplest form.

1.

$$\frac{1}{2} + \frac{3}{8} = \underline{\hspace{1cm}} + \underline{\hspace{1cm}} = \underline{\hspace{1cm}}$$

2.

$$\frac{1}{2} + \frac{2}{5} = \underline{\hspace{1cm}} + \underline{\hspace{1cm}} = \underline{\hspace{1cm}}$$

Name _____

Use fraction strips to find the sum. Write your answer in simplest form.

3.

$$\frac{3}{8} + \frac{1}{4} = \underline{\hspace{1cm}} + \underline{\hspace{1cm}} = \underline{\hspace{1cm}}$$

④ 4.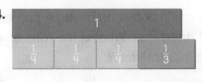

$$\frac{3}{4} + \frac{1}{3} = \underline{\hspace{1cm}} + \underline{\hspace{1cm}} = \underline{\hspace{1cm}}$$

Use fraction strips to find the sum. Write your answer in simplest form.

5. $\frac{2}{5} + \frac{3}{10} = \underline{\hspace{1cm}}$

6. $\frac{1}{4} + \frac{1}{12} = \underline{\hspace{1cm}}$

④ 7. $\frac{1}{2} + \frac{3}{10} = \underline{\hspace{1cm}}$

8. $\frac{2}{3} + \frac{1}{6} = \underline{\hspace{1cm}}$

9. $\frac{5}{8} + \frac{1}{4} = \underline{\hspace{1cm}}$

10. $\frac{1}{2} + \frac{1}{5} = \underline{\hspace{1cm}}$

11. $\frac{3}{4} + \frac{1}{6} = \underline{\hspace{1cm}}$

12. $\frac{1}{2} + \frac{2}{3} = \underline{\hspace{1cm}}$

13. $\frac{7}{8} + \frac{1}{4} = \underline{\hspace{1cm}}$

14. **Write Math** ▶ **Explain** how using fraction strips with like denominators makes it possible to add fractions with unlike denominators.

_____

_____

_____

Problem Solving ....

15. Maya makes trail mix by combining $\frac{1}{3}$ cup of mixed nuts and $\frac{1}{4}$ cup of dried fruit. What is the total amount of ingredients in her trail mix?

$$\frac{1}{3} + \frac{1}{4} = \frac{7}{12}$$

Maya uses $\frac{7}{12}$ cup of ingredients.

Write a new problem using different amounts for each ingredient.
Each amount should be a fraction with a denominator of 2, 3, or 4.
Then use fraction strips to solve your problem.

Pose a problem.

Solve your problem. Draw a picture of the fraction strips you use to solve the problem.

_____

_____

_____

_____

_____

_____

_____

_____

- **Explain** why you chose the amounts you did for your problem.

_____

_____

_____

_____

FOR MORE PRACTICE:
Standards Practice Book, pp. P121–P122

Name _____

Subtraction with Unlike Denominators

Essential Question How can you use models to subtract fractions that have different denominators?

Investigate

Mario fills a hummingbird feeder with $\frac{3}{4}$ cup of sugar water on Friday. On Monday, Mario sees that $\frac{1}{8}$ cup of sugar water is left. How much sugar water did the hummingbirds drink?

Materials ■ fraction strips ■ MathBoard

A. Find $\frac{3}{4} - \frac{1}{8}$. Place three $\frac{1}{4}$ strips under the 1-whole strip on your MathBoard. Then place a $\frac{1}{8}$ strip under the $\frac{1}{4}$ strips.

B. Find fraction strips all with the same denominator that fit exactly under the difference $\frac{3}{4} - \frac{1}{8}$.

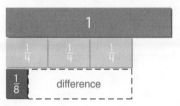

C. Record the difference. $\quad \frac{3}{4} - \frac{1}{8} =$ _____

So, the hummingbirds drank _____ cup of sugar water.

Math Talk MATHEMATICAL PRACTICES
How can you tell if the difference of the fractions is less than 1? **Explain.**

Draw Conclusions . . . .

1. **Describe** how you determined what fraction strips, all with the same denominator, would fit exactly under the difference. What are they?

   _____

   _____

   _____

2. **H.O.T.** **Explain** whether you could have used fraction strips with any other denominator to find the difference. If so, what is the denominator?

   _____

   _____

Make Connections

Sometimes you can use different sets of same-denominator fraction strips to find the difference. All of the answers will be correct.

Solve. $\frac{2}{3} - \frac{1}{6}$

A Find fraction strips, all with the same denominator, that fit exactly under the difference $\frac{2}{3} - \frac{1}{6}$.

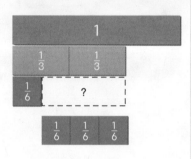

$$\frac{2}{3} - \frac{1}{6} = \frac{3}{6}$$

B Find another set of fraction strips, all with the same denominator, that fit exactly under the difference $\frac{2}{3} - \frac{1}{6}$. Draw the fraction strips you used.

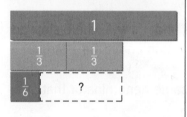

$$\frac{2}{3} - \frac{1}{6} = \underline{\hspace{1cm}}$$

C Find other fraction strips, all with the same denominator, that fit exactly under the difference $\frac{2}{3} - \frac{1}{6}$. Draw the fraction strips you used.

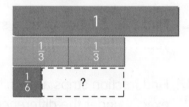

$$\frac{2}{3} - \frac{1}{6} = \underline{\hspace{1cm}}$$

While each answer appears different, all of the answers

can be simplified to _____.

> **Math Talk** MATHEMATICAL PRACTICES
>
> Which other fraction strips with the same denominator could fit exactly in the difference of $\frac{2}{3} - \frac{1}{6}$?

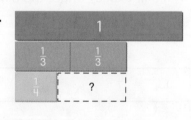

Share and Show MATH BOARD

Use fraction strips to find the difference. Write your answer in simplest form.

1.

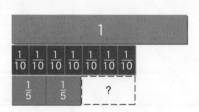

$$\frac{7}{10} - \frac{2}{5} = \underline{\hspace{1cm}}$$

2.

$$\frac{2}{3} - \frac{1}{4} = \underline{\hspace{1cm}}$$

Name _____

Use fraction strips to find the difference. Write your answer in simplest form.

3.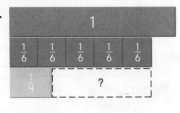

$$\frac{5}{6} - \frac{1}{4} = \underline{\hspace{1cm}}$$

4.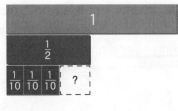

$$\frac{1}{2} - \frac{3}{10} = \underline{\hspace{1cm}}$$

5.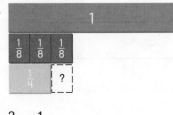

$$\frac{3}{8} - \frac{1}{4} = \underline{\hspace{1cm}}$$

6.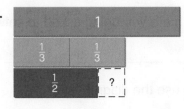

$$\frac{2}{3} - \frac{1}{2} = \underline{\hspace{1cm}}$$

Use fraction strips to find the difference. Write your answer in simplest form.

7. $\frac{3}{5} - \frac{3}{10} = \underline{\hspace{1cm}}$

8. $\frac{5}{12} - \frac{1}{3} = \underline{\hspace{1cm}}$

9. $\frac{1}{2} - \frac{1}{10} = \underline{\hspace{1cm}}$

10. $\frac{3}{5} - \frac{1}{2} = \underline{\hspace{1cm}}$

11. $\frac{7}{8} - \frac{1}{4} = \underline{\hspace{1cm}}$

12. $\frac{5}{6} - \frac{2}{3} = \underline{\hspace{1cm}}$

13. $\frac{3}{4} - \frac{1}{3} = \underline{\hspace{1cm}}$

14. $\frac{5}{6} - \frac{1}{2} = \underline{\hspace{1cm}}$

15. $\frac{3}{4} - \frac{7}{12} = \underline{\hspace{1cm}}$

16. **Write Math** ▶ **Explain** how your model for $\frac{3}{5} - \frac{1}{2}$ is different from your model for $\frac{3}{5} - \frac{3}{10}$.

_____

_____

_____

UNLOCK the Problem REAL WORLD

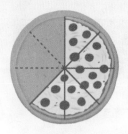

17. The picture at the right shows how much pizza was left over from lunch. Jason eats $\frac{1}{4}$ of the whole pizza for dinner. Which subtraction sentence represents the amount of pizza that is remaining after dinner?

Ⓐ $1 - \frac{1}{4} = \frac{3}{4}$ Ⓒ $\frac{3}{8} - \frac{1}{4} = \frac{2}{8}$

Ⓑ $\frac{5}{8} - \frac{1}{4} = \frac{3}{8}$ Ⓓ $1 - \frac{3}{8} = \frac{5}{8}$

a. What problem are you being asked to solve? _____

_____

b. How will you use the diagram to solve the problem? _____

_____

_____

c. Jason eats $\frac{1}{4}$ of the whole pizza. How many slices does he eat? _____

d. Redraw the diagram of the pizza. Shade the sections of pizza that are remaining after Jason eats his dinner.

e. Write a fraction to represent the amount of pizza that is remaining.

_____

f. Fill in the bubble for the correct answer choice above.

18. The diagram shows what Tina had left from a yard of fabric. She now uses $\frac{2}{3}$ yard of fabric for a project. How much of the original yard of fabric does Tina have left after the project?

Ⓐ $\frac{2}{3}$ yard   Ⓑ $\frac{1}{2}$ yard   Ⓒ $\frac{1}{3}$ yard   Ⓓ $\frac{1}{6}$ yard

Name _____

Estimate Fraction Sums and Differences

Essential Question How can you make reasonable estimates of fraction sums and differences?

🔑 UNLOCK the Problem REAL WORLD

Kimberly will be riding her bike to school this year. The distance from her house to the end of the street is $\frac{1}{6}$ mile. The distance from the end of the street to the school is $\frac{3}{8}$ mile. About how far is Kimberly's house from school?

You can use benchmarks to find reasonable estimates by rounding fractions to 0, $\frac{1}{2}$, or 1.

🔒 One Way Use a number line.

Estimate. $\frac{1}{6} + \frac{3}{8}$

STEP 1 Place a point at $\frac{1}{6}$ on the number line.

The fraction is between _____ and _____.

The fraction $\frac{1}{6}$ is closer to the benchmark _____.

Round to _____.

$$\frac{0}{6} \quad \frac{1}{6} \quad \frac{2}{6} \quad \frac{3}{6} \quad \frac{4}{6} \quad \frac{5}{6} \quad \frac{6}{6}$$
$$0 \qquad\qquad\qquad \frac{1}{2} \qquad\qquad\qquad 1$$

STEP 2 Place a point at $\frac{3}{8}$ on the number line.

The fraction is between _____ and _____.

The fraction $\frac{3}{8}$ is closer to the benchmark _____.

Round to _____.

$$\frac{0}{8} \quad \frac{1}{8} \quad \frac{2}{8} \quad \frac{3}{8} \quad \frac{4}{8} \quad \frac{5}{8} \quad \frac{6}{8} \quad \frac{7}{8} \quad \frac{8}{8}$$
$$0 \qquad\qquad\qquad \frac{1}{2} \qquad\qquad\qquad 1$$

STEP 3 Add the rounded fractions.

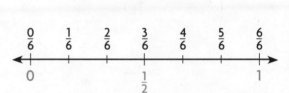

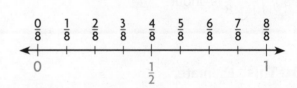

So, Kimberly's house is about _____ mile from the school.

🔑 Another Way Use mental math.

You can compare the numerator and the denominator to round a fraction and find a reasonable estimate.

Estimate. $\frac{9}{10} - \frac{5}{8}$

STEP 1 Round $\frac{9}{10}$. **Think:** The numerator is about the same as the denominator.

Round the fraction $\frac{9}{10}$ to _____ .

STEP 2 Round $\frac{5}{8}$. **Think:** The numerator is about half the denominator.

Round the fraction $\frac{5}{8}$ to _____ .

STEP 3 Subtract.

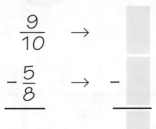

$$\frac{9}{10} \rightarrow$$

$$-\frac{5}{8} \rightarrow -$$

So, $\frac{9}{10} - \frac{5}{8}$ is about _____ .

Remember

A fraction with the same numerator and denominator, such as $\frac{2}{2}$, $\frac{5}{5}$, $\frac{12}{12}$, or $\frac{96}{96}$, is equal to 1.

Math Talk   MATHEMATICAL PRACTICES

Explain another way you could use benchmarks to estimate $\frac{9}{10} - \frac{5}{8}$.

Try This! Estimate.

Ⓐ $2\frac{7}{8} - \frac{2}{5}$

Ⓑ $1\frac{8}{9} + 4\frac{8}{10}$

Name _____

Share and Show

Estimate the sum or difference.

1. $\frac{5}{6} + \frac{3}{8}$

   a. Round $\frac{5}{6}$ to its closest benchmark. _____

   b. Round $\frac{3}{8}$ to its closest benchmark. _____

   c. Add to find the estimate. _____ + _____ = _____

2. $\frac{5}{9} - \frac{3}{8}$

3. $\frac{6}{7} + 2\frac{4}{5}$

✓ 4. $\frac{5}{6} + \frac{2}{5}$

5. $3\frac{9}{10} - 1\frac{2}{9}$

6. $\frac{4}{6} + \frac{1}{9}$

✓ 7. $\frac{9}{10} - \frac{1}{9}$

Math Talk   MATHEMATICAL PRACTICES

Explain how you know whether your estimate for $\frac{9}{10} + 3\frac{6}{7}$ would be greater than or less than the actual sum.

On Your Own

Estimate the sum or difference.

8. $\frac{5}{8} - \frac{1}{5}$

9. $\frac{1}{6} + \frac{3}{8}$

10. $\frac{6}{7} - \frac{1}{5}$

11. $\frac{11}{12} + \frac{6}{10}$

12. $\frac{9}{10} - \frac{1}{2}$

13. $\frac{3}{6} + \frac{4}{5}$

14. $\frac{5}{6} - \frac{3}{8}$

15. $\frac{1}{7} + \frac{8}{9}$

16. $3\frac{5}{12} - 3\frac{1}{10}$

Problem Solving REAL WORLD

17. Lisa and Valerie are picnicking in Trough Creek State Park in Pennsylvania. Lisa has brought a salad that she made with $\frac{3}{4}$ cup of strawberries, $\frac{7}{8}$ cup of peaches, and $\frac{1}{6}$ cup of blueberries. About how many total cups of fruit are in the salad?

_____

18. At Trace State Park in Mississippi, there is a 25-mile mountain bike trail. If Tommy rode $\frac{1}{2}$ of the trail on Saturday and $\frac{1}{5}$ of the trail on Sunday, about what fraction of the trail did he ride?

_____

19. 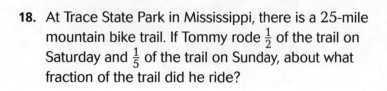 **Explain** how you know that $\frac{5}{8} + \frac{6}{10}$ is greater than 1.

_____

_____

_____

_____

_____

20. **Write Math** ▶ Nick estimated that $\frac{5}{8} + \frac{4}{7}$ is about 2. **Explain** how you know his estimate is not reasonable.

_____

_____

_____

21. **Test Prep** Jake added $\frac{1}{8}$ cup of sunflower seeds and $\frac{4}{5}$ cup of banana chips to his sundae. Which is the best estimate of the total amount of toppings Jake added to his sundae?

Ⓐ about $\frac{1}{2}$ cup

Ⓑ about 1 cup

Ⓒ about $1\frac{1}{2}$ cups

Ⓓ about 2 cups

Common Denominators and Equivalent Fractions

Essential Question How can you rewrite a pair of fractions so that they have a common denominator?

UNLOCK the Problem REAL WORLD

Sarah planted two 1-acre gardens. One had three sections of flowers and the other had 4 sections of flowers. She plans to divide both gardens into more sections so that they have the same number of equal-sized sections. How many sections will each garden have?

You can use a **common denominator** or a common multiple of two or more denominators to write fractions that name the same part of a whole.

One Way Multiply the denominators.

THINK

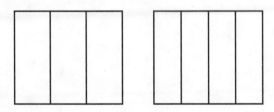

Divide each $\frac{1}{3}$ into fourths and divide each $\frac{1}{4}$ into thirds, each of the wholes will be divided into the same size parts, twelfths.

RECORD

- Multiply the denominators to find a common denominator.

  A common denominator of $\frac{1}{3}$ and $\frac{1}{4}$ is _____.

- Write $\frac{1}{3}$ and $\frac{1}{4}$ as equivalent fractions using the common denominator.

  $\dfrac{1}{3} =$ ▢ $\dfrac{1}{4} =$ ▢

So, both gardens will have _____ sections.

Another Way Use a list.

- Make a list of the first eight nonzero multiples of 3 and 4.

  Multiples of 3: 3, 6, 9, _____, _____, _____, _____, _____

  Multiples of 4: 4, 8, _____, _____, _____, _____, _____, _____

- Circle the common multiples.

- Use one of the common multiples as a common denominator to write equivalent fractions for $\frac{1}{3}$ and $\frac{1}{4}$.

  $\dfrac{1}{3} = \dfrac{}{}$ $\dfrac{1}{4} = \dfrac{}{}$

So, both gardens can have _____ , or _____ sections.

Least Common Denominator Find the least common denominator of two or more fractions by finding the least common multiple of two or more numbers.

🔒 Example   Use the least common denominator.

Find the least common denominator of $\frac{3}{4}$ and $\frac{1}{6}$. Use the least common denominator to write an equivalent fraction for each fraction.

STEP 1   List nonzero multiples of the denominators. Find the least common multiple.

Multiples of 4: _____

Multiples of 6: _____

So, the least common denominator of $\frac{3}{4}$ and $\frac{1}{6}$ is _____.
_____

STEP 2   Using the least common denominator, write an equivalent fraction for each fraction.

Think: What number multiplied by the denominator of the fraction will result in the least common denominator?

$$\frac{3}{4} = \frac{?}{12} = \frac{3 \times 3}{4 \times 3} = \frac{}{}$$ ← least common denominator

$$\frac{1}{6} = \frac{?}{12} = \frac{1 \times }{6 \times } = \frac{}{}$$ ← least common denominator

$\frac{3}{4}$ can be rewritten as _____ and $\frac{1}{6}$ can be rewritten as _____.

Share and Show

Math Talk MATHEMATICAL PRACTICES
Explain two methods for finding a common denominator of two fractions.

1. Find a common denominator of $\frac{1}{6}$ and $\frac{1}{9}$. Rewrite the pair of fractions using the common denominator.

   • Multiply the denominators.
     A common denominator of $\frac{1}{6}$ and $\frac{1}{9}$ is _____.

   • Rewrite the pair of fractions using the common denominator.

   $$\frac{1}{6} = \frac{}{} \qquad \frac{1}{9} = \frac{}{}$$

Use a common denominator to write an equivalent fraction for each fraction.

2. $\frac{1}{3}, \frac{1}{5}$ common
   denominator: _____

3. $\frac{2}{3}, \frac{5}{9}$ common
   denominator: _____

✅ 4. $\frac{2}{9}, \frac{1}{15}$ common
   denominator: _____

_____   _____   _____

Name _____

Use the least common denominator to write an equivalent fraction for each fraction.

5. $\frac{1}{4}, \frac{3}{8}$ least common denominator: _____

_____

6. $\frac{11}{12}, \frac{5}{8}$ least common denominator: _____

_____

7. $\frac{4}{5}, \frac{1}{6}$ least common denominator: _____

_____

Math Talk MATHEMATICAL PRACTICES

Explain what a common denominator of two fractions represents.

On Your Own ·

Use a common denominator to write an equivalent fraction for each fraction.

8. $\frac{3}{5}, \frac{1}{4}$ common denominator: _____

_____

9. $\frac{5}{8}, \frac{1}{5}$ common denominator: _____

_____

10. $\frac{1}{12}, \frac{1}{2}$ common denominator: _____

_____

Practice: Copy and Solve Use the least common denominator to write an equivalent fraction for each fraction.

11. $\frac{1}{6}, \frac{4}{9}$

12. $\frac{7}{9}, \frac{8}{27}$

13. $\frac{7}{10}, \frac{3}{8}$

14. $\frac{1}{3}, \frac{5}{11}$

15. $\frac{5}{9}, \frac{4}{15}$

16. $\frac{1}{6}, \frac{4}{21}$

17. $\frac{5}{14}, \frac{8}{42}$

18. $\frac{7}{12}, \frac{5}{18}$

 Algebra Write the unknown number for each ■.

19. $\frac{1}{5}, \frac{1}{8}$ least common denominator: ■

■ = _____

20. $\frac{2}{5}, \frac{1}{■}$ least common denominator: 15

■ = _____

21. $\frac{3}{■}, \frac{5}{6}$ least common denominator: 42

■ = _____

UNLOCK the Problem REAL WORLD

22. Katie made two pies for the bake sale. One was cut into three equal slices and the other into 5 equal slices. She will continue to cut the pies so each one has the same number of equal-sized slices. What is the least number of equal-sized slices each pie could have?

a. What information are you given? _____

_____

b. What problem are you being asked to solve? _____

_____

c. When Katie cuts the pies more, can she cut each pie the same number

of times and have all the slices the same size? **Explain**. _____

_____

_____

d. Use the diagram to show the steps you use to solve the problem.

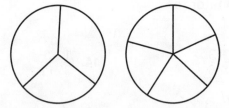

e. Complete the sentences.

The least common denominator of

$\frac{1}{3}$ and $\frac{1}{5}$ is _____.

Katie can cut each piece of the first pie

into _____ and each piece of the

second pie into _____.
That means that Katie can cut each pie into

pieces that are _____ of the whole pie.

23. A cookie recipe calls for $\frac{1}{3}$ cup of brown sugar and $\frac{1}{8}$ cup of walnuts. Find the least common denominator of the fractions used in the recipe.

_____

24. **Test Prep** Which fractions use the least common denominator and are equivalent to $\frac{5}{8}$ and $\frac{7}{10}$?

Ⓐ $\frac{10}{40}$ and $\frac{14}{40}$ Ⓒ $\frac{25}{80}$ and $\frac{21}{80}$

Ⓑ $\frac{25}{40}$ and $\frac{28}{40}$ Ⓓ $\frac{50}{80}$ and $\frac{56}{80}$

Name _____

Add and Subtract Fractions

Essential Question How can you use a common denominator to add and subtract fractions with unlike denominators?

CONNECT You can use what you have learned about common denominators to add or subtract fractions with unlike denominators.

🔑 UNLOCK the Problem ⟩ REAL WORLD

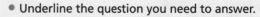

Malia bought shell beads and glass beads to weave into designs in her baskets. She bought $\frac{1}{4}$ pound of shell beads and $\frac{3}{8}$ pound of glass beads. How many pounds of beads did she buy?

- Underline the question you need to answer.
- Draw a circle around the information you will use.

 Add. $\frac{1}{4} + \frac{3}{8}$ **Write your answer in simplest form.**

One Way

Find a common denominator by multiplying the denominators.

$4 \times 8 =$ _____ ← common denominator

Use the common denominator to write equivalent fractions with like denominators. Then add, and write your answer in simplest form.

$$\frac{1}{4} = \frac{1 \times }{4 \times } = $$

$$+ \frac{3}{8} = + \frac{3 \times }{8 \times } = + $$

$$= $$

Another Way

Find the least common denominator.

The least common denominator

of $\frac{1}{4}$ and $\frac{3}{8}$ is _____.

$$\frac{1}{4} = \frac{1 \times }{4 \times } = $$

$$+ \frac{3}{8} \qquad\qquad + $$

So, Malia bought _____ pound of beads.

1. **Explain** how you know whether your answer is reasonable. _____

_____

 **Example**

When subtracting two fractions with unlike denominators, follow the
same steps you follow when adding two fractions. However, instead of
adding the fractions, subtract.

Subtract. $\frac{9}{10} - \frac{2}{5}$ **Write your answer in
simplest form.**

$$\frac{9}{10} =$$

$$-\frac{2}{5} =$$

_____

**Describe the steps you took to solve
the problem.**

_____

_____

_____

_____

2. **Explain** how you know whether your answer is reasonable.

_____

_____

Share and Show . . . .

Find the sum or difference. Write your answer in simplest form.

1. $\frac{5}{12} + \frac{1}{3}$

2. $\frac{2}{5} + \frac{3}{7}$

✓ 3. $\frac{1}{6} + \frac{3}{4}$

4. $\frac{3}{4} - \frac{1}{8}$

5. $\frac{1}{4} - \frac{1}{7}$

✓ 6. $\frac{9}{10} - \frac{1}{4}$

Math Talk
Explain why it is
important to check your answer
for reasonableness.

Name _____

On Your Own ....

Find the sum or difference. Write your answer in simplest form.

7. $\frac{3}{8} + \frac{1}{4}$

8. $\frac{7}{8} + \frac{1}{10}$

9. $\frac{2}{7} + \frac{3}{10}$

10. $\frac{5}{6} + \frac{1}{8}$

11. $\frac{5}{12} + \frac{5}{18}$

12. $\frac{7}{16} - \frac{1}{4}$

13. $\frac{5}{6} - \frac{3}{8}$

14. $\frac{3}{4} - \frac{1}{2}$

15. $\frac{5}{12} - \frac{1}{4}$

Practice: Copy and Solve Find the sum or difference. Write your answer in simplest form.

16. $\frac{1}{3} + \frac{4}{18}$

17. $\frac{3}{5} + \frac{1}{3}$

18. $\frac{3}{10} + \frac{1}{6}$

19. $\frac{1}{2} + \frac{4}{9}$

20. $\frac{1}{2} - \frac{3}{8}$

21. $\frac{5}{7} - \frac{2}{3}$

22. $\frac{4}{9} - \frac{1}{6}$

23. $\frac{11}{12} - \frac{7}{15}$

 Algebra Find the unknown number.

24. $\frac{9}{10} - \blacksquare = \frac{1}{5}$

25. $\frac{5}{12} + \blacksquare = \frac{1}{2}$

$\blacksquare = $ _____

$\blacksquare = $ _____

Problem Solving REAL WORLD

Use the picture for 26–27.

26. Sara is making a key chain using the bead design shown. What fraction of the beads in her design are either blue or red?

_____

27. 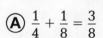 In making the key chain, Sara uses the pattern of beads 3 times. After the key chain is complete, what fraction of the beads in the key chain are either white or blue?

_____

28. **Write Math** ▶ Jamie had $\frac{4}{5}$ of a spool of twine. He then used $\frac{1}{2}$ of a spool of twine to make friendship knots. He claims to have $\frac{3}{10}$ of the original spool of twine left over. **Explain** how you know whether Jamie's claim is reasonable.

_____

_____

_____

_____

29. **Test Prep** Which equation represents the fraction of beads that are green or yellow?

Ⓐ $\frac{1}{4} + \frac{1}{8} = \frac{3}{8}$

Ⓑ $\frac{1}{2} + \frac{1}{4} = \frac{3}{4}$

Ⓒ $\frac{1}{2} + \frac{1}{8} = \frac{5}{8}$

Ⓓ $\frac{3}{4} + \frac{2}{8} = 1$

Name _____

▶ **Vocabulary**

Choose the best term from the box.

Vocabulary
equivalent fractions
common denominator
common multiple

1. A _____ is a number that is a multiple of two or more numbers. **(p. 255)**

2. A _____ is a common multiple of two or more denominators. **(p. 255)**

▶ **Concepts and Skills**

Estimate the sum or difference.

3. $\frac{8}{9} + \frac{4}{7}$

4. $3\frac{2}{5} - \frac{5}{8}$

5. $1\frac{5}{6} + 2\frac{2}{11}$

Use a common denominator to write an equivalent fraction for each fraction.

6. $\frac{1}{6}, \frac{1}{9}$ common denominator: _____

7. $\frac{3}{8}, \frac{3}{10}$ common denominator: _____

8. $\frac{1}{9}, \frac{5}{12}$ common denominator: _____

_____

_____

_____

Use the least common denominator to write an equivalent fraction for each fraction.

9. $\frac{2}{5}, \frac{1}{10}$ least common denominator: _____

10. $\frac{5}{6}, \frac{3}{8}$ least common denominator: _____

11. $\frac{1}{3}, \frac{2}{7}$ least common denominator: _____

_____

_____

_____

Find the sum or difference. Write your answer in simplest form.

12. $\frac{11}{18} - \frac{1}{6}$

13. $\frac{2}{7} + \frac{2}{5}$

14. $\frac{3}{4} - \frac{3}{10}$

Fill in the bubble completely to show your answer.

15. Mrs. Michaels bakes a pie for her book club meeting. The shaded part of the diagram below shows the amount of pie left after the meeting. That evening, Mr. Michaels eats $\frac{1}{4}$ of the whole pie. Which fraction represents the amount of pie remaining?

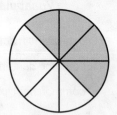

Ⓐ $\frac{1}{4}$

Ⓑ $\frac{3}{8}$

Ⓒ $\frac{5}{8}$

Ⓓ $\frac{3}{4}$

16. Keisha bakes a pan of brownies for a family picnic. She takes $\frac{1}{2}$ of the brownies to the picnic. At the picnic, her family eats $\frac{3}{8}$ of the whole pan of brownies. Which fraction of the whole pan of brownies does Keisha bring back from the picnic?

Ⓐ $\frac{1}{8}$

Ⓑ $\frac{1}{4}$

Ⓒ $\frac{2}{5}$

Ⓓ $\frac{1}{2}$

17. Mario is mixing paint for his walls. He mixes $\frac{1}{6}$ gallon blue paint and $\frac{5}{8}$ gallon green paint in a large container. Which fraction represents the total amount of paint Mario mixes?

Ⓐ $\frac{2}{3}$ gallon

Ⓑ $\frac{3}{7}$ gallon

Ⓒ $\frac{9}{12}$ gallon

Ⓓ $\frac{19}{24}$ gallon

Name _____

Add and Subtract Mixed Numbers

Essential Question How can you add and subtract mixed numbers with unlike denominators?

🔑 UNLOCK the Problem · REAL WORLD

Denise mixed $1\frac{4}{5}$ ounces of blue paint with $2\frac{1}{10}$ ounces of yellow paint. How many ounces of paint did Denise mix?

- What operation should you use to solve the problem?

  _____

- Do the fractions have the same denominator?

  _____

 Add. $1\frac{4}{5} + 2\frac{1}{10}$

To find the sum of mixed numbers with unlike denominators, you can use a common denominator.

STEP 1   Estimate the sum. _____

_____

STEP 2   Find a common denominator. Use the common denominator to write equivalent fractions with like denominators.

_____

STEP 3   Add the fractions. Then add the whole numbers. Write the answer in simplest form.

$$1\frac{4}{5} =$$

$$+\ 2\frac{1}{10} = +$$

_____

So, Denise mixed _____ ounces of paint.

Math Talk   MATHEMATICAL PRACTICES
Did you use the least common denominator? Explain.

---

1. Explain how you know whether your answer is reasonable. _____

_____

_____

2. What other common denominator could you have used? _____

_____

 **Example**

Subtract. $4\frac{5}{6} - 2\frac{3}{4}$

You can also use a common denominator to find the difference of mixed numbers with unlike denominators.

STEP 1	Estimate the difference. _____	

$4\frac{5}{6} = \boxed{}$

STEP 2 Find a common denominator. Use the common denominator to write equivalent fractions with like denominators.

$-2\frac{3}{4} = -$

STEP 3 Subtract the fractions. Subtract the whole numbers. Write the answer in simplest form.

3. **Explain** how you know whether your answer is reasonable. _____

_____

_____

Share and Show · · · · · · · · · · · · · · · · · ·

1. Use a common denominator to write equivalent fractions with like denominators and then find the sum. Write your answer in simplest form.

$7\frac{2}{5} = \boxed{}$

$+4\frac{3}{4} = + \boxed{}$

$\boxed{}$

Find the sum. Write your answer in simplest form.

2. $2\frac{3}{4} + 3\frac{3}{10}$

3. $5\frac{3}{4} + 1\frac{1}{3}$

✓ 4. $3\frac{4}{5} + 2\frac{3}{10}$

Find the difference. Write your answer in simplest form.

5. $9\frac{5}{6} - 2\frac{1}{3}$

6. $10\frac{5}{9} - 9\frac{1}{6}$

☑ **7.** $7\frac{2}{3} - 3\frac{1}{6}$

On Your Own ·····································

> **MATHEMATICAL PRACTICES**
> **Math Talk** **Explain** why you need to write equivalent fractions with common denominators to add $4\frac{5}{6}$ and $1\frac{1}{8}$.

Find the sum or difference. Write your answer in simplest form.

8. $1\frac{3}{10} + 2\frac{2}{5}$

9. $3\frac{4}{9} + 3\frac{1}{2}$

10. $2\frac{1}{2} + 2\frac{1}{3}$

11. $5\frac{1}{4} + 9\frac{1}{3}$

12. $8\frac{1}{6} + 7\frac{3}{8}$

13. $14\frac{7}{12} - 5\frac{1}{4}$

14. $12\frac{3}{4} - 6\frac{1}{6}$

15. $2\frac{5}{8} - 1\frac{1}{4}$

16. $10\frac{1}{2} - 2\frac{1}{5}$

Practice: Copy and Solve Find the sum or difference. Write your answer in simplest form.

17. $1\frac{5}{12} + 4\frac{1}{6}$

18. $8\frac{1}{2} + 6\frac{3}{5}$

19. $2\frac{1}{6} + 4\frac{5}{9}$

20. $3\frac{5}{8} + \frac{5}{12}$

21. $3\frac{2}{3} - 1\frac{1}{6}$

22. $5\frac{6}{7} - 1\frac{2}{3}$

23. $2\frac{7}{8} - \frac{1}{2}$

24. $4\frac{7}{12} - 1\frac{2}{9}$

Problem Solving REAL WORLD

Use the table to solve 25–28.

25. Gavin is mixing a batch of Sunrise Orange paint for an art project. How much paint does Gavin mix?

_____

26. Gavin plans to mix a batch of Tangerine paint. He expects to have a total of $5\frac{3}{10}$ ounces of paint after he mixes the amounts of red and yellow. **Explain** how you can tell if Gavin's expectation is reasonable.

_____

_____

_____

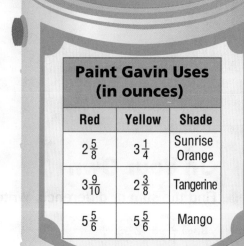

Paint Gavin Uses (in ounces)

Red	Yellow	Shade
$2\frac{5}{8}$	$3\frac{1}{4}$	Sunrise Orange
$3\frac{9}{10}$	$2\frac{3}{8}$	Tangerine
$5\frac{5}{6}$	$5\frac{5}{6}$	Mango

27. **H.O.T.** For a special project, Gavin mixes the amount of red from one shade of paint with the amount of yellow from a different shade. He mixes the batch so he will have the greatest possible amount of paint. What amounts of red and yellow from which shades are used in the mixture for the special project? **Explain** your answer.

_____

_____

_____

_____

SHOW YOUR WORK

28. Gavin needs to make 2 batches of Mango paint. **Explain** how you could find the total amount of paint Gavin mixed.

_____

_____

_____

29. **Test Prep** Yolanda walked $3\frac{6}{10}$ miles. Then she walked $4\frac{1}{2}$ more miles. How many miles did Yolanda walk?

(A) $7\frac{1}{10}$ miles

(C) $8\frac{1}{10}$ miles

(B) $7\frac{7}{10}$ miles

(D) $8\frac{7}{10}$ miles

FOR MORE PRACTICE:
Standards Practice Book, pp. P131–P132

Name _____

Subtraction with Renaming

Essential Question How can you use renaming to find the difference of two mixed numbers?

🔑 UNLOCK the Problem REAL WORLD

To practice for a race, Kara is running $2\frac{1}{2}$ miles. When she reaches the end of her street, she knows that she has already run $1\frac{5}{6}$ miles. How many miles does Kara have left to run?

- Underline the sentence that tells you what you need to find.
- What operation should you use to solve the problem?

_____

🔑 One Way Rename the first mixed number.

Subtract. $2\frac{1}{2} - 1\frac{5}{6}$

STEP 1 Estimate the difference. _____

STEP 2 Find a common denominator. Use the common denominator to write equivalent fractions with like denominators.

STEP 3 Rename $2\frac{6}{12}$ as a mixed number with a fraction greater than 1.

Think: $2\frac{6}{12} = 1 + 1 + \frac{6}{12} = 1 + \frac{12}{12} + \frac{6}{12} = 1\frac{18}{12}$

$2\frac{6}{12} =$ _____

STEP 4 Find the difference of the fractions. Then find the difference of the whole numbers. Write the answer in simplest form. Check to make sure your answer is reasonable.

$$2\frac{1}{2} = \quad 2\frac{6}{12} = \boxed{}$$

$$-1\frac{5}{6} = -1\frac{10}{12} = -1\frac{10}{12}$$

$$\boxed{} = \boxed{}$$

So, Kara has _____ mile left to run.

- **Explain** why it is important to write equivalent fractions before renaming. _____

_____

Another Way Rename both mixed numbers as fractions greater than 1.

Subtract. $2\frac{1}{2} - 1\frac{5}{6}$

STEP 1 Write equivalent fractions, using a common denominator.

A common denominator of $\frac{1}{2}$ and $\frac{5}{6}$ is 6.

$2\frac{1}{2} \longrightarrow$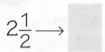

$1\frac{5}{6} \longrightarrow$

STEP 2 Rename both mixed numbers as fractions greater than 1.

$2\frac{3}{6} =$ Think: $\frac{6}{6} + \frac{6}{6} + \frac{3}{6}$

$1\frac{5}{6} =$ Think: $\frac{6}{6} + \frac{5}{6}$

STEP 3 Find the difference of the fractions. Then write the answer in simplest form.

 $-$ $=$

$=$

$2\frac{1}{2} - 1\frac{5}{6} =$ _____

Share and Show

Estimate. Then find the difference and write it in simplest form.

1. Estimate: _____

$1\frac{3}{4} - \frac{7}{8}$

2. Estimate: _____

$12\frac{1}{9} - 7\frac{1}{3}$

Name _____

Estimate. Then find the difference and write it in simplest form.

✓ **3.** Estimate: _____

$$4\frac{1}{2} - 3\frac{4}{5}$$

✓ **4.** Estimate: _____

$$9\frac{1}{6} - 2\frac{3}{4}$$

Math Talk MATHEMATICAL PRACTICES
Explain the strategy you could use to solve $3\frac{1}{9} - 2\frac{1}{3}$.

On Your Own . . . .

Estimate. Then find the difference and write it in simplest form.

5. Estimate: _____

$$3\frac{2}{3} - 1\frac{11}{12}$$

6. Estimate: _____

$$4\frac{1}{4} - 2\frac{1}{3}$$

7. Estimate: _____

$$5\frac{2}{5} - 1\frac{1}{2}$$

8. Estimate: _____

$$7\frac{5}{9} - 2\frac{5}{6}$$

9. Estimate: _____

$$7 - 5\frac{2}{3}$$

10. Estimate: _____

$$2\frac{1}{5} - 1\frac{9}{10}$$

Practice: Copy and Solve Find the difference and write it in simplest form.

11. $11\frac{1}{9} - 3\frac{2}{3}$

12. $6 - 3\frac{1}{2}$

13. $4\frac{3}{8} - 3\frac{1}{2}$

14. $9\frac{1}{6} - 3\frac{5}{8}$

15. $1\frac{1}{5} - \frac{1}{2}$

16. $13\frac{1}{6} - 3\frac{4}{5}$

17. $12\frac{2}{5} - 5\frac{3}{4}$

18. $7\frac{3}{8} - 2\frac{7}{9}$

Connect to Reading

Summarize

An amusement park in Sandusky, Ohio, offers 17 amazing roller coasters for visitors to ride. One of the roller coasters runs at 60 miles per hour and has 3,900 feet of twisting track. This coaster also has 3 trains with 8 rows per train. Riders stand in rows of 4, for a total of 32 riders per train.

The operators of the coaster recorded the number of riders on each train during a run. On the first train, the operators reported that $7\frac{1}{4}$ rows were filled. On the second train, all 8 rows were filled, and on the third train, $5\frac{1}{2}$ rows were filled. How many more rows were filled on the first train than on the third train?

When you *summarize*, you restate the most important information in a shortened form to more easily understand what you have read.

Summarize the information given.

_____

_____

_____

_____

_____

Use the summary to solve.

19. Solve the problem above.

_____

_____

20. **H.O.T.** How many rows were empty on the third train? How many additional riders would it take to fill the empty rows? **Explain** your answer.

_____

_____

_____

FOR MORE PRACTICE:
Standards Practice Book, pp. P133–P134

Patterns with Fractions

Essential Question How can you use addition or subtraction to describe
a pattern or create a sequence with fractions?

🔑 UNLOCK the Problem REAL WORLD

Mr. Patrick wants to develop a new chili recipe for his restaurant. Each
batch he makes uses a different amount of chili powder. The first batch
uses $3\frac{1}{2}$ ounces, the second batch uses $4\frac{5}{6}$ ounces, the third uses
$6\frac{1}{6}$ ounces, and the fourth uses $7\frac{1}{2}$ ounces. If this pattern continues,
how much chili powder will he use in the sixth batch?

You can find the pattern in a sequence by comparing one term with
the next term.

STEP 1   Write the terms in the sequence as equivalent fractions with a
common denominator. Then examine the sequence and compare
the consecutive terms to find the rule used to make the sequence
of fractions.

$$+1\frac{2}{6}$$

difference between terms

$$3\frac{1}{2}, 4\frac{5}{6}, 6\frac{1}{6}, 7\frac{1}{2}, \cdots \rightarrow \underline{\quad} \text{oz}, \underline{\quad} \text{oz}, \underline{\quad} \text{oz}, \underline{\quad} \text{oz}$$

terms with common denominator

batch 1   batch 2   batch 3   batch 4

STEP 2   Write a rule that describes the pattern in the sequence.

- Is the sequence increasing or decreasing from one term to the
next? **Explain.**

_____

_____

Rule: _____

STEP 3   Extend the sequence to solve the problem.

$$3\frac{1}{2}, 4\frac{5}{6}, 6\frac{1}{6}, 7\frac{1}{2}, \underline{\quad}, \underline{\quad}$$

So, Mr. Patrick will use _____ ounces of chili powder in the sixth batch.

🔑 Example Find the unknown terms in the sequence.

$1\frac{3}{4}$, $1\frac{9}{16}$, $1\frac{3}{8}$, $1\frac{3}{16}$, _____ , _____ , _____ , $\frac{7}{16}$, $\frac{1}{4}$

STEP 1 Write the terms in the sequence as equivalent fractions with a common denominator.

_____ , _____ , _____ , _____ , $\overset{?}{\_\_\_\_\_}$, $\overset{?}{\_\_\_\_\_}$, $\overset{?}{\_\_\_\_\_}$, _____ , _____

_____

STEP 2 Write a rule describing the pattern in the sequence.

• What operation can be used to describe a sequence that increases?

_____

• What operation can be used to describe a sequence that decreases?

_____

Rule: _____

_____

STEP 3 Use your rule to find the unknown terms. Then complete the sequence above.

MATHEMATICAL PRACTICES

Math Talk Explain how you know whether your rule for a sequence would involve addition or subtraction.

Try This!

A Write a rule for the sequence. Then find the unknown term.

$1\frac{1}{12}$, $\frac{5}{6}$, _____ , $\frac{1}{3}$, $\frac{1}{12}$

Rule: _____

B Write the first four terms of the sequence.

Rule: start at $\frac{1}{4}$, add $\frac{3}{8}$

_____ , _____ , _____ , _____

274

Name _____

Share and Show

Write a rule for the sequence.

1. $\frac{1}{4}, \frac{1}{2}, \frac{3}{4}, \cdots$

    Think: Is the sequence increasing or decreasing?

    Rule: _____

2. $\frac{1}{9}, \frac{1}{3}, \frac{5}{9}, \cdots$

    Rule: _____

Write a rule for the sequence. Then, find the unknown term.

3. $\frac{3}{10}, \frac{2}{5},$ _____ $, \frac{3}{5}, \frac{7}{10}$

    Rule: _____

4. $10\frac{2}{3}, 9\frac{11}{18}, 8\frac{5}{9},$ _____ $, 6\frac{4}{9}$

    Rule: _____

5. $1\frac{1}{6},$ _____ $, 1, \frac{11}{12}, \frac{5}{6}$

    Rule: _____

6. $2\frac{3}{4}, 4, 5\frac{1}{4}, 6\frac{1}{2},$ _____

    Rule: _____

On Your Own

Write a rule for the sequence. Then, find the unknown term.

7. $\frac{1}{8}, \frac{1}{2},$ _____ $, 1\frac{1}{4}, 1\frac{5}{8}$

    Rule: _____

8. $1\frac{2}{3}, 1\frac{3}{4}, 1\frac{5}{6}, 1\frac{11}{12},$ _____

    Rule: _____

9. $12\frac{7}{8}, 10\frac{3}{4},$ _____ $, 6\frac{1}{2}, 4\frac{3}{8}$

    Rule: _____

10. $9\frac{1}{3},$ _____ $, 6\frac{8}{9}, 5\frac{2}{3}, 4\frac{4}{9}$

    Rule: _____

Write the first four terms of the sequence.

11. Rule: start at $5\frac{3}{4}$, subtract $\frac{5}{8}$

    _____ , _____ , _____ , _____

12. Rule: start at $\frac{3}{8}$, add $\frac{3}{16}$

    _____ , _____ , _____ , _____

13. Rule: start at $2\frac{1}{3}$, add $2\frac{1}{4}$

    _____ , _____ , _____ , _____

14. Rule: start at $\frac{8}{9}$, subtract $\frac{1}{18}$

    _____ , _____ , _____ , _____

Problem Solving REAL WORLD

15. When Bill bought a marigold plant, it was $\frac{1}{4}$ inch tall. After the first week, it measured $1\frac{1}{12}$ inches tall. After the second week, it was $1\frac{11}{12}$ inches. After week 3, it was $2\frac{3}{4}$ inches tall. Assuming the growth of the plant was constant, what was the height of the plant at the end of week 4?

_____

16. **H.O.T.** **What if** Bill's plant grew at the same rate but was $1\frac{1}{2}$ inches when he bought it? How tall would the plant be after 3 weeks?

_____

17. **Write Math** ▶ Vicki wanted to start jogging. The first time she ran, she ran $\frac{3}{16}$ mile. The second time, she ran $\frac{3}{8}$ mile, and the third time, she ran $\frac{9}{16}$ mile. If she continued this pattern, when was the first time she ran more than 1 mile? **Explain.**

_____

_____

_____

18. Mr. Conners drove $78\frac{1}{3}$ miles on Monday, $77\frac{1}{12}$ miles on Tuesday, and $75\frac{5}{6}$ miles on Wednesday. If he continues this pattern on Thursday and Friday, how many miles will he drive on Friday?

_____

19. **Test Prep** Zack watered his garden with $1\frac{3}{8}$ gallons of water the first week he planted it. He watered it with $1\frac{3}{4}$ gallons the second week, and $2\frac{1}{8}$ gallons the third week. If he continued watering in this pattern, how much water did he use on the fifth week?

(A) $2\frac{1}{2}$ gallons

(C) $3\frac{1}{4}$ gallons

(B) $2\frac{7}{8}$ gallons

(D) $6\frac{7}{8}$ gallons

. . . . **SHOW YOUR WORK** . . . . . .

Name _____

Problem Solving
Practice Addition and Subtraction

Essential Question How can the strategy *work backward* help you solve
a problem with fractions that involves addition and subtraction?

UNLOCK the Problem REAL WORLD

The Diaz family is cross-country skiing the Big Tree trails,
which have a total length of 4 miles. Yesterday, they skied
the $\frac{7}{10}$ mile Oak Trail. Today, they skied the $\frac{3}{5}$ mile Pine Trail.
If they plan to ski all of the Big Tree trails, how many more
miles do they have left to ski?

Use the graphic organizer to help you solve the problem.

Read the Problem

What do I need to find?	**What information do I need to use?**	**How will I use the information?**
I need to find the distance _____.	I need to use the distance _____ and the total distance _____.	I can work backward by starting with the _____ and _____ each distance they have already skied to find amount they have left.

Solve the Problem

Addition and subtraction are inverse operations. By working backward and
using the same numbers, one operation undoes the other.

- Write an equation.

miles skied yesterday	+	miles skied today	+	miles they need to ski	=	total distance
↓		↓		↓		↓
_____	+	_____	+	m	=	4

- Then work backward to find m.

_____ – _____ – _____ = m

_____ = m

So, the family has _____ miles left to ski.

- **Explain** how you know your answer is reasonable. _____

_____

🔓 Try Another Problem

As part of their study of Native American basket weaving, Lia's class is making wicker baskets. Lia starts with a strip of wicker 36 inches long. From the strip, she first cuts one piece but does not know its length, and then cuts a piece that is $6\frac{1}{2}$ inches long. The piece left is $7\frac{3}{4}$ inches long. What is the length of the first piece she cut from the strip?

Read the Problem

What do I need to find?	What information do I need to use?	How will I use the information?

Solve the Problem

So, the length of the first piece cut was _____ inches.

Math Talk MATHEMATICAL PRACTICES
What other strategy could you use to solve the problem?

278

© Houghton Mifflin Harcourt Publishing Company

Name _____

Share and Show

✓ Plan your solution by deciding on the steps you will use.

✓ Check your exact answer by comparing it with your estimate.

✓ Check your answer for reasonableness.

1. Caitlin has $4\frac{3}{4}$ pounds of clay. She uses $1\frac{1}{10}$ pounds to make a cup, and another 2 pounds to make a jar. How many pounds are left?

   First, write an equation to model the problem.

   _____

   Next, work backwards and rewrite the equation to find x.

   _____

   Solve.

   _____

   So, _____ pounds of clay remain.

2. **H.O.T.** **What if** Caitlin had used more than 2 pounds of clay to make a jar? Would the amount remaining have been more or less than your answer to Exercise 1?

   _____

3. A pet store donated 50 pounds of food for adult dogs, puppies, and cats to an animal shelter. $19\frac{3}{4}$ pounds was adult dog food and $18\frac{7}{8}$ pounds was puppy food. How many pounds of cat food did the pet store donate?

   _____

4. Thelma spent $\frac{1}{6}$ of her weekly allowance on dog toys, $\frac{1}{4}$ on a dog collar, and $\frac{1}{3}$ on dog food. What fraction of her weekly allowance is left?

   _____

· · · · · · **SHOW YOUR WORK** · · · · · · ·

On Your Own.....................

MATHEMATICAL
PRACTICES

Model • Reason • Make Sense

**Choose a
STRATEGY**

Act It Out
Draw a Diagram
Make a Table
Solve a Simpler Problem
Work Backward
Guess, Check, and Revise

5. Martin is making a model of a Native American canoe. He has $5\frac{1}{2}$ feet of wood. He uses $2\frac{3}{4}$ feet for the hull and $1\frac{1}{4}$ feet for the paddles and struts. How much wood does he have left?

_____

6. ⚡H.O.T.⚡ **What if** Martin makes a hull and two sets of paddles and struts? How much wood does he have left?

_____

7. Beth's summer vacation lasted 87 days. At the beginning of her vacation, she spent 3 weeks at soccer camp, 5 days at her grandmother's house, and 13 days visiting Glacier National Park with her parents. How many vacation days remained?

_____

8. **Write Math** ▶ You can buy 2 DVDs for the same price you would pay for 3 CDs selling for $13.20 apiece. **Explain** how you could find the price of 1 DVD.

_____

_____

_____

_____

SHOW YOUR WORK

9. **Test Prep** During the 9 hours between 8 A.M. and 5 P.M., Bret spent $5\frac{3}{4}$ hours in class and $1\frac{1}{2}$ hours at band practice. How much time did he spend on other activities?

(A) $\frac{3}{4}$ hour   (C) $1\frac{1}{2}$ hours

(B) $1\frac{1}{4}$ hours   (D) $1\frac{3}{4}$ hours

Use Properties of Addition

Essential Question How can properties help you add fractions with unlike denominators?

CONNECT You can use properties of addition to help you add fractions with unlike denominators.

Commutative Property: $\frac{1}{2} + \frac{3}{5} = \frac{3}{5} + \frac{1}{2}$

Associative Property: $\left(\frac{2}{9} + \frac{1}{8}\right) + \frac{3}{8} = \frac{2}{9} + \left(\frac{1}{8} + \frac{3}{8}\right)$

Remember
Parentheses ( ) tell which operation to do first.

🔑 UNLOCK the Problem REAL WORLD

Jane and her family are driving to Big Lagoon State Park. On the first day, they travel $\frac{1}{3}$ of the total distance. On the second day, they travel $\frac{1}{3}$ of the total distance in the morning and then $\frac{1}{6}$ of the total distance in the afternoon. How much of the total distance has Jane's family driven by the end of the second day?

 Use the Associative Property.

Day 1 + Day 2

$$\frac{1}{3} + \left(\frac{1}{3} + \frac{1}{6}\right) = \left(\boxed{} + \boxed{}\right) + \boxed{}$$

$$= \boxed{} + \boxed{}$$

$$= \boxed{} + \boxed{}$$

$$= \boxed{}$$

Write the number sentence to represent the problem. Use the Associative Property to group fractions with like denominators together.

Use mental math to add the fractions with like denominators.

Write equivalent fractions with like denominators. Then add.

So, Jane's family has driven _____ of the total distance by the end of the second day.

Math Talk MATHEMATICAL PRACTICES
Explain why grouping the fractions differently makes it easier to find the sum.

 **Example** Add. $\left(2\frac{5}{8} + 1\frac{2}{3}\right) + 1\frac{1}{8}$

Use the Commutative Property and the Associative Property.

$\left(2\dfrac{5}{8} + 1\dfrac{2}{3}\right) + 1\dfrac{1}{8} = \left(\quad + \quad\right) +$

Use the Commutative Property to put fractions with like denominators next to each other.

$= \quad + \left(\quad + \quad\right)$

Use the Associative Property to group fractions with like denominators together.

$= \quad + $

Use mental math to add the fractions with like denominators.

$= \quad + $

Write equivalent fractions with like denominators.
Then add.

$= \quad = $

Rename and simplify.

Try This! Use properties to solve. Show each step and name the property used.

A $5\frac{1}{4} + \left(\frac{3}{4} + 1\frac{5}{12}\right)$

B $\left(\frac{1}{5} + \frac{3}{10}\right) + \frac{2}{5}$

Name _____

Share and Show

Use the properties and mental math to solve. Write your answer in simplest form.

1. $\left(2\frac{5}{8} + \frac{5}{6}\right) + 1\frac{1}{8}$

✅ 2. $\frac{5}{12} + \left(\frac{5}{12} + \frac{3}{4}\right)$

✅ 3. $\left(3\frac{1}{4} + 2\frac{5}{6}\right) + 1\frac{3}{4}$

Math Talk MATHEMATICAL PRACTICES
Explain how solving Exercise 3 is different from solving Exercise 1.

On Your Own

Use the properties and mental math to solve. Write your answer in simplest form.

4. $\left(\frac{2}{7} + \frac{1}{3}\right) + \frac{2}{3}$

5. $\left(\frac{1}{5} + \frac{1}{2}\right) + \frac{2}{5}$

6. $\left(\frac{1}{6} + \frac{3}{7}\right) + \frac{2}{7}$

7. $\left(2\frac{5}{12} + 4\frac{1}{4}\right) + \frac{1}{4}$

8. $1\frac{1}{8} + \left(5\frac{1}{2} + 2\frac{3}{8}\right)$

9. $\frac{5}{9} + \left(\frac{1}{9} + \frac{4}{5}\right)$

Problem Solving REAL WORLD

Use the map to solve 10–12.

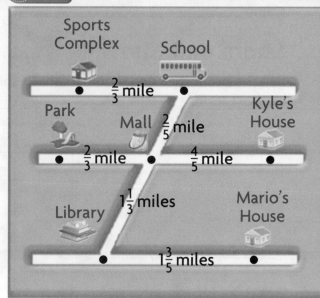

Sports Complex — School

Park — Mall $\frac{2}{5}$ mile — Kyle's House

$\frac{2}{3}$ mile

$\frac{2}{3}$ mile — $\frac{4}{5}$ mile

Library — $1\frac{1}{3}$ miles — Mario's House

$1\frac{3}{5}$ miles

10. In the morning, Julie rides her bike from the sports complex to the school. In the afternoon, she rides from the school to the mall, and then to Kyle's house. How far does Julie ride her bike?

_____

11. On one afternoon, Mario walks from his house to the library. That evening, Mario walks from the library to the mall, and then to Kyle's house. **Describe** how you can use the properties to find how far Mario walks.

_____

_____

_____

_____

SHOW YOUR WORK

12. **H.O.T. Pose a Problem** Write and solve a new problem that uses the distances between four locations.

_____

_____

_____

_____

13. **Test Prep** Which property or properties does the problem below use?

$$\frac{1}{9} + \left(\frac{4}{9} + \frac{1}{6}\right) = \left(\frac{1}{9} + \frac{4}{9}\right) + \frac{1}{6}$$

(A) Commutative Property

(B) Associative Property

(C) Commutative Property and Associative Property

(D) Distributive Property

Name _____

Chapter Review/Test

▶ Vocabulary

Choose the best term from the box.

1. A _____ is a number that is a
   common multiple of two or more denominators. **(p. 255)**

▶ Concepts and Skills

**Use a common denominator to write an equivalent
fraction for each fraction.**

2. $\frac{2}{5}, \frac{1}{8}$ common
   denominator: _____

   _____

3. $\frac{3}{4}, \frac{1}{2}$ common
   denominator: _____

   _____

4. $\frac{2}{3}, \frac{1}{6}$ common
   denominator: _____

   _____

Find the sum or difference. Write your answer in simplest form

5. $\frac{5}{6} + \frac{7}{8}$

6. $2\frac{2}{3} - 1\frac{2}{5}$

7. $7\frac{3}{4} + 3\frac{7}{20}$

Estimate. Then find the difference and write it in simplest form.

8. Estimate: _____

   $1\frac{2}{5} - \frac{2}{3}$

9. Estimate: _____

   $7 - \frac{3}{7}$

10. Estimate: _____

    $5\frac{1}{9} - 3\frac{5}{6}$

**Use the properties and mental math to solve. Write your answer in
simplest form.**

11. $\left(\frac{3}{8} + \frac{2}{3}\right) + \frac{1}{3}$

12. $1\frac{4}{5} + \left(2\frac{3}{20} + \frac{3}{5}\right)$

13. $3\frac{5}{9} + \left(1\frac{7}{9} + 2\frac{5}{12}\right)$

© Houghton Mifflin Harcourt Publishing Company

**GO
Online** Assessment Options
Chapter Test

Fill in the bubble completely to show your answer.

14. Ursula mixed $3\frac{1}{8}$ cups of dry ingredients with $1\frac{2}{5}$ cups of liquid ingredients. Which answer represents the best estimate of the total amount of ingredients Ursula mixed?

   Ⓐ about 4 cups

   Ⓑ about $4\frac{1}{2}$ cups

   Ⓒ about 5 cups

   Ⓓ about $5\frac{1}{2}$ cups

15. Samuel walks in the Labor Day parade. He walks $3\frac{1}{4}$ miles along the parade route and $2\frac{5}{6}$ miles home. How many miles does Samuel walk?

   Ⓐ $\frac{5}{10}$ mile

   Ⓑ $5\frac{1}{12}$ miles

   Ⓒ $5\frac{11}{12}$ miles

   Ⓓ $6\frac{1}{12}$ miles

16. A gardener has a container with $6\frac{1}{5}$ ounces of liquid plant fertilizer. On Sunday, the gardener uses $2\frac{1}{2}$ ounces on a flower garden. How many ounces of liquid plant fertilizer are left?

   Ⓐ $3\frac{7}{10}$ ounces

   Ⓑ $5\frac{7}{10}$ ounces

   Ⓒ $6\frac{7}{10}$ ounces

   Ⓓ $8\frac{7}{10}$ ounces

17. Aaron is practicing for a triathlon. On Sunday, he bikes $12\frac{5}{8}$ miles and swims $5\frac{2}{3}$ miles. On Monday, he runs $6\frac{3}{8}$ miles. How many total miles does Aaron cover on the two days?

   Ⓐ $23\frac{1}{6}$ miles   Ⓒ $24\frac{2}{3}$ miles

   Ⓑ $24\frac{7}{12}$ miles   Ⓓ $25\frac{7}{12}$ miles

Name _____

Fill in the bubble completely to show your answer.

18. Mrs. Friedmon baked a walnut cake for her class. The pictures below show how much cake she brought to school and how much she had left at the end of the day.

Before School

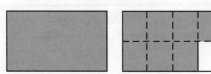

After School

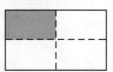

Which fraction represents the difference between the amounts of cake Mrs. Friedmon had before school and after school?

(A) $\frac{5}{8}$

(B) $1\frac{1}{2}$

(C) $1\frac{5}{8}$

(D) $2\frac{1}{2}$

19. Cody is designing a pattern for a wood floor. The length of the pieces of wood are $1\frac{1}{2}$ inches, $1\frac{13}{16}$ inches, and $2\frac{1}{8}$ inches. What is the length of the 5th piece of wood if the pattern continues?

(A) $2\frac{7}{16}$ inches

(B) $2\frac{3}{4}$ inches

(C) $3\frac{1}{2}$ inches

(D) 4 inches

20. Julie spends $\frac{3}{4}$ hour studying on Monday and $\frac{1}{6}$ hour studying on Tuesday. How many hours does Julie study on those two days?

(A) $\frac{1}{3}$ hour

(B) $\frac{2}{5}$ hour

(C) $\frac{5}{6}$ hour

(D) $\frac{11}{12}$ hour

▶ Constructed Response

21. A class uses $8\frac{5}{6}$ sheets of white paper and $3\frac{1}{12}$ sheets of red paper for a project. How much more white paper is used than red paper? Show your work using words, pictures, or numbers. **Explain** how you know your answer is reasonable.

_____

_____

▶ Performance Task

22. For a family gathering, Marcos uses the recipe below to make a lemon-lime punch.

> **Lemon-Lime Punch**
>
> $\frac{1}{4}$ gallon lime juice
>
> $\frac{2}{3}$ gallon lemon juice
>
> $1\frac{1}{4}$ gallons carbonated water

A How would you decide the size of a container you need for one batch of the Lemon-Lime Punch?

_____

_____

B If Marcos needs to make two batches of the recipe, how much of each ingredient will he need? How many gallons of punch will he have? Show your math solution and explain your thinking when you solve both questions.

_____

_____

C Marcos had $1\frac{1}{3}$ gallons of punch left over. He poured all of it into several containers for family members to take home. Use fractional parts of a gallon to suggest a way he could have shared the punch in three different-sized containers.

_____

_____

© Houghton Mifflin Harcourt Publishing Company

Chapter 7 Multiply Fractions

Show What You Know ✓

Check your understanding of important skills.

Name _____

▶ **Part of a Group** Write a fraction that names the shaded part.

1. 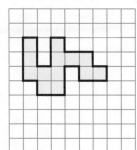   shaded parts _____

   total parts _____

   fraction _____

2. 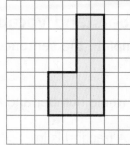   shaded parts _____

   total parts _____

   fraction _____

▶ **Area** Write the area of each shape.

3.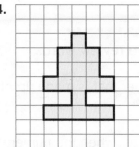

   _____ square units

4.

   _____ square units

5.

   _____ square units

▶ **Equivalent Fractions** Write an equivalent fraction.

6. $\frac{3}{4}$ _____

7. $\frac{9}{15}$ _____

8. $\frac{24}{40}$ _____

9. $\frac{5}{7}$ _____

Carmen recovered 2 gold bars that were stolen from a safe. The first bar weighed $2\frac{2}{5}$ pounds. The second bar weighed $1\frac{2}{3}$ times as much as the first bar. Be a Math Detective and find out how much gold was recovered.

Vocabulary Builder

▶ **Visualize It** ●

Match the review words with their examples.

What is it?   **What are some examples?**

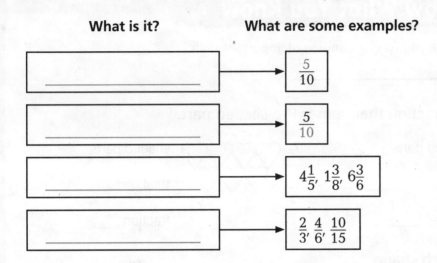

_____	$\frac{5}{10}$
_____	$\frac{5}{10}$
_____	$4\frac{1}{5}, 1\frac{3}{8}, 6\frac{3}{6}$
_____	$\frac{2}{3}, \frac{4}{6}, \frac{10}{15}$

▶ **Understand Vocabulary** ●

Complete the sentences by using the review words.

1. A _____ is a number that is made up of a whole number and a fraction.

2. A fraction is in _____ when the numerator and denominator have only 1 as a common factor.

3. The number below the bar in a fraction that tells how many equal parts are in the whole or in the group is the

   _____.

4. The _____ is the answer to a multiplication problem.

5. Fractions that name the same amount or part are called

   _____.

6. The _____ is the number above the bar in a fraction that tells how many equal parts of the whole are being considered.

GO Online • eStudent Edition • Multimedia eGlossary

Name _____

Find Part of a Group

Essential Question How can you find a fractional part of a group?

🔑 UNLOCK the Problem REAL WORLD

Maya collects stamps. She has 20 stamps in her collection. Four-fifths of her stamps have been canceled. How many of the stamps in Maya's collection have been canceled?

 Find $\frac{4}{5}$ of 20.

▲ The post office cancels stamps to keep them from being reused.

- Put 20 counters on your MathBoard.

  Since you want to find $\frac{4}{5}$ of the stamps, you should arrange the 20 counters in _____ equal groups.

- Draw the counters in equal groups below. How many counters are in each group? _____

- Each group represents _____ of the stamps. Circle $\frac{4}{5}$ of the counters.

  How many groups did you circle? _____

  How many counters did you circle? _____

  $\frac{4}{5}$ of 20 = _____, or $\frac{4}{5} \times 20$ = _____

So, _____ of the stamps have been canceled.

Math Talk MATHEMATICAL PRACTICES
How many groups would you circle if $\frac{3}{5}$ of the stamps were canceled? Explain.

🔑 Example

Max's stamp collection has stamps from different countries. He has 12 stamps from Canada. Of those twelve, $\frac{2}{3}$ of them have pictures of Queen Elizabeth II. How many stamps have the queen on them?

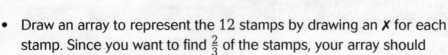

- Draw an array to represent the 12 stamps by drawing an **X** for each stamp. Since you want to find $\frac{2}{3}$ of the stamps, your array should

  show _____ rows of equal size.

- Circle _____ of the 3 rows to show $\frac{2}{3}$ of 12. Then count the number of **X**s in the circle.

  There are _____ **X**s circled.

- Complete the number sentences.

  $\frac{2}{3}$ of 12 = _____, or $\frac{2}{3} \times 12 =$ _____

So, there are _____ stamps with a picture of Queen Elizabeth II.

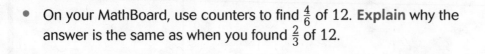

- On your MathBoard, use counters to find $\frac{4}{6}$ of 12. **Explain** why the answer is the same as when you found $\frac{2}{3}$ of 12.

_____

_____

Try This! Draw an array.

Susan has 16 stamps. In her collection, $\frac{3}{4}$ of the stamps are from the United States. How many of her stamps are from the United States and how many are not?

So, _____ of Susan's stamps are from the United States and _____ stamps are not.

292

Name _____

Share and Show

1. Complete the model to solve.

   $\frac{7}{8}$ of 16, or $\frac{7}{8} \times 16$

   • How many rows of counters are there? _____

   • How many counters are in each row? _____

   • Circle _____ rows to solve the problem.

   • How many counters are circled? _____

   $\frac{7}{8}$ of 16 = _____, or $\frac{7}{8} \times 16 =$ _____

Use a model to solve.

2. $\frac{2}{3} \times 18 =$ _____

3. $\frac{2}{5} \times 15 =$ _____

4. $\frac{2}{3} \times 6 =$ _____

On Your Own

Use a model to solve.

5. $\frac{5}{8} \times 24 =$ _____

6. $\frac{3}{4} \times 24 =$ _____

7. $\frac{4}{7} \times 21 =$ _____

8. $\frac{2}{9} \times 27 =$ _____

9. $\frac{3}{5} \times 20 =$ _____

10. $\frac{7}{11} \times 22 =$ _____

Problem Solving REAL WORLD

Use the table for 11–12.

Stamps Collected	
Name	**Number of Stamps**
Zack	30
Teri	18
Paco	24

11. Four-fifths of Zack's stamps have pictures of animals. How many stamps with pictures of animals does Zack have? Use a model to solve.

_____

12. **H.O.T.** **Write Math** ▸ Zack, Teri, and Paco combined the foreign stamps from their collections for a stamp show. Out of their collections, $\frac{3}{10}$ of Zack's stamps, $\frac{5}{6}$ of Teri's stamps, and $\frac{3}{8}$ of Paco's stamps were from foreign countries. How many stamps were in their display? **Explain** how you solved the problem.

SHOW YOUR WORK

_____

_____

_____

_____

_____

13. Paula has 24 stamps in her collection. Among her stamps, $\frac{1}{3}$ have pictures of animals. Out of her stamps with pictures of animals, $\frac{3}{4}$ of those stamps have pictures of birds. How many stamps have pictures of birds on them?

_____

14. **Test Prep** Barry bought 21 stamps from a hobby shop. He gave $\frac{3}{7}$ of them to his sister. How many stamps did he have left?

Ⓐ 3 stamps

Ⓑ 6 stamps

Ⓒ 9 stamps

Ⓓ 12 stamps

FOR MORE PRACTICE:
Standards Practice Book, pp. P145–P146

Name _____

Multiply Fractions and Whole Numbers

Essential Question How can you use a model to show the product of a fraction and a whole number?

Investigate

Martin is planting a vegetable garden. Each row is two meters long. He wants to plant carrots along $\frac{3}{4}$ of each row. How many meters of each row will he plant with carrots?

 Multiply. $\frac{3}{4} \times 2$

Materials ■ fraction strips ■ MathBoard

A. Place two 1-whole fraction strips side-by-side to represent the length of the garden.

B. Find 4 fraction strips all with the same denominator that fit exactly under the two wholes.

C. Draw a picture of your model.

1	1

D. Circle $\frac{3}{4}$ of 2 on the model you drew.

E. Complete the number sentence. $\frac{3}{4} \times 2 =$ _____

So, Martin will plant carrots along _____ meters of each row.

Draw Conclusions ·····················

1. **Explain** why you placed four fraction strips with the same denominator under the two 1-whole strips.

_____

_____

2. **Explain** how you would model $\frac{3}{10}$ of 2?

_____

_____

Make Connections ....

In the Investigate, you multiplied a whole number by a fraction. You can also use a model to multiply a fraction by a whole number.

Margo was helping clean up after a class party. There were 3 boxes remaining with pizza in them. Each box had $\frac{3}{8}$ of a pizza left. How much pizza was left in all?

Materials ▪ fraction circles

STEP 1 Find $3 \times \frac{3}{8}$. Model three 1-whole fraction circles to represent the number of boxes containing pizza.

STEP 2 Place $\frac{1}{8}$ fraction circle pieces on each circle to represent the amount of pizza that was left in each box.

• Shade the fraction circles below to show your model.

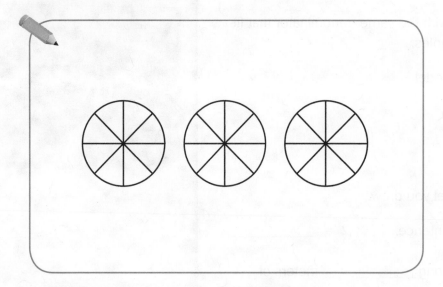

Each circle shows _____ eighths of a whole.

The 3 circles show _____ eighths of a whole.

STEP 3 Complete the number sentences.

$\frac{3}{8} + \frac{3}{8} + \frac{3}{8} =$ _____

$3 \times \frac{3}{8} =$ _____

So, Margo had _____ boxes of pizza left.

Math Talk MATHEMATICAL PRACTICES
Explain how you would know there is more than one pizza left.

Name _____

Share and Show

Use the model to find the product.

1. $\dfrac{5}{6} \times 3 =$ _____

1		1		1	
$\frac{1}{2}$	$\frac{1}{2}$	$\frac{1}{2}$	$\frac{1}{2}$	$\frac{1}{2}$	$\frac{1}{2}$

2. $2 \times \dfrac{5}{6} =$ _____

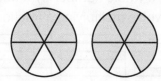

Find the product.

3. $\dfrac{5}{12} \times 3 =$ _____

4. $9 \times \dfrac{1}{3} =$ _____

5. $\dfrac{7}{8} \times 4 =$ _____

6. $4 \times \dfrac{3}{5} =$ _____

7. $\dfrac{7}{8} \times 2 =$ _____

8. $7 \times \dfrac{2}{5} =$ _____

9. $\dfrac{3}{8} \times 4 =$ _____

10. $11 \times \dfrac{3}{4} =$ _____

11. $\dfrac{4}{15} \times 5 =$ _____

12. **Write Math** ▶ Matt has a 5-pound bag of apples. To make a pie, he needs to use $\dfrac{3}{5}$ of the bag. How many pounds of apples will he use for the pie? **Explain** what a model for this problem might look like.

_____

_____

_____

Problem Solving

 Pose a Problem

13. Tarique drew the model below for a problem. Write 2 problems
that can be solved using this model. One of your problems should
involve multiplying a whole number by a fraction and the other
problem should involve multiplying a fraction by a whole number.

| | | | | | |

Pose problems.   **Solve your problems.**

• How could you change the model to give you an answer of $4\frac{4}{5}$?
Explain and write a new equation.

FOR MORE PRACTICE:
Standards Practice Book, pp. P147–P148

Name _____

Fraction and Whole Number Multiplication

Essential Question How can you find the product of a fraction and a whole number without using a model?

 UNLOCK the Problem REAL WORLD

Charlene has five 1-pound bags of different color sands. For an art project, she will use $\frac{3}{8}$ pound of each bag of sand to create a colorful sand-art jar. How much sand will be in Charlene's sand-art jar?

- How much sand is in each bag?

  _____

- Will Charlene use all of the sand in each bag? Explain.

  _____

🔒 **Multiply a fraction by a whole number.**

MODEL

- Shade the model to show 5 groups of $\frac{3}{8}$.

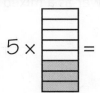
$$5 \times \;\;\;\; =$$

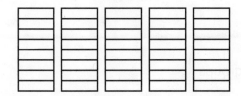

- Rearrange the shaded pieces to fill as many wholes as possible.

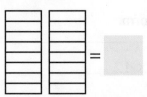

RECORD

- Write an expression to represent the problem.

  $$5 \times \frac{3}{8}$$ **Think:** I need to find 5 groups of 3 eighth-size pieces.

- Multiply the number of eighth-size pieces in each whole by 5. Then write the answer as the total number of eighth-size pieces.

  $$\frac{\times}{8} = \frac{}{}$$

- Write the answer as a mixed number in simplest form.

  $$\frac{}{} = \frac{}{}$$

So, there are _____ pounds of sand in Charlene's sand-art jar.

Math Talk MATHEMATICAL PRACTICES
Explain how you can find how much sand Charlene has left.

🔓 Example Multiply a whole number by a fraction.

Kirsten brought in 4 loaves of bread to make sandwiches for the class picnic. Her classmates used $\frac{2}{3}$ of the bread. How many loaves of bread were used?

MODEL

- Shade the model to show $\frac{2}{3}$ of 4.

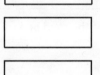

Think: I can cut the loaves into thirds and show $\frac{2}{3}$ of them being used.

- Rearrange the shaded pieces to fill as many wholes as possible.

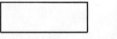
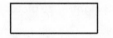

So, _____ loaves of bread were used.

RECORD

- Write an expression to represent the problem.

$$\frac{2}{3} \times 4$$

Think: I need to find $\frac{2}{3}$ of 4 wholes.

- Multiply 4 by the number of third-size pieces in each whole. Then write the answer as the total number of third-size pieces.

$$\frac{ \times }{} = \frac{}{}$$

- Write the answer as a mixed number.

$$\frac{}{} = \frac{}{}$$

- Would we have the same amount of bread if we had 4 groups of $\frac{2}{3}$ of a loaf? **Explain.**

_____

Try This! Find the product. Write the product in simplest form.

 $4 \times \frac{7}{8}$

 $\frac{5}{9} \times 12$

Name _____

Share and Show

Find the product. Write the product in simplest form.

1. $3 \times \dfrac{2}{5} =$ _____

- Multiply the numerator by the whole number. Write the product over the denominator.

- Write the answer as a mixed number in simplest form.

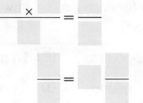

☑ 2. $\dfrac{2}{3} \times 5 =$ _____   **☑ 3.** $6 \times \dfrac{2}{3} =$ _____   **4.** $\dfrac{5}{7} \times 4 =$ _____

On Your Own

Find the product. Write the product in simplest form.

5. $5 \times \dfrac{2}{3} =$ _____   **6.** $\dfrac{1}{4} \times 3 =$ _____   **7.** $7 \times \dfrac{7}{8} =$ _____

8. $2 \times \dfrac{4}{5} =$ _____   **9.** $4 \times \dfrac{3}{4} =$ _____   **10.** $\dfrac{7}{9} \times 2 =$ _____

Practice: Copy and Solve. Find the product. Write the product in simplest form.

11. $\dfrac{3}{5} \times 11$ **12.** $3 \times \dfrac{3}{4}$ **13.** $\dfrac{5}{8} \times 3$

H.O.T. **Algebra** Find the unknown digit.

14. $\dfrac{\blacksquare}{2} \times 8 = 4$ **15.** $\blacksquare \times \dfrac{5}{6} = \dfrac{20}{6}$, or $3\dfrac{1}{3}$ **16.** $\dfrac{1}{\blacksquare} \times 18 = 3$

$\blacksquare =$ _____ $\blacksquare =$ _____ $\blacksquare =$ _____

UNLOCK the Problem   REAL WORLD

17. The caterer wants to have enough turkey to feed 24 people. If he wants to provide $\frac{3}{4}$ of a pound of turkey for each person, how much turkey does he need?

Ⓐ 72 pounds   Ⓒ 18 pounds

Ⓑ 24 pounds   Ⓓ 6 pounds

a. What do you need to find? _____

_____

b. What operation will you use? _____

c. What information are you given? _____

_____

d. Solve the problem.

e. Complete the sentences.

The caterer wants to serve 24 people

_____ of a pound of turkey each.

He will need _____ × _____, or

_____ pounds of turkey.

f. Fill in the bubble for the correct answer choice.

18. Patty wants to run $\frac{5}{6}$ of a mile every day for 5 days. How far will she run in that time?

Ⓐ 25 miles

Ⓑ 5 miles

Ⓒ $4\frac{1}{6}$ miles

Ⓓ $1\frac{2}{3}$ miles

19. Doug has 33 feet of rope. He wants to use $\frac{2}{3}$ of it for his canoe. How many feet of rope will he use for his canoe?

Ⓐ 11 feet

Ⓑ 22 feet

Ⓒ 33 feet

Ⓓ 66 feet

© Houghton Mifflin Harcourt Publishing Company

Name _____

Multiply Fractions

Essential Question How can you use an area model to show the product of two fractions?

Investigate

Jane is making reusable grocery bags and lunch bags. She needs $\frac{3}{4}$ yard of material to make a grocery bag. A lunch bag needs $\frac{2}{3}$ of the amount of material a grocery bag needs. How much material does she need to make a lunch bag?

 Find $\frac{2}{3}$ of $\frac{3}{4}$. **Materials** ■ color pencils

A. Fold a sheet of paper vertically into 4 equal parts. Using the vertical folds as a guide, shade $\frac{3}{4}$ yellow.

B. Fold the paper horizontally into 3 equal parts. Using the horizontal folds as a guide, shade $\frac{2}{3}$ of the yellow sections blue.

C. Count the number of sections into which the whole paper is folded.

- How many rectangles are formed by all

  the folds in the paper? _____

- What fraction of the whole sheet of paper

  does one rectangle represent? _____

D. Count the sections that are shaded twice and record

the answer. $\frac{2}{3} \times \frac{3}{4} =$ _____

So, Jane needs _____ yard of material to make a lunch bag.

Draw Conclusions ·····································

1. **Explain** why you shaded $\frac{2}{3}$ of the yellow sections blue rather than shading $\frac{2}{3}$ of the whole.

   _____

   _____

2. **Analyze** what you are finding if a model shows $\frac{1}{2}$ of a sheet of paper shaded yellow and $\frac{1}{3}$ of the yellow section shaded blue?

   _____

Make Connections

You can find a part of a part in different ways. Margo and James both correctly solved the problem $\frac{1}{3} \times \frac{3}{4}$ using the steps shown.

Use the steps to show how each person found $\frac{1}{3} \times \frac{3}{4}$.

Margo

- Shade the model to show $\frac{3}{4}$ of the whole.

- How many $\frac{1}{4}$ pieces did you shade?

  _____ one-fourth pieces

- To find $\frac{1}{3}$ of $\frac{3}{4}$, circle $\frac{1}{3}$ of the three $\frac{1}{4}$ pieces that are shaded.

- What part of the whole is $\frac{1}{3}$ of the shaded

  pieces? _____ of the whole

So, $\frac{1}{3} \times \frac{3}{4}$ is _____.

James

- Shade the model to show $\frac{3}{4}$ of the whole.

- Divide each $\frac{1}{4}$ piece into thirds.

- What part of the whole is each

  small piece? _____

- To find $\frac{1}{3}$ of $\frac{3}{4}$, circle $\frac{1}{3}$ of each of the three $\frac{1}{4}$ pieces that are shaded.

- How many $\frac{1}{12}$ pieces are circled?

  _____ one-twelfth pieces

So, $\frac{1}{3} \times \frac{3}{4}$ is _____.

- **Pose a Problem** that can be solved using the equation above.

_____

_____

Share and Show

Use the model to find the product.

1.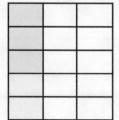

   $\frac{3}{5} \times \frac{1}{3} =$ _____

2.

   Circle $\frac{2}{3}$ of $\frac{3}{5}$.

   $\frac{2}{3} \times \frac{3}{5} =$ _____

Name _____

Find the product. Draw a model.

3. $\frac{2}{3} \times \frac{1}{5} =$ _____

4. $\frac{1}{2} \times \frac{5}{6} =$ _____

5. $\frac{3}{5} \times \frac{1}{3} =$ _____

6. $\frac{3}{4} \times \frac{1}{6} =$ _____

7. $\frac{2}{5} \times \frac{5}{6} =$ _____

8. $\frac{5}{6} \times \frac{3}{5} =$ _____

Problem Solving REAL WORLD

H.O.T. What's the Error?

9. Cheryl and Marcus are going to make a two-tiered cake. The smaller tier is $\frac{2}{3}$ the size of the larger tier. The recipe for the bottom tier calls for $\frac{3}{5}$ cup of water. How much water will they need to make the smaller tier?

They made a model to represent the problem. Cheryl says they need $\frac{6}{9}$ cup of water. Marcus says they need $\frac{2}{5}$ cup water. Who is correct? **Explain**.

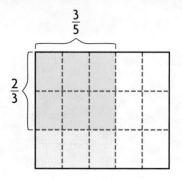

Cheryl's answer

Marcus' answer

FOR MORE PRACTICE:
Standards Practice Book, pp. P151–P152

Name _____

Compare Fraction Factors and Products

Essential Question How does the size of the product compare to the size of one factor when multiplying fractions?

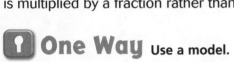 **UNLOCK the Problem** REAL WORLD

Multiplication can be thought of as resizing one number by another number. For example, 2×3 will result in a product that is 2 times as great as 3.

What happens to the size of a product when a number is multiplied by a fraction rather than a whole number?

 One Way Use a model.

A **During the week, the Smith family ate $\frac{3}{4}$ of a box of cereal.**

- Shade the model to show $\frac{3}{4}$ of a box of cereal.

- Write an expression for $\frac{3}{4}$ of 1 box of cereal. $\frac{3}{4} \times$ _____

- Will the product be *equal to, greater than,* or *less than* 1?

_____

B **The Ling family has 4 boxes of cereal. They ate $\frac{3}{4}$ of all the cereal during the week.**

- Shade the model to show $\frac{3}{4}$ of 4 boxes of cereal.

- Write an expression for $\frac{3}{4}$ of 4 boxes of cereal. $\frac{3}{4} \times$ _____

- Will the product be *equal to, greater than,* or *less than* 4?

_____

C **The Carter family has only $\frac{1}{2}$ of a box of cereal at the beginning of the week. They ate $\frac{3}{4}$ of the $\frac{1}{2}$ box of cereal.**

- Shade the model to show $\frac{3}{4}$ of $\frac{1}{2}$ box of cereal.

- Write an expression to show $\frac{3}{4}$ of $\frac{1}{2}$ box of cereal. $\frac{3}{4} \times$ _____

- Will the product be *equal to, greater than,* or *less than* $\frac{1}{2}$? than $\frac{3}{4}$?

_____

🔑 Another Way Use a diagram.

You can use a diagram to show the relationship between the products when a fraction is multiplied or scaled (resized) by a number.

Graph a point to show $\frac{3}{4}$ scaled by 1, $\frac{1}{2}$, and 4.

Ⓐ $1 \times \frac{3}{4}$

Think: Locate $\frac{3}{4}$ on the diagram and shade that distance from 0. Then graph a point to show 1 of $\frac{3}{4}$.

Ⓑ $\frac{1}{2} \times \frac{3}{4}$

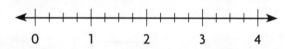

Think: Locate $\frac{3}{4}$ on the diagram and shade that distance from 0. Then graph a point to show $\frac{1}{2}$ of $\frac{3}{4}$.

Ⓒ $4 \times \frac{3}{4}$

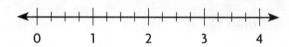

Think: Locate $\frac{3}{4}$ on the diagram and shade that distance from 0. Then graph a point to show 4 times $\frac{3}{4}$.

Complete each statement with *equal to, greater than,* or *less than*.

- The product of 1 and $\frac{3}{4}$ will be _____ $\frac{3}{4}$.

- The product of a number less than 1 and $\frac{3}{4}$ will be

  _____ $\frac{3}{4}$ and _____ the other factor.

- The product of a number greater than 1 and $\frac{3}{4}$ will

  be _____ $\frac{3}{4}$ and _____ the other factor.

Math Talk

MATHEMATICAL PRACTICES
What if $\frac{3}{5}$ was multiplied by $\frac{1}{6}$ or by the whole number 7? Would the products be equal to, greater than, or less than $\frac{3}{5}$? **Explain.**

Name _____

Share and Show

Complete the statement with *equal to, greater than,* or *less than.*

1. $4 \times \frac{7}{8}$ will be _____ $\frac{7}{8}$.

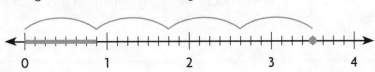

2. $\frac{3}{5} \times \frac{2}{7}$ will be _____ $\frac{3}{5}$.

3. $\frac{5}{8} \times 6$ will be _____ $\frac{5}{8}$.

4. $\frac{2}{3} \times \frac{5}{5}$ will be _____ $\frac{2}{3}$.

5. $8 \times \frac{7}{8}$ will be _____ 8.

On Your Own

Complete the statement with *equal to, greater than,* or *less than.*

6. $\frac{4}{9} \times \frac{3}{8}$ will be _____ $\frac{3}{8}$.

7. $7 \times \frac{9}{10}$ will be _____ $\frac{9}{10}$.

8. $5 \times \frac{1}{3}$ will be _____ $\frac{1}{3}$.

9. $\frac{6}{11} \times 1$ will be _____ $\frac{6}{11}$.

10. $\frac{1}{6} \times \frac{7}{7}$ will be _____ 1.

11. $4 \times \frac{3}{5}$ will be _____ $\frac{3}{5}$.

Problem Solving REAL WORLD

12. Lola is making cookies. She plans to multiply the recipe by 3 so she can make enough cookies for the whole class. If the recipe calls for $\frac{2}{3}$ cup of sugar, will she need more than $\frac{2}{3}$ or less than $\frac{2}{3}$ cup of sugar to make all the cookies?

13. Peter is planning on spending $\frac{2}{3}$ as many hours watching television this week as he did last week. Is Peter going to spend more hours or fewer hours watching television this week?

14. **Test Prep** Rochelle saves $\frac{1}{4}$ of her allowance. If she decides to start saving $\frac{1}{2}$ as much, which statement below is true?

(A) She will be saving the same amount.

(C) She will be saving less.

(B) She will be saving more.

(D) She will be saving twice as much.

Connect to Art

A scale model is a representation of an object with the same shape as the real object. Models can be larger or smaller than the actual object but are often smaller.

Architects often make scale models of the buildings or structures they plan to build. Models can give them an idea of how the structure will look when finished. Each measurement of the building is scaled up or down by the same factor.

Bob is building a scale model of his bike. He wants his model to be $\frac{1}{5}$ as long as his bike.

15. If Bob's bike is 60 inches long, how long will his model be? _____

16. H.O.T. If one wheel on Bob's model is 4 inches across, how many inches across is the actual wheel on his bike? **Explain.**

_____

_____

Fraction Multiplication

Essential Question How do you multiply fractions?

 UNLOCK the Problem REAL WORLD

Sasha has $\frac{3}{5}$ of a scarf left to knit. If she finishes $\frac{1}{2}$ of that today, how much of the scarf will Sasha knit today?

Multiply. $\frac{1}{2} \times \frac{3}{5}$

🔑 One Way Use a model.

- Shade $\frac{3}{5}$ of the model yellow.

- Draw a horizontal line across the rectangle to show 2 equal parts.

- Shade $\frac{1}{2}$ of the yellow sections blue.

- Count the sections that are shaded twice and write a fraction for the parts of the whole that are shaded twice.

$$\frac{1}{2} \times \frac{3}{5} = \underline{\hspace{1cm}}$$

- Compare the numerator and denominator of the product with the numerators and denominators of the factors. **Describe** what you notice.

_____

_____

_____

- How much of the scarf does Sasha have left to knit?

_____

- Of the fraction that is left, how much will she finish today?

_____

🔑 Another Way Use paper and pencil.

You can multiply fractions without using a model.

- Multiply the numerators.

- Multiply the denominators.

$$\frac{1}{2} \times \frac{3}{5} = \frac{1 \times \boxed{}}{2 \times \boxed{}}$$

$$= \frac{\boxed{}}{\boxed{}}$$

So, Sasha will knit _____ of the scarf today.

CONNECT Remember you can write a whole number as a fraction with a denominator of 1.

 Example

Find $4 \times \frac{5}{12}$. **Write the product in simplest form.**

$$4 \times \frac{5}{12} = \frac{4}{\boxed{}} \times \frac{5}{12}$$

Write the whole number as a fraction.

$$= \frac{4 \times \boxed{}}{\boxed{} \times \boxed{}} = \frac{\boxed{}}{\boxed{}}$$

Multiply the numerators.
Multiply the denominators.

$$= \frac{\boxed{} \div \boxed{}}{12 \div \boxed{}} = \frac{\boxed{}}{\boxed{}} , \text{ or } \boxed{}$$

Write the product as a fraction or a mixed number in simplest form.

So, $4 \times \frac{5}{12} =$ _____, or _____.

Math Talk MATHEMATICAL PRACTICES Is the answer reasonable? **Explain.**

Try This! Evaluate $c \times \frac{4}{5}$ **for** $c = \frac{5}{8}$.

- What number does c represent? _____

- Replace c in the expression with _____.

- Multiply the numerators.

- Multiply the denominators.

- Write the product in simplest form.

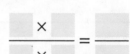

So, $c \times \frac{4}{5}$ is equal to _____ for $c = \frac{5}{8}$.

- Since $\frac{4}{5}$ is being multiplied by a number less than one, should the

  product be *greater than* or *less than* $\frac{4}{5}$? **Explain.** _____

  _____

  _____

312

© Houghton Mifflin Harcourt Publishing Company

Name _____

Share and Show

Find the product. Write the product in simplest form.

1. $6 \times \frac{3}{8}$

$$\frac{6}{1} \times \frac{3}{8} = \underline{\quad}$$

2. $\frac{3}{8} \times \frac{8}{9}$

3. $\frac{2}{3} \times 27$

4. $\frac{5}{12} \times \frac{3}{5}$

5. $\frac{1}{2} \times \frac{3}{5}$

6. $\frac{2}{3} \times \frac{4}{5}$

7. $\frac{1}{3} \times \frac{5}{8}$

8. $4 \times \frac{1}{5}$

MATHEMATICAL PRACTICES

Math Talk **Explain** how to find the product $\frac{1}{6} \times \frac{2}{3}$ in simplest form.

On Your Own

Find the product. Write the product in simplest form.

9. $2 \times \frac{1}{8}$

10. $\frac{4}{9} \times \frac{4}{5}$

11. $\frac{1}{12} \times \frac{2}{3}$

12. $\frac{1}{7} \times 30$

13. Of the pets in the pet show, $\frac{5}{6}$ are cats. $\frac{4}{5}$ of the cats are calico cats. What fraction of the pets are calico cats?

14. Five cats each ate $\frac{1}{4}$ cup of food. How much food did they eat altogether?

Algebra Evaluate for the given value.

15. $\frac{2}{5} \times c$ for $c = \frac{4}{7}$

16. $m \times \frac{4}{5}$ for $m = \frac{7}{8}$

17. $\frac{2}{3} \times t$ for $t = \frac{1}{8}$

18. $y \times \frac{4}{5}$ for $y = 5$

Problem Solving **REAL WORLD**

Speedskating is a popular sport in the Winter Olympics. Many young athletes in the U.S. participate in speedskating clubs and camps.

19. At a camp in Green Bay, Wisconsin, $\frac{7}{9}$ of the participants were from Wisconsin. Of that group, $\frac{3}{5}$ were 12 years old. What fraction of the group was from Wisconsin and 12 years old?

_____

20. **H.O.T.** Maribel wants to skate $1\frac{1}{2}$ miles on Monday. If she skates $\frac{9}{10}$ mile Monday morning and $\frac{2}{3}$ of that distance Monday afternoon, will she reach her goal? **Explain**.

_____

_____

SHOW YOUR WORK

21. **Write Math** ► On the first day of camp, $\frac{5}{6}$ of the skaters were beginners. Of the beginners, $\frac{1}{3}$ were girls. What fraction of the skaters were girls and beginners? **Explain** why your answer is reasonable.

_____

_____

_____

_____

22. **Test Prep** On Wednesday, Danielle skated $\frac{2}{3}$ of the way around the track in 2 minutes. Her younger brother skated $\frac{3}{4}$ of Danielle's distance in 2 minutes. What fraction of the track did Danielle's brother finish in 2 minutes?

(A) $\frac{1}{3}$ (C) $\frac{5}{7}$

(B) $\frac{1}{2}$ (D) $\frac{3}{4}$

Name _____

Mid-Chapter Checkpoint

▶ **Concepts and Skills**

1. Explain how you would model $5 \times \frac{2}{3}$.

_____

_____

2. When you multiply $\frac{2}{3}$ by a fraction less than one, how does the product compare to the factors. **Explain**.

_____

_____

_____

Find the product. Write the product in simplest form.

3. $\frac{2}{3} \times 6$

4. $\frac{4}{5} \times 7$

5. $8 \times \frac{5}{7}$

_____ | _____ | _____

6. $\frac{7}{8} \times \frac{3}{8}$

7. $\frac{1}{2} \times \frac{3}{4}$

8. $\frac{7}{8} \times \frac{4}{7}$

_____ | _____ | _____

9. $2 \times \frac{3}{11}$

10. $\frac{5}{8} \times \frac{2}{3}$

11. $\frac{7}{12} \times 8$

_____ | _____ | _____

Complete the statement with *equal to*, *greater than*, or *less than*.

12. $3 \times \frac{2}{3}$ will be _____ 3.

13. $\frac{5}{7} \times 3$ will be _____ $\frac{5}{7}$.

Fill in the bubble completely to show your answer.

14. There is $\frac{5}{6}$ of an apple pie left from dinner. Tomorrow, Victor plans to eat $\frac{1}{6}$ of the pie that was left. How much of the whole pie will he eat tomorrow?

   Ⓐ $\frac{1}{36}$

   Ⓑ $\frac{5}{36}$

   Ⓒ $\frac{1}{3}$

   Ⓓ $\frac{2}{3}$

15. Everett and Marie are going to make brownies for their family reunion. They want to make 4 times the amount the recipe makes. If the recipe calls for $\frac{2}{3}$ cup of oil, how much oil will they need?

   Ⓐ 8 cups

   Ⓑ $4\frac{2}{3}$ cups

   Ⓒ $2\frac{2}{3}$ cups

   Ⓓ 2 cups

16. Matt made the model below to help him solve his math problem. Which of the expressions could he have been working on?

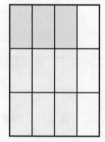

   Ⓐ $\frac{3}{12} \times \frac{3}{4}$

   Ⓑ $\frac{3}{4} \times 3$

   Ⓒ $\frac{3}{12} \times \frac{3}{12}$

   Ⓓ $\frac{1}{3} \times \frac{3}{4}$

Name _____

Area and Mixed Numbers

Essential Question How can you use a unit tile to find the area of a rectangle with fractional side lengths?

Investigate

You can use square tiles with side lengths that are unit fractions to find the area of a rectangle.

Sonja wants to cover the rectangular floor of her closet with tile. The floor is $2\frac{1}{2}$ feet by $3\frac{1}{2}$ feet. She wants to use the fewest tiles possible and doesn't want to cut any tiles. The tiles come in three sizes: 1 foot by 1 foot, $\frac{1}{2}$ foot by $\frac{1}{2}$ foot, and $\frac{1}{4}$ foot by $\frac{1}{4}$ foot. Choose the tile that Sonja should use. What is the area of the closet floor?

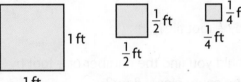

A. Choose the largest tile Sonja can use to tile the floor of the closet and avoid gaps or overlaps.

- Which square tile should Sonja choose? **Explain.** _____

_____

_____

B. On the grid, let each square represent the dimensions of the tile you chose. Then draw a diagram of the floor.

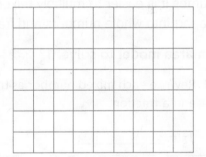

C. Count the squares in your diagram.

- How many squares cover the diagram?

  _____ × _____ , or _____ squares

- What is the area of the tile you chose? _____

- Since 1 square on your diagram represents an area of _____ square foot,

  the area represented by _____ squares is _____ × _____ ,

  or _____ square feet.

So, the area of the floor written as a mixed number

is _____ square feet.

MATHEMATICAL PRACTICES

Math Talk **Explain** how you found the area of the tile you chose.

Chapter 7 317

Draw Conclusions

1. Write a number sentence for the area of the floor using fractions greater than 1. **Explain** how you knew which operation to use in your number sentence.

   _____

   _____

2. **Explain** how using fractions greater than 1 could help you multiply mixed numbers.

   _____

   _____

3. How many $\frac{1}{4}$ foot by $\frac{1}{4}$ foot tiles would Sonja need to cover one $\frac{1}{2}$ foot by $\frac{1}{2}$ foot tile? _____

   $\frac{1}{2}$ foot

   $\frac{1}{2}$ foot

4. How could you find the number of $\frac{1}{4}$ foot by $\frac{1}{4}$ foot tiles needed to cover the same closet floor?

   _____

   _____

Make Connections

Sometimes it is easier to multiply mixed numbers if you break them apart into whole numbers and fractions.

Use an area model to solve. $1\frac{3}{5} \times 2\frac{3}{4}$

STEP 1 Rewrite each mixed number as the sum of a whole number and a fraction.

   $1\frac{3}{5} =$ _____ $2\frac{3}{4} =$ _____

STEP 2 Draw an area model to show the original multiplication problem.

STEP 3 Draw dashed lines and label each section to show how you broke apart the mixed numbers in Step 1.

STEP 4 Find the area of each section.

STEP 5 Add the area of each section to find the total area of the rectangle.

   _____

So, the product of $1\frac{3}{5} \times 2\frac{3}{4}$ is _____.

318

© Houghton Mifflin Harcourt Publishing Company

Name _____

Share and Show

Use the grid to find the area. Let each square represent
$\frac{1}{3}$ meter by $\frac{1}{3}$ meter.

1. $1\frac{2}{3} \times 1\frac{1}{3}$

- Draw a diagram to represent the dimensions.

- How many squares cover the diagram? _____

- What is the area of each square? _____

- What is the area of the diagram? _____

Use the grid to find the area. Let each square represent
$\frac{1}{4}$ foot by $\frac{1}{4}$ foot.

2. $1\frac{3}{4} \times 1\frac{2}{4} =$ _____

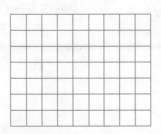

The area is _____ square feet.

3. $1\frac{1}{4} \times 1\frac{1}{2} =$ _____

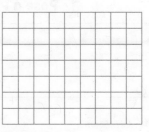

The area is _____ square feet.

Use an area model to solve.

4. $1\frac{3}{4} \times 2\frac{1}{2}$

_____

5. $1\frac{3}{8} \times 2\frac{1}{2}$

_____

6. $1\frac{1}{9} \times 1\frac{2}{3}$

_____

7. **Write Math** ▶ **Explain** how finding the area of a rectangle with whole-number side lengths compares to finding the area of a rectangle with fractional side lengths.

_____

_____

Problem Solving REAL WORLD

 Pose a Problem

8. Terrance is designing a garden. He drew the following diagram of his garden. Pose a problem using mixed numbers that can be solved using his diagram.

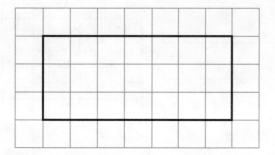

Pose a Problem.

_____

_____

_____

_____

_____

_____

_____

Solve your problem.

- **Describe** how you decided on the dimensions of Terrance's garden.

_____

_____

_____

Name _____

Compare Mixed Number Factors and Products

Essential Question How does the size of the product compare to the size of one factor when multiplying fractions greater than one?

🔓 UNLOCK the Problem REAL WORLD

You can make generalizations about the relative size of a product when one factor is equal to 1, less than 1, or greater than 1.

🔑 One Way Use a model.

Jane has a recipe that calls for $1\frac{1}{4}$ cups of flour. She wants to know how much flour she would need if she made the recipe as written, if she made half the recipe, and if she made $1\frac{1}{2}$ times the recipe.

Shade the models to show $1\frac{1}{4}$ scaled by 1, by $\frac{1}{2}$, and by $1\frac{1}{2}$.

Ⓐ $1 \times 1\frac{1}{4}$

Think: I can use what I know about the Identity Property.

- What can you say about the product when $1\frac{1}{4}$ is multiplied by 1?

_____

Ⓑ $\frac{1}{2} \times 1\frac{1}{4}$

Think: The product will be half of what I started with.

- What can you say about the product when $1\frac{1}{4}$ is multiplied by a

fraction less than 1? _____

Ⓒ $1\frac{1}{2} \times 1\frac{1}{4} = \left(1 \times 1\frac{1}{4}\right) + \left(\frac{1}{2} \times 1\frac{1}{4}\right)$

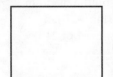 +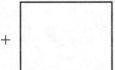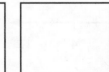

Think: The product will be what I started with and $\frac{1}{2}$ more.

- What can you say about the product when $1\frac{1}{4}$ is multiplied by a number greater than 1?

_____

Math Talk MATHEMATICAL PRACTICES
Which expression has the greatest product? Which has the least product?

CONNECT You can also use a diagram to show the relationship between the products when a fraction greater than one is multiplied or scaled (resized) by a number.

Another Way Use a diagram.

Jake wants to train for a road race. He plans to run $2\frac{1}{2}$ miles on the first day. On the second day, he plans to run $\frac{3}{5}$ of the distance he runs on the first day. On the third day, he plans to run $1\frac{2}{5}$ of the distance he runs on the first day. Which distance is greater: the distance on day 2 when he runs $\frac{3}{5}$ of $2\frac{1}{2}$ miles, or the distance on day 3 when he runs $1\frac{2}{5}$ of $2\frac{1}{2}$ miles?

Graph a point on the diagram to show the size of the product. Then complete the statement with *equal to*, *greater than*, or *less than*.

A $1 \times 2\frac{1}{2}$

Think: Locate $2\frac{1}{2}$ on the diagram and shade that distance. Then graph a point to show 1 of $2\frac{1}{2}$.

- The product of 1 and $2\frac{1}{2}$ will be _____ $2\frac{1}{2}$.

B $\frac{3}{5} \times 2\frac{1}{2}$

Think: Locate $2\frac{1}{2}$ on the diagram and shade that distance. Then graph a point to show $\frac{3}{5}$ of $2\frac{1}{2}$.

- The product of a number less than 1 and $2\frac{1}{2}$

  is _____ $2\frac{1}{2}$.

C $1\frac{2}{5} \times 2\frac{1}{2} = \left(1 \times 2\frac{1}{2}\right) + \left(\frac{2}{5} \times 2\frac{1}{2}\right)$

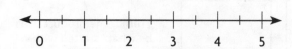

Think: Locate $2\frac{1}{2}$ on the diagram and shade that distance. Then graph a point to show 1 of $2\frac{1}{2}$ and $\frac{2}{5}$ more of $2\frac{1}{2}$.

- The product of a number greater than 1 and $2\frac{1}{2}$ will

  be _____ $2\frac{1}{2}$ and _____ the other factor.

So, _____ of _____ miles is a greater distance than _____ of _____ miles.

Name _____

Share and Show

Complete the statement with *equal to, greater than,* or *less than*.

1. $\frac{5}{6} \times 2\frac{1}{5}$ will be _____ $2\frac{1}{5}$.

Shade the model to show $\frac{5}{6} \times 2\frac{1}{5}$.

2. $1\frac{1}{5} \times 2\frac{2}{3}$ will be _____ $2\frac{2}{3}$.

3. $\frac{4}{5} \times 2\frac{2}{5}$ will be _____ $2\frac{2}{5}$.

On Your Own

Complete the statement with *equal to, greater than,* or *less than*.

4. $\frac{2}{2} \times 1\frac{1}{2}$ will be _____ $1\frac{1}{2}$.

5. $\frac{2}{3} \times 3\frac{1}{6}$ will be _____ $3\frac{1}{6}$.

6. $2 \times 2\frac{1}{4}$ will be _____ $2\frac{1}{4}$.

7. $4 \times 1\frac{3}{7}$ will be _____ $1\frac{3}{7}$.

H.O.T. **Algebra** Tell whether the unknown factor is *less than 1* or *greater than 1*.

8. ▇ $\times 1\frac{2}{3} = \frac{5}{6}$

9. ▇ $\times 1\frac{1}{4} = 2\frac{1}{2}$

The unknown factor is _____ 1.

The unknown factor is _____ 1.

Problem Solving REAL WORLD

10. Kyle is making a scale drawing of his math book. The dimensions of his drawing will be $\frac{1}{3}$ the dimensions of his book. If the width of his book is $8\frac{1}{2}$ inches, will the width of his drawing be equal to, greater than, or less than $8\frac{1}{2}$ inches?

_____

11. **Write Math** ▸ **Sense or Nonsense?**
Penny wants to make a model of a beetle that is larger than life-size. Penny says she is going to use a scaling factor of $\frac{7}{12}$. Does this make sense or is it nonsense? **Explain**.

_____

_____

_____

12. **H.O.T.** Shannon, Mary, and John earn a weekly allowance. Shannon earns an amount that is $\frac{2}{3}$ of what John earns. Mary earns an amount that is $1\frac{2}{3}$ of what John earns. John earns $20 a week. Who earns the greatest allowance? Who earns the least?

_____

_____

13. **Test Prep** Addie's puppy weighs $1\frac{2}{3}$ times what it weighed when it was born. It weighed $1\frac{1}{3}$ pounds at birth. Which statement below is true?

Ⓐ The puppy weighs the same as it did at birth.

Ⓑ The puppy weighs less than it did at birth.

Ⓒ The puppy weighs more than it did at birth.

Ⓓ The puppy weighs twice what it did at birth.

© Houghton Mifflin Harcourt Publishing Company

> SHOW YOUR WORK

Name _____

Multiply Mixed Numbers

Essential Question How do you multiply mixed numbers?

🔑 UNLOCK the Problem REAL WORLD

One-third of a $1\frac{1}{4}$ acre park has been set aside as a dog park. Find the number of acres that are used as a dog park.

Multiply. $\frac{1}{3} \times 1\frac{1}{4}$

> • Is the area of the dog park less than or greater than the area of the $1\frac{1}{4}$ acre park?
>
> _____

🔑 One Way Use a model.

STEP 1 Shade the model to represent the whole park.

Think: The whole park is _____ acres.

_____

STEP 2 Double-shade the model to represent the part of the park that is a dog park.

Think: The dog park is _____ of the park.

Draw horizontal lines across each rectangle to show _____.

• How many parts does each rectangle show? _____

• What fraction of each rectangle is shaded twice?

  _____ and _____

• What fraction represents all the parts which are shaded twice?

  _____ + _____ = _____

So, _____ acre has been set aside.

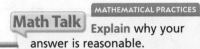

🔑 Another Way Rename the mixed number as a fraction.

STEP 1 Write the mixed number as a fraction greater than 1.

_____

STEP 2 Multiply the fractions.

$$\frac{1}{3} \times 1\frac{1}{4} = \frac{1}{3} \times \frac{\boxed{}}{4}$$

$$= \frac{1 \times \boxed{}}{3 \times 4} = \frac{\boxed{}}{\boxed{}}$$

So, $\frac{1}{3} \times 1\frac{1}{4} = $ _____.

Math Talk MATHEMATICAL PRACTICES **Explain** why your answer is reasonable.

🔑 Example 1 Rename the whole number.

Multiply. $12 \times 2\frac{1}{6}$ **Write the product in simplest form.**

STEP 1 Determine how the product will compare to the greater factor.

$12 \times 2\frac{1}{6}$ will be _____ 12.

STEP 2 Write the whole number and mixed number as fractions.

STEP 3 Multiply the fractions.

STEP 4 Write the product in simplest form.

So, $12 \times 2\frac{1}{6} =$ _____.

$$12 \times 2\frac{1}{6} = \frac{}{1} \times \frac{}{6}$$

$$= \frac{}{} = \frac{}{}, or $$

🔑 Example 2 Use the Distributive Property.

Multiply. $16 \times 4\frac{1}{8}$ **Write the product in simplest form.**

STEP 1 Rewrite the expression by using the Distributive Property.

STEP 2 Multiply 16 by each number.

STEP 3 Add.

So, $16 \times 4\frac{1}{8} =$ _____.

$$16 \times 4\frac{1}{8} = 16 \times \left(\underline{} + \frac{1}{8} \right)$$

$$= (16 \times 4) + \left(16 \times \frac{}{} \right)$$

$$= \underline{} + 2 = \underline{}$$

Math Talk
Explain how you know that your answers to both examples are reasonable.

1. **Explain** why you might choose to use the Distributive Property to solve Example 2.

_____

2. When you multiply two factors greater than 1, is the product less than, between, or greater than the two factors? **Explain.**

_____

_____

326

Name _____

Share and Show

Find the product. Write the product in simplest form.

1. $1\frac{2}{3} \times 3\frac{4}{5} = \dfrac{\boxed{}}{3} \times \dfrac{\boxed{}}{5}$

   $= \dfrac{\boxed{}}{\boxed{}}$

   $= \underline{}$

2. $\frac{1}{2} \times 1\frac{1}{3}$

Shade the model to find the product.

_____

3. $1\frac{1}{8} \times 2\frac{1}{3}$

_____

4. $\frac{3}{4} \times 6\frac{5}{6}$

_____

5. $1\frac{2}{7} \times 1\frac{3}{4}$

_____

6. $\frac{3}{4} \times 1\frac{1}{4}$

_____

Use the Distributive Property to find the product.

7. $16 \times 2\frac{1}{2}$

_____

8. $1\frac{4}{5} \times 15$

_____

> **Math Talk** MATHEMATICAL PRACTICES
>
> **Explain** how multiplying a mixed number by a whole number is similar to multiplying two mixed numbers.

On Your Own

Find the product. Write the product in simplest form.

9. $\frac{3}{4} \times 1\frac{1}{2}$

_____

10. $4\frac{2}{5} \times 1\frac{1}{2}$

_____

11. $5\frac{1}{3} \times \frac{3}{4}$

_____

12. $2\frac{1}{2} \times 1\frac{1}{5}$

_____

13. $12\frac{3}{4} \times 2\frac{2}{3}$

_____

14. $3 \times 4\frac{1}{2}$

_____

15. $2\frac{3}{8} \times \frac{4}{9}$

_____

16. $1\frac{1}{3} \times 1\frac{1}{4} \times 1\frac{1}{5}$

_____

Use the Distributive Property to find the product.

17. $10 \times 2\frac{3}{5}$

_____

18. $3\frac{3}{4} \times 12$

_____

Changing Recipes

You can make a lot of recipes more healthful by reducing the amounts of fat, sugar, and salt.

Kelly has a muffin recipe that calls for $1\frac{1}{2}$ cups of sugar. She wants to use $\frac{1}{2}$ that amount of sugar and more cinnamon and vanilla. How much sugar will she use?

Multiply $1\frac{1}{2}$ by $\frac{1}{2}$ to find what part of the original amount of sugar to use.

Write the mixed number as a fraction greater than 1.

$$\frac{1}{2} \times 1\frac{1}{2} = \frac{1}{2} \times \frac{}{2}$$

Multiply.

$$= \frac{}{}$$

So, Kelly will use _____ cup of sugar.

19. Michelle has a recipe that calls for $2\frac{1}{2}$ cups of vegetable oil. She wants to use $\frac{2}{3}$ that amount of oil and use applesauce to replace the rest. How much vegetable oil will she use?

_____

20. Tony's recipe for soup calls for $1\frac{1}{4}$ teaspoons of salt. He wants to use $\frac{1}{2}$ that amount. How much salt will he use?

_____

21. Jeffrey's recipe for oatmeal muffins calls for $2\frac{1}{4}$ cups of oatmeal and makes one dozen muffins. If he makes $1\frac{1}{2}$ dozen muffins for a club meeting, how much oatmeal will he use?

_____

22. **H.O.T.** Cara's muffin recipe calls for $1\frac{1}{2}$ cups of flour for the muffins and $\frac{1}{4}$ cup of flour for the topping. If she makes $\frac{1}{2}$ of the original recipe, how much flour will she use?

_____

FOR MORE PRACTICE:
Standards Practice Book, pp. P161–P162

Name _____

Problem Solving • Find Unknown Lengths

Essential Question How can you use the strategy *guess, check, and revise* to solve problems with fractions?

🔑 UNLOCK the Problem REAL WORLD

Sarah wants to design a rectangular garden with a section for flowers that attract butterflies. She wants the area of this section to be $\frac{3}{4}$ square yard. If she wants the width to be $\frac{1}{3}$ the length, what will the dimensions of the butterfly section be?

Read the Problem

What do I need to find?	**What information do I need to use?**	**How will I use the information?**
I need to find _____ _____ _____ _____ _____.	The part of the garden for butterflies has an area of _____ square yard and the width is _____ the length.	I will _____ the sides of the butterfly area. Then I will _____ my guess and _____ it if it is not correct.

Solve the Problem

I can try different lengths and calculate the widths by finding $\frac{1}{3}$ the length. For each length and width, I find the area and then compare. If the product is less than or greater than $\frac{3}{4}$ square yard, I need to revise the length.

Guess		Check	Revise
Length (in yards)	**Width (in yards) ($\frac{1}{3}$ of the length)**	**Area of Butterfly Garden (in square yards)**	
$\frac{3}{4}$	$\frac{1}{3} \times \frac{3}{4} = \frac{1}{4}$	$\frac{3}{4} \times \frac{1}{4} = \frac{3}{16}$ too low	Try a longer length.
$2\frac{1}{4}$, or $\frac{9}{4}$			

So, the dimensions of Sarah's butterfly garden will be _____ yard by _____ yards.

🔓 Try Another Problem

Marcus is building a rectangular box for his kitten to sleep in. He wants the area of the bottom of the box to be 360 square inches and the length of one side to be $1\frac{3}{5}$ the length of the other side. What should the dimensions of the bottom of the bed be?

Read the Problem

What do I need to find?	What information do I need to use?	How will I use the information?

Solve the Problem

So, the dimensions of the bottom of the kitten's bed will be _____ by _____.

• **What if** the longer side was still $1\frac{3}{5}$ the length of the shorter side and the shorter side was 20 inches long? What would the area of

the bottom of the bed be then? _____

_____

330

© Houghton Mifflin Harcourt Publishing Company

Share and Show

1. When Pascal built a dog house, he knew he wanted the floor of the house to have an area of 24 square feet. He also wanted the width to be $\frac{2}{3}$ the length. What are the dimensions of the dog house?

   First, choose two numbers that have a product of 24.

   Guess: _____ feet and _____ feet

   Then, check those numbers. Is the greater number $\frac{2}{3}$ of the other number?

   Check: $\frac{2}{3} \times$ _____ = _____

   My guess is _____.

   Finally, if the guess is not correct, revise it and check again. Continue until you find the correct answer.

   So, the dimensions of the dog house are _____.

2. **What if** Pascal wanted the area of the floor to be 54 square feet and the width still to be $\frac{2}{3}$ the length? What would the dimensions of the floor be?

3. Leo wants to paint a mural that covers a wall with an area of 1,440 square feet. The height of the wall is $\frac{2}{5}$ of its length. What is the length and the height of the wall?

On Your Own.....

Choose a STRATEGY

Act It Out
Draw a Diagram
Make a Table
Solve a Simpler Problem
Work Backward
Guess, Check, and Revise

4. Barry wants to make a drawing that is $\frac{1}{4}$ the size of the original. If a tree in the original drawing is 14 inches tall, how tall will the tree in Barry's drawing be?

_____

5. **H.O.T.** A blueprint is a scale drawing of a building. The dimensions of the blueprint for Penny's doll house are $\frac{1}{4}$ of the measurements of the actual doll house. The floor of the doll house has an area of 864 square inches. If the width of the doll house is $\frac{2}{3}$ the length, what are the dimensions of the floor on the blueprint of the doll house?

_____

6. **Write Math** ▶ **Pose a Problem** Look back at Exercise 4. Write a similar problem using a different measurement and a different fraction. Then solve your problem.

_____

_____

_____

_____

SHOW YOUR WORK

7. **Test Prep** Albert's photograph has an area of 80 square inches. The length of the photo is $1\frac{1}{4}$ the width. Which of the following could be the dimensions of the photograph?

(**A**) 5 inches by 16 inches

(**B**) 12 inches by 10 inches

(**C**) 6 inches by 5 inches

(**D**) 10 inches by 8 inches

FOR MORE PRACTICE:
Standards Practice Book, pp. P163–P164

✓ 🔲 **Chapter Review/Test**

▶ **Concepts and Skills**

1. When you multiply $3\frac{1}{4}$ by a number greater than one, how does the product compare to $3\frac{1}{4}$? **Explain**.

_____

_____

Use a model to solve.

2. $\frac{2}{3} \times 6$

3. $\frac{3}{7} \times 14$

4. $\frac{5}{8} \times 24$

Find the product. Write the product in simplest form.

5. $\frac{3}{5} \times 8 =$ _____

6. $\frac{1}{4} \times 10 =$ _____

7. $\frac{5}{7} \times 15 =$ _____

8. $\frac{5}{6} \times \frac{2}{3} =$ _____

9. $\frac{1}{5} \times \frac{5}{7} =$ _____

10. $\frac{3}{8} \times \frac{1}{6} =$ _____

Complete the statement with *equal to, greater than,* or *less than*.

11. $\frac{7}{8} \times \frac{6}{6}$ will be _____ $\frac{7}{8}$.

12. $\frac{1}{2} \times \frac{8}{9}$ will be _____ $\frac{8}{9}$.

GO
Online

Assessment Options
Chapter Test

Fill in the bubble completely to show your answer.

13. Wolfgang wants to enlarge a picture he developed. Which factor listed below would scale up (enlarge) his picture the most if he used it to multiply its current dimensions?

Ⓐ $\frac{7}{8}$

Ⓑ $\frac{14}{14}$

Ⓒ $1\frac{4}{9}$

Ⓓ $\frac{3}{2}$

14. Rachel wants to reduce the size of her photo. Which factor listed below would scale down (reduce) the size of her picture the most?

Ⓐ $\frac{5}{8}$

Ⓑ $\frac{11}{16}$

Ⓒ $\frac{3}{4}$

Ⓓ $\frac{8}{5}$

15. Marteen wants to paint $\frac{2}{3}$ of her room today. She wants to paint $\frac{1}{4}$ of that before lunch. How much of her room will she paint today before lunch?

Ⓐ $\frac{1}{12}$

Ⓑ $\frac{1}{6}$

Ⓒ $\frac{5}{12}$

Ⓓ $\frac{11}{12}$

Name _____

Fill in the bubble completely to show your answer.

16. Gia's bus route to school is $5\frac{1}{2}$ miles. The bus route home is $1\frac{3}{5}$ times as long. How long is Gia's bus route home?

Ⓐ $5\frac{3}{10}$ miles

Ⓑ 8 miles

Ⓒ $8\frac{4}{5}$ miles

Ⓓ $17\frac{3}{5}$ miles

17. Carl's dog weighs $2\frac{1}{3}$ times what Judy's dog weighs. If Judy's dog weighs $35\frac{1}{2}$ pounds, how much does Carl's dog weigh?

Ⓐ $88\frac{3}{4}$ pounds

Ⓑ $82\frac{5}{6}$ pounds

Ⓒ $81\frac{2}{3}$ pounds

Ⓓ 71 pounds

18. In a fifth grade class, $\frac{4}{5}$ of the girls have brown hair. Of the brown-haired girls, $\frac{3}{4}$ of the girls have long hair. What fraction of the girls in the class have long brown hair?

Ⓐ $\frac{1}{20}$

Ⓑ $\frac{1}{5}$

Ⓒ $\frac{3}{5}$

Ⓓ $\frac{1}{4}$

▶ Constructed Response

19. Tasha plans to tile the floor in her room with square tiles that are $\frac{1}{4}$ foot long. Will she use more or fewer tiles if she is only able to purchase square tiles that are $\frac{1}{3}$ foot long? **Explain**.

_____

_____

_____

▶ Performance Task

20. For a bake sale, Violet wants to use the recipe at the right.

> ### Sugar Cookies
> $2\frac{3}{4}$ cups flour
> 1 tsp baking soda
> $\frac{1}{2}$ tsp baking powder
> $1\frac{1}{2}$ cups sugar
> 1 cup butter
> 1 egg
> 1 tsp vanilla

A If she wants to double the recipe, how much flour will she need?

_____

B Baxter wants to make $1\frac{1}{2}$ times the recipe. Will he need more or less sugar than Violet needs if she doubles the recipe? **Explain**.

_____

_____

_____

C As shown, the recipe makes 60 cookies. Jorge wants to bring 150 cookies. How much flour will he need to make 150 cookies? **Explain** how you got your answer. (Hint: what can you multiply 60 by to get 150?)

_____

_____

_____

_____

Chapter 8

Divide Fractions

Show What You Know

Check your understanding of important skills.

Name _____

▶ **Part of a Group** Write a fraction that names the shaded part.

1. total counters _____

   shaded counters _____

   fraction _____

2. total groups _____

   shaded groups _____

   fraction _____

▶ **Relate Multiplication and Division** Use inverse operations and fact families to solve.

3. Since $6 \times 4 = 24$,

   then _____ $\div 4 = 6$.

4. Since _____ $\times 8 = 56$,

   then _____ $\div 7 = 8$.

5. Since $9 \times 3 =$ _____,

   then _____ $\div 3 = 9$.

6. Since _____ $\div 4 = 10$,

   then $4 \times 10 =$ _____.

▶ **Equivalent Fractions** Write an equivalent fraction.

7. $\frac{16}{20}$ _____

8. $\frac{3}{8}$ _____

9. $\frac{5}{12}$ _____

10. $\frac{25}{45}$ _____

Emily spent $\frac{1}{2}$ of her money at the grocery store.
Then, she spent $\frac{1}{2}$ of what was left at the bakery.
Next, at the music store, she spent $\frac{1}{2}$ of what was left
on a CD that was on sale. She spent the remaining
$6.00 on lunch at the diner. Be a Math Detective and
find how much money Emily started with.

Vocabulary Builder

▶ **Visualize It** ●

Complete the flow map using the review words.

Inverse Operations

Multiplication

factor		factor		product
$\frac{1}{3}$	×	6	=	2

Division

2	÷	$\frac{1}{3}$	=	6

▶ **Understand Vocabulary** ●

Complete the sentences using the review words.

1. The number that divides the dividend is the

   _____.

2. An algebraic or numerical sentence that shows that two

   quantities are equal is an _____.

3. A number that names a part of a whole or a part of a group

   is called a _____.

4. The _____ is the number that is to be divided
   in a division problem.

5. The _____ is the number, not including the
   remainder, that results from dividing.

GO Online • eStudent Edition • Multimedia eGlossary

Name _____

Divide Fractions and Whole Numbers

Essential Question How do you divide a whole number by a fraction and divide a fraction by a whole number?

Investigate

Materials ■ fraction strips

A. Mia walks a 2-mile fitness trail. She stops to exercise every $\frac{1}{5}$ mile. How many times does Mia stop to exercise?

- Draw a number line from 0 to 2. Divide the number line into fifths. Label each fifth on your number line.

- Skip count by fifths from 0 to 2 to find $2 \div \frac{1}{5}$.

There are _____ one-fifths in 2 wholes.

You can use the relationship between multiplication and division to explain and check your solution.

- Record and check the quotient.

$2 \div \frac{1}{5} =$ _____ because _____ $\times \frac{1}{5} = 2$.

So, Mia stops to exercise _____ times.

B. Roger has 2 yards of string. He cuts the string into pieces that are $\frac{1}{3}$ yard long. How many pieces of string does Roger have?

- Model 2 using 2 whole fraction strips.

- Then place enough $\frac{1}{3}$ strips to fit exactly under the 2 wholes. There are _____ one-third-size pieces in 2 wholes.

- Record and check the quotient.

$2 \div \frac{1}{3} =$ _____ because _____ $\times \frac{1}{3} = 2$.

So, Roger has _____ pieces of string.

© Houghton Mifflin Harcourt Publishing Company

Draw Conclusions

1. When you divide a whole number by a fraction, how does the quotient compare to the dividend? **Explain.**

   _____

   _____

2. **Explain** how knowing the number of fifths in 1 could help you find the number of fifths in 2.

   _____

   _____

3. **Describe** how you would find $4 \div \frac{1}{5}$.

   _____

   _____

Make Connections

You can use fraction strips to divide a fraction by a whole number.

Calia shares half of a package of clay equally among herself and each of 2 friends. What fraction of the whole package of clay will each friend get?

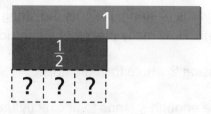

STEP 1 Place a $\frac{1}{2}$ strip under a 1-whole strip to show the $\frac{1}{2}$ package of clay.

STEP 2 Find 3 fraction strips, all with the same denominator, that fit exactly under the $\frac{1}{2}$ strip.

Each piece is _____ of the whole.

Think: How much of the whole is each piece when $\frac{1}{2}$ is divided into 3 equal pieces?

STEP 3 Record and check the quotient.

$\frac{1}{2} \div 3 =$ _____ because _____ $\times 3 = \frac{1}{2}$.

So, each friend will get _____ of the whole package of clay.

Math Talk

MATHEMATICAL PRACTICES

When you divide a fraction by a whole number, how does the quotient compare to the dividend? **Explain.**

Name _____

Share and Show ·····································

Divide and check the quotient.

1.

$3 \div \frac{1}{3} =$ _____ because _____ $\times \frac{1}{3} = 3.$

2.

0   1   2   3

Think: What label should I write for each tick mark?

$3 \div \frac{1}{6} =$ _____ because

_____ $\times \frac{1}{6} = 3.$

3.

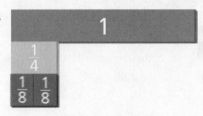

$\frac{1}{4} \div 2 =$ _____ because

_____ $\times 2 = \frac{1}{4}.$

Divide. Draw a number line or use fraction strips.

4. $1 \div \frac{1}{3} =$ _____

⊘ 5. $3 \div \frac{1}{4} =$ _____

⊘ 6. $\frac{1}{5} \div 2 =$ _____

7. $2 \div \frac{1}{2} =$ _____

8. $\frac{1}{4} \div 3 =$ _____

9. $5 \div \frac{1}{2} =$ _____

10. $4 \div \frac{1}{2} =$ _____

11. $\frac{1}{6} \div 2 =$ _____

12. $3 \div \frac{1}{5} =$ _____

Problem Solving ....

Sense or Nonsense?

13. Emilio and Julia used different ways to find $\frac{1}{2} \div 4$. Emilio used a model to find the quotient. Julia used a related multiplication equation to find the quotient. Whose answer makes sense? Whose answer is nonsense? **Explain** your reasoning.

Emilio's Work

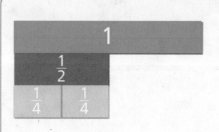

$\frac{1}{2} \div 4 = \frac{1}{4}$

Julia's Work

If $\frac{1}{2} \div 4 = $ ■, then ■ $\times 4 = \frac{1}{2}$.

I know that $\frac{1}{8} \times 4 = \frac{1}{2}$.

So, $\frac{1}{2} \div 4 = \frac{1}{8}$ because $\frac{1}{8} \times 4 = \frac{1}{2}$.

• For the answer that is nonsense, describe how to find the correct answer.

_____

_____

• If you were going to find $\frac{1}{2} \div 5$, **explain** how you would find the

quotient using fraction strips. _____

_____

_____

FOR MORE PRACTICE:
Standards Practice Book, pp. P169–P170

Name _____

Problem Solving • Use Multiplication

Essential Question How can the strategy *draw a diagram* help you solve
fraction division problems by writing a multiplication sentence?

🔑 UNLOCK the Problem REAL WORLD

Erica makes 6 submarine sandwiches and cuts each
sandwich into thirds. How many $\frac{1}{3}$-size sandwich
pieces does she have?

Read the Problem	Solve the Problem
What do I need to find? I need to find _____ _____ _____ .	Since Erica cuts 6 submarine sandwiches, my diagram needs to show 6 rectangles to represent the sandwiches. I can divide each of the 6 rectangles into thirds.

Read the Problem

What do I need to find?

I need to find _____

_____

_____ .

What information do I need to use?

I need to use the size of each _____ of

sandwich and the number of _____ she
cuts.

How will I use the information?

I can _____ to
organize the information from the problem. Then
I can use the organized information to find

_____

_____ .

Solve the Problem

Since Erica cuts 6 submarine sandwiches,
my diagram needs to show 6 rectangles to
represent the sandwiches. I can divide each
of the 6 rectangles into thirds.

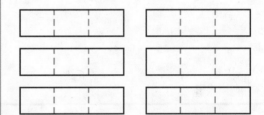

To find the total number of thirds in the
6 rectangles, I can multiply the number of
thirds in each rectangle by the number of
rectangles.

$$6 \div \frac{1}{3} = 6 \times \underline{} = \underline{}$$

So, Erica has _____ one-third-size sandwich pieces.

Math Talk MATHEMATICAL PRACTICES
Explain how you
can use multiplication to check
your answer.

🔑 Try Another Problem

Roberto is cutting 3 blueberry pies into halves to give to his neighbors. How many neighbors will get a $\frac{1}{2}$-size pie piece?

Read the Problem	Solve the Problem
What do I need to find?	
What information do I need to use?	
How will I use the information?	

So, _____ neighbors will get a $\frac{1}{2}$-size pie piece.

• **Explain** how the diagram you drew for the division problem helps you write a multiplication sentence.

_____

_____

344

Share and Show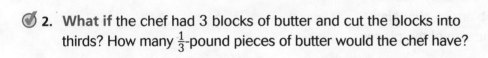

1. A chef has 5 blocks of butter. Each block weighs 1 pound. She cuts each block into fourths. How many $\frac{1}{4}$-pound pieces of butter does the chef have?

    First, draw rectangles to represent the blocks of butter.

    Then, divide each rectangle into fourths.

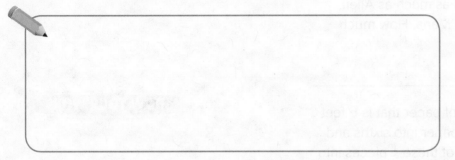

    Finally, multiply the number of fourths in each block by the number of blocks.

    So, the chef has _____ one-fourth-pound pieces of butter.

· · · · · SHOW YOUR WORK · · · · ·

2. **What if** the chef had 3 blocks of butter and cut the blocks into thirds? How many $\frac{1}{3}$-pound pieces of butter would the chef have?

    _____

3. Jason has 2 pizzas that he cuts into fourths. How many $\frac{1}{4}$-size pizza slices does he have?

    _____

4. Thomas makes 5 sandwiches that he cuts into thirds. How many $\frac{1}{3}$-size sandwich pieces does he have?

    _____

5. Holly cuts 3 pans of brownies into eighths. How many $\frac{1}{8}$-size brownie pieces does she have?

    _____

On Your Own . . . .

6. Julie wants to make a drawing that is $\frac{1}{4}$ the size of the original. If a tree in the original drawing is 8 inches tall, how tall will the tree in Julie's drawing be?

_____

7. Three friends go to a book fair. Allen spends $2.60. Maria spends 4 times as much as Allen. Akio spends $3.45 less than Maria. How much does Akio spend?

_____

8. **H.O.T.** Brianna has a sheet of paper that is 6 feet long. She cuts the length of paper into sixths and then cuts the length of each of these $\frac{1}{6}$ pieces into thirds. How many pieces does she have? How many inches long is each piece?

_____

9. **Write Math** ▶ **Pose a Problem** Look back at Problem 8. Write a similar problem by changing the length of the paper and the size of the pieces.

_____

_____

_____

_____

_____

10. **Test Prep** Adrian made 3 carrot cakes. He cut each cake into fourths. How many $\frac{1}{4}$-size cake pieces does he have?

(A) 16

(C) $1\frac{1}{3}$

(B) 12

(D) 1

Choose a STRATEGY

Act It Out

Draw a Diagram

Make a Table

Solve a Simpler Problem

Work Backward

Guess, Check, and Revise

· · · · · **SHOW YOUR WORK** · · · · ·

© Houghton Mifflin Harcourt Publishing Company

FOR MORE PRACTICE:
Standards Practice Book, pp. P171–P172

Name _____

Connect Fractions to Division

Essential Question How does a fraction represent division?

CONNECT A fraction can be written as a division problem.

$$\frac{3}{4} = 3 \div 4 \qquad \frac{12}{2} = 12 \div 2$$

🔑 UNLOCK the Problem REAL WORLD

There are 3 students in a crafts class and 2 sheets of construction paper for them to share equally. What part of the construction paper will each student get?

- Circle the dividend.
- Underline the divisor.

 Use a drawing.

Divide. 2 ÷ 3

STEP 1 Draw lines to divide each piece of paper into 3 equal pieces.

 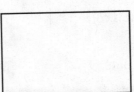

Each student's share of one sheet of construction paper is _____.

STEP 2 Count the number of thirds each student gets. Since there are 2 sheets of construction paper, each student will

get 2 of the _____, or 2 × _____.

STEP 3 Complete the number sentence.

2 ÷ 3 = ▢/▢

STEP 4 Check your answer.

Since _____ × _____ = _____, the quotient is correct.
       quotient   divisor   dividend

So, each student will get _____ of a sheet of construction paper.

Math Talk MATHEMATICAL PRACTICES Describe a division problem where each student gets $\frac{3}{4}$ of a sheet of construction paper.

🔑 Example

Four friends share 6 granola bars equally. How many granola bars does each friend get?

Divide. 6 ÷ 4

STEP 1 Draw lines to divide each of the 6 bars into fourths.

Each friend's share of 1 granola bar is_____.

STEP 2 Count the number of fourths each friend gets. Since there are 6 granola bars, each friend will

get _____ of the fourths, or ——.

STEP 3 Complete the number sentence. Write the fraction as a mixed number in simplest form.

6 ÷ 4 = ——, or ☐ ——

STEP 4 Check your answer.

Since _____ × 4 = _____, the quotient is correct.

So, each friend will get _____ granola bars.

Math Talk **Describe** a different way the granola bars could have been divided into 4 equal shares.

Try This!

Ms. Ruiz has a piece of string that is 125 inches long. For a science experiment, she divides the string equally among 8 groups of students. How much string will each group get?

You can represent this problem as a division equation or a fraction.

- Divide. Write the remainder as a fraction. 125 ÷ 8 = _____

- Write $\frac{125}{8}$ as a mixed number in simplest form. $\frac{125}{8}$ = _____

So, each group will get _____ inches of string.

- **Explain** why 125 ÷ 8 gives the same result as $\frac{125}{8}$.

_____

_____

Name _____

Share and Show ....

Draw lines on the model to complete the number sentence.

1. Six friends share 4 pizzas equally.

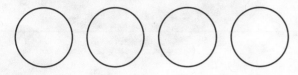

$4 \div 6 =$ _____

Each friend's share is _____ of a pizza.

2. Four brothers share 5 sandwiches equally.

$5 \div 4 =$ _____

Each brother's share is _____ sandwiches.

Complete the number sentence to solve.

3. Twelve friends share 3 pies equally. What fraction of a pie does each friend get?

$3 \div 12 =$ _____

Each friend's share is _____ of a pie.

4. Three students share 8 blocks of clay equally. How much clay does each student get?

$8 \div 3 =$ _____

Each student's share is _____ blocks of clay.

Math Talk MATHEMATICAL PRACTICES
Explain how you can check your answer.

On Your Own ....

Complete the number sentence to solve.

5. Four students share 7 oranges equally. How many oranges does each student get?

$7 \div 4 =$ _____

Each student's share is _____ oranges.

6. Eight girls share 5 fruit bars equally. What fraction of a fruit bar does each girl get?

$5 \div 8 =$ _____

Each girl's share is _____ of a fruit bar.

7. Nine friends share 6 pizzas equally. What fraction of a pizza does each friend get?

$6 \div 9 =$ _____

Each friend's share is _____ of a pizza.

8. Two boys share 9 feet of rope equally. How many feet of rope does each boy get?

$9 \div 2 =$ _____

Each boy's share is _____ feet of rope.

Problem Solving REAL WORLD

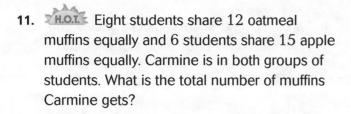

9. Shawna has 3 adults and 2 children coming over for dessert. She is going to serve 2 small apple pies. If she plans to give each person, including herself, an equal amount of pie, how much pie will each person get?

_____

10. There are 36 members in the math club. Addison brought 81 brownies to share with all the members. How many brownies does each member get?

_____

SHOW YOUR WORK

11. **H.O.T.** Eight students share 12 oatmeal muffins equally and 6 students share 15 apple muffins equally. Carmine is in both groups of students. What is the total number of muffins Carmine gets?

_____

12. **Write Math** ▶ Nine friends order 4 large pizzas. Four of the friends share 2 pizzas equally and the other 5 friends share 2 pizzas equally. In which group does each member get a greater amount of pizza? **Explain** your reasoning.

_____

_____

_____

13. **Test Prep** Jason baked 5 cherry pies. He wants to share them equally among 3 of his neighbors. How many pies will each neighbor get?

(A) $\frac{3}{8}$ (C) $1\frac{2}{3}$

(B) $\frac{3}{5}$ (D) $2\frac{2}{3}$

Name _____

▶ **Concepts and Skills**

1. **Explain** how you can tell, without computing, whether the quotient $\frac{1}{2} \div 6$ is greater than 1 or less than 1.

_____

_____

Divide. Draw a number line or use fraction strips.

2. $3 \div \frac{1}{2} =$ _____

3. $1 \div \frac{1}{4} =$ _____

4. $\frac{1}{2} \div 2 =$ _____

5. $\frac{1}{3} \div 4 =$ _____

6. $2 \div \frac{1}{6} =$ _____

7. $\frac{1}{4} \div 3 =$ _____

Complete the number sentence to solve.

8. Two students share 3 granola bars equally. How many granola bars does each student get?

$3 \div 2 =$ _____

Each student's share is _____ granola bars.

9. Five girls share 4 sandwiches equally. What fraction of a sandwich does each girl get?

$4 \div 5 =$ _____

Each girl's share is _____ of a sandwich.

10. Nine boys share 4 pizzas equally. What fraction of a pizza does each boy get?

$4 \div 9 =$ _____

Each boy's share is _____ of a pizza.

11. Four friends share 10 cookies equally. How many cookies does each friend get?

$10 \div 4 =$ _____

Each friend's share is _____ cookies.

Fill in the bubble completely to show your answer.

12. Carmine has 8 liters of punch for a party. Each glass holds $\frac{1}{5}$ liter of punch. How many glasses can Carmine fill with punch?

Ⓐ $\frac{5}{8}$

Ⓒ 13

Ⓑ $1\frac{3}{5}$

Ⓓ 40

13. Four friends share 3 fruit bars equally. What fraction of a fruit bar does each friend get?

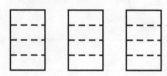

Ⓐ $\frac{3}{7}$

Ⓒ $1\frac{1}{3}$

Ⓑ $\frac{3}{4}$

Ⓓ $2\frac{1}{3}$

14. Caleb and 2 friends are sharing $\frac{1}{2}$ quart of milk equally. What fraction of a quart of milk does each of the 3 friends get?

Ⓐ $\frac{3}{2}$ quarts

Ⓑ $\frac{2}{3}$ quart

Ⓒ $\frac{1}{4}$ quart

Ⓓ $\frac{1}{6}$ quart

15. Makayla has 3 yards of ribbon to use for a craft project. She cuts the ribbon into pieces that are $\frac{1}{4}$ yard long. How many pieces of ribbon does Makayla have?

Ⓐ $1\frac{1}{4}$

Ⓒ 7

Ⓑ $2\frac{1}{3}$

Ⓓ 12

Name _____

Fraction and Whole-Number Division

Essential Question How can you divide fractions by solving a related multiplication sentence?

 UNLOCK the Problem REAL WORLD

Three friends share a $\frac{1}{4}$-pound block of fudge equally. What fraction of a pound of fudge does each friend get?

Divide. $\frac{1}{4} \div 3$

- Let the rectangle represent a 1-pound block of fudge. Divide the rectangle into fourths and then divide each fourth into three equal parts.

  The rectangle is now divided into _____ equal parts.

- When you divide one fourth into 3 equal parts, you are finding one of three equal parts or $\frac{1}{3}$ of $\frac{1}{4}$. Shade $\frac{1}{3}$ of $\frac{1}{4}$.

  The shaded part is _____ of the whole rectangle.

- Complete the number sentence.

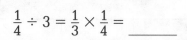

$$\frac{1}{4} \div 3 = \frac{1}{3} \times \frac{1}{4} = \text{_____}$$

So, each friend gets _____ of a pound of fudge.

Example

Brad has 9 pounds of ground turkey to make turkey burgers for a picnic. How many $\frac{1}{3}$-pound turkey burgers can he make?

Divide. $9 \div \frac{1}{3}$

- Will the number of turkey burgers be less than or greater than 9?

  _____

- Draw 9 rectangles to represent each pound of ground turkey. Divide each rectangle into thirds.

- When you divide the _____ rectangles into thirds, you are finding the number of thirds in 9 rectangles or

  finding 9 groups of _____. There are _____ thirds.

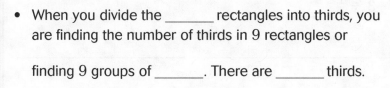

- Complete the number sentence.

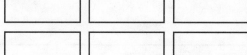

$$9 \div \frac{1}{3} = \text{_____} \times \text{_____} = \text{_____}$$

So, Brad can make _____ one-third-pound turkey burgers.

CONNECT You have learned how to use a model and write a multiplication sentence to solve a division problem.

🔑 Examples

A $\frac{1}{4} \div 2 = \frac{1}{8}$ $\frac{1}{2} \times \frac{1}{4} = \frac{1}{8}$

B $4 \div \frac{1}{2} = 8$ $4 \times 2 = 8$

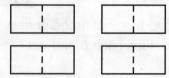

1. Look at Example A. **Describe** how the model shows that dividing by 2 is the same as multiplying by $\frac{1}{2}$.

_____

_____

2. Look at Example B. **Describe** how the model shows that dividing by $\frac{1}{2}$ is the same as multiplying by 2.

_____

_____

When you divide whole numbers, the quotient is always less than the dividend. For example, the quotient for $6 \div 2$ is less than 6 and the quotient for $2 \div 3$ is less than 2. Complete the Try This! to learn how the quotient compares to the dividend when you divide fractions and whole numbers.

Try This!

For the two expressions below, which will have a quotient that is greater than its dividend? **Explain.**

$$\frac{1}{2} \div 3 \qquad\qquad 3 \div \frac{1}{2}$$

_____

_____

So, when I divide a fraction by a whole number, the quotient is _____ the dividend. When I divide a whole number by a fraction less than 1,

the quotient is _____ the dividend.

Name _____

Share and Show ·

1. Use the model to complete the number sentence.

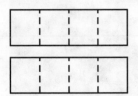

$2 \div \frac{1}{4} = 2 \times$ _____ = _____

2. Use the model to complete the number sentence.

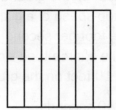

$\frac{1}{6} \div 2 =$ _____ $\times \frac{1}{6} =$ _____

Write a related multiplication sentence to solve.

3. $3 \div \frac{1}{4}$

4. $\frac{1}{5} \div 4$

5. $\frac{1}{9} \div 3$

6. $7 \div \frac{1}{2}$

On Your Own ·

Write a related multiplication sentence to solve.

7. $5 \div \frac{1}{3}$

8. $8 \div \frac{1}{2}$

9. $\frac{1}{7} \div 4$

10. $\frac{1}{2} \div 9$

11. $\frac{1}{3} \div 4$

12. $\frac{1}{4} \div 12$

13. $6 \div \frac{1}{5}$

14. H.O.T. $\frac{2}{3} \div 3$

🔑 UNLOCK the Problem REAL WORLD

15. The slowest mammal is the three-toed sloth. The top speed of a three-toed sloth on the ground is about $\frac{1}{4}$ foot per second. The top speed of a giant tortoise on the ground is about $\frac{1}{3}$ foot per second. How much longer would it take a three-toed sloth than a giant tortoise to travel 10 feet on the ground?

(A) 10 seconds (C) 40 seconds

(B) 30 seconds (D) 70 seconds

a. What do you need to find? _____

_____

b. What operations will you use to solve the problem? _____

_____

c. Show the steps you used to solve the problem.

d. Complete the sentences.

A three-toed sloth would travel 10 feet in

_____ seconds.

A giant tortoise would travel 10 feet in

_____ seconds.

Since _____ – _____ = _____, it would take a three-toed sloth

_____ seconds longer to travel 10 feet.

e. Fill in the bubble for the correct answer choice.

16. Robert divides 8 cups of almonds into $\frac{1}{8}$-cup servings. How many servings does he have?

(A) 1   (C) 8

(B) 16   (D) 64

17. Tina cuts $\frac{1}{3}$ yard of fabric into 4 equal parts. What is the length of each part?

(A) 12 yards   (C) $\frac{3}{4}$ yard

(B) $1\frac{1}{3}$ yards   (D) $\frac{1}{12}$ yard

Interpret Division with Fractions

Essential Question How can you use diagrams, equations, and story problems to represent division?

 UNLOCK the Problem REAL WORLD

Elizabeth has 6 cups of raisins. She divides the raisins into $\frac{1}{4}$-cup servings. How many servings does she have?

You can use diagrams, equations, and story problems to represent division.

- How many $\frac{1}{4}$-cups are in 1 cup?

_____

- How many cups does Elizabeth have?

_____

 Draw a diagram to solve.

- Draw 6 rectangles to represent the cups of raisins. Draw lines to divide each rectangle into fourths.

- To find $6 \div \frac{1}{4}$, count the total number of fourths in the 6 rectangles.

$6 \div$ _____ = _____

So, Elizabeth has _____ servings.

Example 1 Write an equation to solve.

Four friends share $\frac{1}{4}$ of a gallon of orange juice. What fraction of a gallon of orange juice does each friend get?

STEP 1

Write an equation.

$\frac{1}{4} \div$ _____ $= n$

STEP 2

Write a related multiplication equation. Then solve.

$\frac{1}{4} \times$ _____ $= n$

_____ $= n$

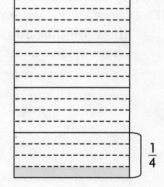

So, each friend will get _____ of a gallon of orange juice.

🔑 Example 2 Write a story problem. Then draw a diagram to solve.

$4 \div \dfrac{1}{3}$

STEP 1 Choose the item you want to divide.

> **Think:** Your problem should be about how many groups of $\frac{1}{3}$ are in 4 wholes.

Possible items: 4 sandwiches, 4 feet of ribbon, 4 pies

STEP 2 Write a story problem to represent $4 \div \frac{1}{3}$ using the item you chose. Describe how it is divided into thirds. Then ask how many thirds there are.

_____

_____

_____

STEP 3 Draw a diagram to solve.

$4 \div \dfrac{1}{3} = $ _____

🔑 Example 3 Write a story problem. Then draw a diagram to solve.

$\dfrac{1}{2} \div 5$

STEP 1 Choose the item you want to divide.

> **Think:** Your problem should describe $\frac{1}{2}$ of an item that can be divided into 5 equal parts.

Possible items: $\frac{1}{2}$ of a pizza, $\frac{1}{2}$ of a yard of rope, $\frac{1}{2}$ of a gallon of milk

STEP 2 Write a story problem to represent $\frac{1}{2} \div 5$ using the item you chose. Describe how it is divided into 5 equal parts. Then ask about the size of each part.

_____

_____

_____

STEP 3 Draw a diagram to solve.

$\dfrac{1}{2} \div 5 = $ _____

Math Talk MATHEMATICAL PRACTICES
Explain how you decided what type of diagram to draw for your problem.

Name _____

Share and Show

1. Complete the story problem to represent $3 \div \frac{1}{4}$.

   Carmen has a roll of paper that is _____ feet long. She cuts

   the paper into pieces that are each _____ foot long. How many
   pieces of paper does Carmen have?

2. Draw a diagram to represent the problem. Then solve.

   April has 6 fruit bars. She cuts the bars into halves. How many $\frac{1}{2}$-size bar pieces does she have?

3. Write an equation to represent the problem. Then solve.

   Two friends share $\frac{1}{4}$ of a large peach pie. What fraction of the whole pie does each friend get?

   _____

On Your Own

4. Write an equation to represent the problem. Then solve.

   Benito has $\frac{1}{3}$-kilogram of grapes. He divides the grapes equally into 3 bags. What fraction of a kilogram of grapes is in each bag?

   _____

   _____

5. Draw a diagram to represent the problem. Then solve.

   Sonya has 5 sandwiches. She cuts each sandwich into fourths. How many $\frac{1}{4}$-size sandwich pieces does she have?

   _____

   _____

6. Write a story problem to represent $2 \div \frac{1}{8}$. Then solve.

   _____

   _____

   _____

Problem Solving REAL WORLD

H.O.T. Pose a Problem

7. Amy wrote the following problem to represent $4 \div \frac{1}{6}$.

Jacob has a board that is 4 feet long. He cuts the board into pieces that are each $\frac{1}{6}$ foot long. How many pieces does Jacob have now?

Then Amy drew this diagram to solve her problem.

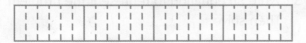

So, Jacob has 24 pieces.

Write a new problem using a different item to be divided and different fractional pieces. Then draw a diagram to solve your problem.

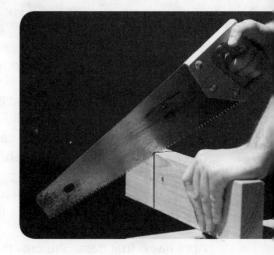

Pose a problem.

_____

_____

_____

_____

_____

_____

Draw a diagram to solve your problem.

8. **Test Prep** Melvin has $\frac{1}{4}$ of a gallon of fruit punch. He shares the punch equally with each of 2 friends and himself. Which equation represents the fraction of a gallon of punch that each of the friends get?

(A) $\frac{1}{4} \div \frac{1}{3} = n$ (C) $3 \div \frac{1}{4} = n$

(B) $\frac{1}{4} \div 3 = n$ (D) $3 \div 4 = n$

Name _____

Chapter Review/Test

▶ **Concepts and Skills**

Divide. Draw a number line or use fraction strips.

1. $2 \div \frac{1}{3} =$ _____

2. $1 \div \frac{1}{5} =$ _____

3. $\frac{1}{4} \div 3 =$ _____

Complete the number sentence to solve.

4. Three students share 4 sandwiches equally. How many sandwiches does each student get?

$4 \div 3 =$ _____

Each student's share is _____ sandwiches.

5. Six girls share 5 pints of milk equally. What fraction of a pint of milk does each girl get?

$5 \div 6 =$ _____

Each girl's share is _____ pint of milk.

Write a related multiplication sentence to solve.

6. $\frac{1}{4} \div 5$

7. $\frac{1}{3} \div 9$

8. $8 \div \frac{1}{2}$

9. $5 \div \frac{1}{6}$

_____ _____ _____ _____

10. Write a story problem to represent $\frac{1}{2} \div 3$. Then solve.

_____

_____

11. Write a story problem to represent $3 \div \frac{1}{2}$. Then solve.

_____

_____

_____

GO Online Assessment Options
Chapter Test

Fill in the bubble completely to show your answer.

12. Michelle cuts $\frac{1}{4}$ yard of ribbon into 4 equal pieces. What is the length of each piece?

    (A) $\frac{1}{16}$ yard

    (B) $\frac{1}{8}$ yard

    (C) 1 yard

    (D) 16 yards

13. Ashton picked 6 pounds of pecans. He wants to share the pecans equally among 5 of his neighbors. How many pounds of pecans will each neighbor get?

    (A) $\frac{5}{11}$ pound

    (B) $\frac{5}{6}$ pound

    (C) $1\frac{1}{5}$ pounds

    (D) $2\frac{1}{5}$ pounds

14. Isabella has 5 pounds of trail mix. She divides the mix into $\frac{1}{4}$-pound servings. How many $\frac{1}{4}$-pound servings does she have?

    (A) $1\frac{1}{4}$

    (B) 9

    (C) 16

    (D) 20

15. Melvin has $\frac{1}{2}$ of a cake. He shares the cake equally with each of 2 friends and himself. Which equation represents the fraction of the whole cake that each of the friends get?

    (A) $\frac{1}{2} \div \frac{1}{3} = n$

    (B) $\frac{1}{2} \div 3 = n$

    (C) $2 \div \frac{1}{3} = n$

    (D) $2 \div 3 = n$

Name _____

Fill in the bubble completely to show your answer.

16. Camille has 8 feet of rope. She cuts the rope into $\frac{1}{3}$-foot pieces for a science project. How many $\frac{1}{3}$-foot pieces of rope does she have?

Ⓐ 24

Ⓑ 8

Ⓒ 3

Ⓓ $2\frac{2}{3}$

17. Awan makes 3 sandwiches and cuts each sandwich into sixths. How many $\frac{1}{6}$-size sandwich pieces does he have?

Ⓐ $\frac{1}{2}$

Ⓑ 2

Ⓒ 9

Ⓓ 18

18. Eight students share 5 blocks of modeling clay equally. What fraction of one block of modeling clay does each student get?

Ⓐ $\frac{1}{40}$

Ⓑ $\frac{1}{8}$

Ⓒ $\frac{5}{8}$

Ⓓ $1\frac{3}{5}$

19. The diagram below represents which division problem?

Ⓐ $5 \div \frac{1}{3}$

Ⓑ $\frac{1}{3} \div 5$

Ⓒ $5 \div \frac{1}{4}$

Ⓓ $\frac{1}{4} \div 5$

► Constructed Response

20. Dora buys one package each of the 1-pound, 2-pound, and 4-pound packages of ground beef to make hamburgers. How many $\frac{1}{4}$-pound hamburgers can she make? Show your work using words, pictures, or numbers.

_____

Explain how you found your answer.

_____

_____

_____

► Performance Task

21. Suppose your teacher gives you the division problem $6 \div \frac{1}{5}$.

A In the space below, draw a diagram to represent $6 \div \frac{1}{5}$.

B Write a story problem to represent $6 \div \frac{1}{5}$.

_____

_____

C Use a related multiplication expression to solve your story problem. Show your work.

_____

D Write a division problem that shows a unit fraction divided by a whole number. Write a story problem to represent your division problem. Then solve.

_____

_____

_____

© Houghton Mifflin Harcourt Publishing Company

Geometry and Measurement

Developing
understanding of volume

A lunar rover is a surface
exploration vehicle used
on the moon. ▶

Space Architecture

NASA's Lunar Architecture Team develops ideas for rovers and space habitats. A space habitat is made up of modules linked by airlocks. Airlocks are double doors that allow people to move between the modules without losing atmosphere.

Get Started

Work with a partner to design a space habitat made up of 3 modules. The Important Facts name some modules that you can choose for your design. Cut out, fold, and tape the patterns for each of the modules that you have selected, and for the measuring cube.

Use a formula to find the volume of the measuring cube in cubic centimeters. Estimate the volume of each module by filling it with rice, then pouring the rice into the measuring cube. Let every cubic centimeter in the measuring cube represent 32 cubic feet. Determine what the volume of your space habitat would be in cubic feet.

Connect the modules to complete your space habitat.

Important Facts

Modules of a Space Habitat

- sleeping room
- kitchen
- exercise room
- bathroom
- work room
- airlock
- life-support room (for air and water supplies)

_____

_____

_____

_____

_____

_____

Completed by _____

Show What You Know ✓

Check your understanding of important skills.

Name _____

▶ **Read and Use a Bar Graph** Use the graph to answer the questions.

1. Which fruit received the most votes?

   _____

2. Which fruit received 5 votes? _____

3. There were _____ votes in all.

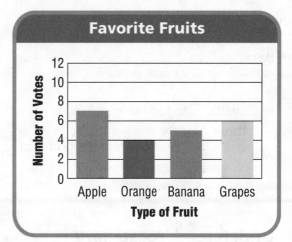

Favorite Fruits

▶ **Extend Patterns** Find the missing numbers. Then write a description for each pattern.

4. 0, 5, 10, 15, _____, _____, _____

   description: _____

5. 70, 60, 50, 40, _____, _____, _____

   description: _____

6. 12, 18, 24, 30, _____, _____, _____

   description: _____

7. 150, 200, 250, 300, _____, _____, _____

   description: _____

8. 200, 180, 160, 140, _____, _____, _____

   description: _____

MATH DETECTIVE

WITH

CARMEN SANDIEGO™

Be a math detective by graphing and connecting the map coordinates to locate the secret documents in the lost briefcase.

(3, 3), (4, 2), (4, 4), (5, 3)

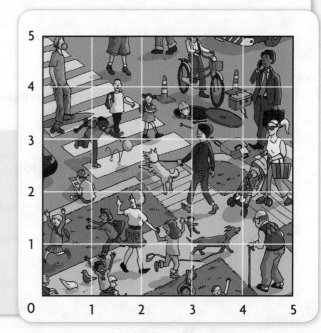

Vocabulary Builder

▶ **Visualize It** ••••••••••••••••••••••••••

Use the checked words to complete the tree map.

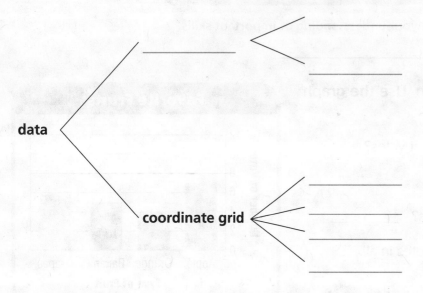

▶ **Understand Vocabulary** •••••••••••••••••••

Complete the sentences using the preview words.

1. A graph that uses line segments to show how data changes over time

   is called a _____.

2. The pair of numbers used to locate points on a grid is

   an _____.

3. The point, (0, 0), also called the _____, is where the
   x-axis and the *y*-axis intersect.

4. On a coordinate grid, the horizontal number line is the _____

   and the vertical number line is the _____.

5. The first number in an ordered pair is the _____ and

   the second number in an ordered pair is the _____.

6. The difference between the values on the scale of a graph

   is an _____.

GO Online • eStudent Edition • Multimedia eGlossary

Line Plots

Essential Question How can a line plot help you find an average with data given in fractions?

🔑 UNLOCK the Problem · REAL WORLD

Students have measured different amounts of water into beakers for an experiment. The amount of water in each beaker is listed below.

$\frac{1}{4}$ cup, $\frac{1}{4}$ cup, $\frac{1}{2}$ cup, $\frac{3}{4}$ cup, $\frac{1}{4}$ cup, $\frac{1}{4}$ cup,

$\frac{1}{4}$ cup, $\frac{1}{2}$ cup, $\frac{1}{4}$ cup, $\frac{3}{4}$ cup, $\frac{1}{4}$ cup, $\frac{3}{4}$ cup

If the total amount of water stayed the same, what would be the average amount of water in a beaker?

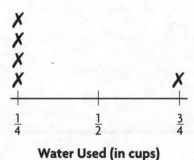

Water Used (in cups)

STEP 1 Count the number of cups for each amount. Draw an ✗ for the number of times each amount is recorded to complete the line plot.

$\frac{1}{4}$: _____ $\frac{1}{2}$: _____ $\frac{3}{4}$: _____

STEP 2 Find the total amount of water in all of the beakers that contain $\frac{1}{4}$ cup of water.

There are _____ beakers with $\frac{1}{4}$ cup of water. So, there are _____ fourths, or

— , or ▢ — cups.

STEP 3 Find the total amount of water in all of the beakers that contain $\frac{1}{2}$ cup of water.

There are _____ beakers with $\frac{1}{2}$ cup of water. So, there are _____ halves, or

— , or 1 cup.

STEP 4 Find the total amount of water in all of the beakers that contain $\frac{3}{4}$ cup of water.

$3 \times \frac{3}{4} = $ — , or ▢ —

STEP 5 Add to find the total amount of water in all of the beakers.

$1\frac{3}{4} + 1 + 2\frac{1}{4} = $ _____

STEP 6 Divide the sum you found in Step 5 by the number of beakers to find the average.

$5 \div 12 = $ —

So, the average amount of water in a beaker is _____ cup.

You can use the order of operations to find the average. Solve the problem as a series of expressions that use parentheses and brackets to separate them. Perform operations from inside the parentheses to the outer brackets.

$\left[\left(7 \times \frac{1}{4}\right) + \left(2 \times \frac{1}{2}\right) + \left(3 \times \frac{3}{4}\right)\right] \div 12$ Perform the operations inside the parentheses.

 $\div 12$ Next, perform the operations in the brackets.

☐ $\div 12$ Divide.

☐   Write the expression as a fraction.

🔑 Example

Raine divides three 2-ounce bags of rice into smaller bags. The first bag is divided into bags weighing $\frac{1}{6}$-ounce each, the second bag is divided into bags weighing $\frac{1}{3}$-ounce each, and the third bag is divided into bags weighing $\frac{1}{2}$-ounce each.

Find the number of $\frac{1}{6}$-, $\frac{1}{3}$-, and $\frac{1}{2}$-ounce rice bags. Then graph the results on the line plot.

STEP 1 Write a title for your line plot. It should describe what you are counting.

STEP 2 Label $\frac{1}{6}$, $\frac{1}{3}$, and $\frac{1}{2}$ on the line plot to show the different amounts into which the three 2-ounce bags of rice are divided.

STEP 3 Use division to find the number of $\frac{1}{6}$-ounce, $\frac{1}{3}$-ounce, and $\frac{1}{2}$-ounce bags that were made from the three original 2-ounce bags of rice.

$2 \div \frac{1}{6}$ $2 \div \frac{1}{3}$ $2 \div \frac{1}{2}$

$2 \times \square = \square$ $2 \times \square = \square$ $2 \times \square = \square$

STEP 4 Draw an **X** above $\frac{1}{6}$, $\frac{1}{3}$, or $\frac{1}{2}$ to show the number of rice bags.

Math Talk MATHEMATICAL PRACTICES
Explain why there are more $\frac{1}{6}$-ounce rice bags than $\frac{1}{2}$-ounce rice bags.

Name _____

Share and Show

Use the data to complete the line plot. Then answer the questions.

Lilly needs to buy beads for a necklace. The beads are sold by mass. She sketches a design to determine what beads are needed, and then writes down their sizes. The sizes are shown below.

$\frac{2}{5}$ g, $\frac{2}{5}$ g, $\frac{4}{5}$ g, $\frac{2}{5}$ g, $\frac{1}{5}$ g, $\frac{1}{5}$ g, $\frac{3}{5}$ g,

$\frac{4}{5}$ g, $\frac{1}{5}$ g, $\frac{2}{5}$ g, $\frac{3}{5}$ g, $\frac{3}{5}$ g, $\frac{2}{5}$ g

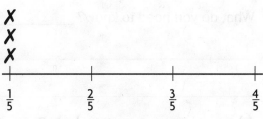

Mass of Beads (in grams)

1. What is the combined mass of the beads with a mass of $\frac{1}{5}$ gram?

   Think: There are _____ Xs above $\frac{1}{5}$ on the line plot, so the combined mass of the beads

   is _____ fifths, or _____ gram.

2. What is the combined mass of all the beads with a mass of $\frac{2}{5}$ gram?

   _____

3. What is the combined mass of all the beads on the necklace?

   _____

4. What is the average weight of the beads on the necklace?

   _____

On Your Own

Use the data to complete the line plot. Then answer the questions.

A breakfast chef used different amounts of milk when making pancakes, depending on the number of pancakes ordered. The results are shown below.

$\frac{1}{2}$ c, $\frac{1}{4}$ c, $\frac{1}{2}$ c, $\frac{3}{4}$ c, $\frac{1}{2}$ c, $\frac{3}{4}$ c, $\frac{1}{2}$ c, $\frac{1}{4}$ c, $\frac{1}{2}$ c, $\frac{1}{2}$ c

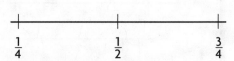

Milk in Pancake Orders (in cups)

5. How much milk combined is used in

   $\frac{1}{4}$-cup amounts? _____

6. How much milk combined is used in

   $\frac{1}{2}$-cup amounts? _____

7. How much milk combined is used in

   $\frac{3}{4}$-cup amounts? _____

8. How much milk is used in all the orders

   of pancakes? _____

9. What is the average amount of milk used

   for an order of pancakes? _____

10. **H.O.T.** **Describe** an amount you could add to the data that would make the average increase.

   _____

UNLOCK the Problem REAL WORLD

11. For 10 straight days, Samantha measured the amount of food that her cat Dewey ate, recording the results, which are shown below. Graph the results on the line plot. What is the average amount of cat food that Dewey ate daily?

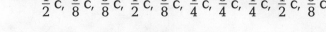

$\frac{1}{2}$ c, $\frac{3}{8}$ c, $\frac{5}{8}$ c, $\frac{1}{2}$ c, $\frac{5}{8}$ c, $\frac{1}{4}$ c, $\frac{3}{4}$ c, $\frac{1}{4}$ c, $\frac{1}{2}$ c, $\frac{5}{8}$ c

a. What do you need to know? _____

_____

_____

b. How can you use a line plot to organize the information?

_____

_____

_____

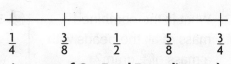

$\frac{1}{4}$ $\frac{3}{8}$ $\frac{1}{2}$ $\frac{5}{8}$ $\frac{3}{4}$

Amount of Cat Food Eaten (in cups)

c. What steps could you use to find the average amount of food that Dewey ate daily?

_____

_____

_____

_____

_____

d. Fill in the blanks for the totals of each amount measured.

$\frac{1}{4}$ cup: _____

$\frac{3}{8}$ cup: _____

$\frac{1}{2}$ cup: _____

$\frac{5}{8}$ cup: _____

$\frac{3}{4}$ cup: _____

e. Find the total amount of cat food eaten over 10 days.

_____ + _____ + _____ + _____ +

_____ = _____

So, the average amount of food Dewey

ate daily was _____.

12. **Test Prep** How many days did Dewey eat the least amount of cat food?

(A) 1 day

(B) 2 days

(C) 3 days

(D) 4 days

FOR MORE PRACTICE:
Standards Practice Book, pp. P183–P184

Name _____

Ordered Pairs

Essential Question How can you identify and plot points on a coordinate grid?

CONNECT Locating a point on a coordinate grid is similar to describing directions using North-South and West-East. The horizontal number line on the grid is the **x-axis**. The vertical number line on the grid is the **y-axis**.

Each point on the coordinate grid can be described by an **ordered pair** of numbers. The **x-coordinate**, the first number in the ordered pair, is the horizontal location, or the distance the point is from 0 in the direction of the x-axis. The **y-coordinate**, the second number in the ordered pair, is the vertical location, or the distance the point is from 0 in the direction of the y-axis.

$$(x, y)$$
x-coordinate ⌐ ⌐ y-coordinate

The x-axis and the y-axis intersect at the point (0, 0), called the **origin**.

🔑 UNLOCK the Problem · REAL WORLD

🔒 **Write the ordered pairs for the locations of the arena and the aquarium.**

Locate the point for which you want to write an ordered pair.

Look below at the x-axis to identify the point's horizontal distance from 0, which is its x-coordinate.

Look to the left at the y-axis to identify the point's vertical distance from 0, which is its y-coordinate.

So, the ordered pair for the arena is (3, 2) and the ordered pair for the aquarium

is (_____, _____).

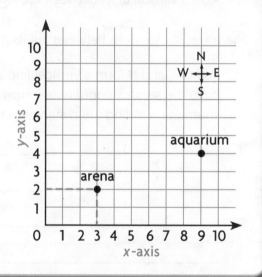

- Describe the path you would take to get from the origin to the aquarium, using horizontal, then vertical movements.

_____

_____

🔑 Example 1 Use the graph.

A point on a coordinate grid can be labeled with an ordered pair, a letter, or both.

Ⓐ Plot the point (5, 7) and label it _J_.

From the origin, move right 5 units and then up 7 units.

Plot and label the point.

Ⓑ Plot the point (8, 0) and label it _S_.

From the origin, move right _____ units and

then up _____ units.

Plot and label the point.

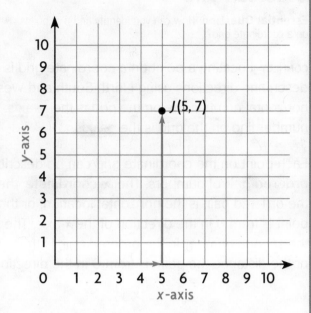

🔑 Example 2 Find the distance between two points.

You can find the distance between two points when the points are along the same horizontal or vertical line.

- Draw a line segment to connect point _A_ and point _B_.

- Count vertical units between the two points.

There are _____ units between points _A_ and _B_.

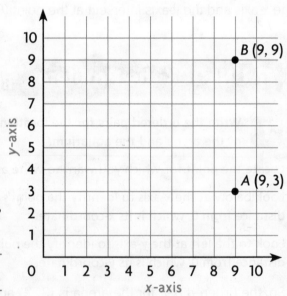

1. Points _A_ and _B_ form a vertical line segment and have the same _x_-coordinates. How can you use subtraction to find the distance between the points?

_____

_____

_____

_____

2. Graph the points (3, 2) and (5, 2). **Explain** how you can use subtraction to find the horizontal distance between these two points.

_____

_____

374

Name _____

Graph Data

Essential Question How can you use a coordinate grid to display data collected in an experiment?

Investigate

Materials ■ paper cup ■ water ■ Fahrenheit thermometer ■ ice cubes ■ stopwatch

When data is collected, it can be organized in a table.

A. Fill the paper cup more than halfway with room-temperature water.

B. Place the Fahrenheit thermometer in the water and find its beginning temperature before adding any ice. Record this temperature in the table at 0 seconds.

C. Place three cubes of ice in the water and start the stopwatch. Find the temperature every 10 seconds for 60 seconds. Record the temperatures in the table.

Water Temperature	
Time (in seconds)	Temperature (in °F)
0	
10	
20	
30	
40	
50	
60	

Draw Conclusions

1. **Explain** why you would record the beginning temperature at 0 seconds.

   _____

   _____

2. **Describe** what happens to the temperature of the water in 60 seconds, during the experiment.

   _____

   _____

3. **H.O.T.** **Analyze** your observations of the temperature of the water during the 60 seconds, and explain what you think would happen to the temperature if the experiment continued for 60 seconds longer.

   _____

   _____

   _____

Make Connections

You can use a coordinate grid to graph and analyze the data you collected in the experiment.

STEP 1 Write the related pairs of data as ordered pairs.

(0, _____)   (40, _____)

(10, _____)   (50, _____)

(20, _____)   (60, _____)

(30, _____)

STEP 2 Construct a coordinate grid and write a title for it. Label each axis.

STEP 3 Plot a point for each ordered pair.

Math Talk MATHEMATICAL PRACTICES
What is the ordered pair that you recorded for the data at 10 seconds? **Explain** what each coordinate represents.

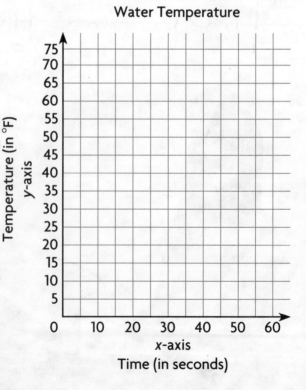

Water Temperature

Temperature (in °F)
y-axis

Time (in seconds)
x-axis

Line Graphs

Essential Question How can you use a line graph to display and analyze real-world data?

UNLOCK the Problem REAL WORLD

A **line graph** is a graph that uses line segments to show how data changes over time. The series of numbers placed at fixed distances that label the graph are the graph's **scale**. The **intervals**, or difference between the values on the scale, should be equal.

 Graph the data. Use the graph to determine the times between which the greatest temperature change occurred.

- Write related number pairs of data as ordered pairs.

  (1:00 , 51) (_____ , _____)

  (_____ , _____) (_____ , _____)

  (_____ , _____) (_____ , _____)

  (_____ , _____)

Recorded Temperatures

Time (A.M.)	1:00	2:00	3:00	4:00	5:00	6:00	7:00
Temperature (in °F)	51	49	47	44	45	44	46

STEP 1 For the vertical axis, choose a scale and an interval that are appropriate for the data. You can show a break in the scale between 0 and 40, since there are no temperatures between 0°F and 44°F.

STEP 2 For the horizontal axis, write the times of day. Write a title for the graph and name each axis. Then graph the ordered pairs. Complete the graph by connecting the points with line segments.

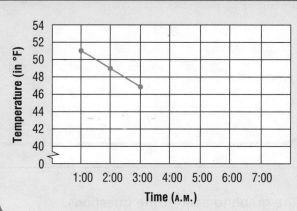

Look at each line segment in the graph. Find the line segment that shows the greatest change in temperature between two consecutive points.

The greatest temperature change occurred between _____ and _____.

Try This! Jill used a rain gauge to collect data on the total rainfall during 6 days at her home in Miami. She read the amount of rain collected in the rain gauge each day and did not pour it out. Her data is shown in the table. Make a line graph to display Jill's data.

STEP 1 Write related pairs of data as ordered pairs.

(Mon , 2) (____, ____) (____, ____)

(____, ____) (____, ____) (____, ____)

STEP 2 Choose a scale and an interval for the data.

STEP 3 Label the horizontal and vertical axes. Write a title for the graph. Graph the ordered pairs. Connect the points with line segments.

Rainfall Collected

Day	Rainfall (in inches)
Mon	2
Tue	2
Wed	3
Thu	6
Fri	8
Sat	9

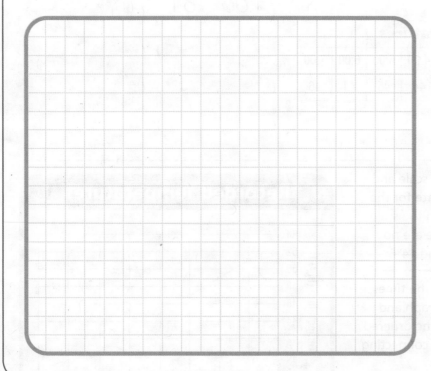

Math Talk MATHEMATICAL PRACTICES
Explain how you could use the graph to identify the two readings between which it did not rain.

Use the graph to answer the questions.

1. On which day was the total rainfall recorded the greatest?

_____

2. On which day did Jill record the greatest increase in rainfall collected from the previous day?

_____

Name _____

Share and Show

Use the table at the right for 1–3.

1. What scale and intervals would be appropriate to make a graph of the data?

   _____

   _____

2. Write the related pairs as ordered pairs.

   _____

   _____

☑ 3. Make a line graph of the data.

☑ 4. Use the graph to determine between which two months the least change in average temperature occurs.

   _____

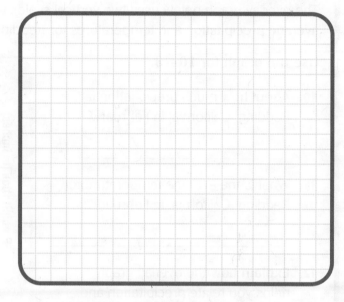

Average Monthly Temperature in Tupelo, Mississippi					
Month	Jan	Feb	Mar	Apr	May
Temperature (in °F)	40	44	54	62	70

On Your Own

Use the table at the right for 5–7.

5. Write the related number pairs for the plant height as ordered pairs.

   _____

6. What scale and intervals would be appropriate to make a graph of the data?

   _____

   _____

7. Make a line graph of the data.

8. Use the graph to find the difference in height between Month 1 and Month 2.

   _____

9. 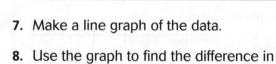 Use the graph to estimate the height at $1\frac{1}{2}$ months.

   _____

Plant Height				
Month	1	2	3	4
Height (in inches)	20	25	29	32

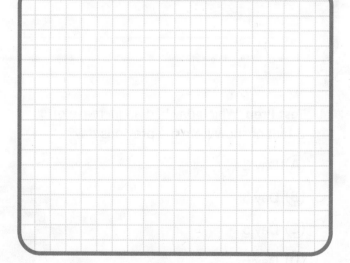

Connect to Science

Evaporation changes water on Earth's surface into water vapor. Water vapor condenses in the atmosphere and returns to the surface as precipitation. This process is called the water cycle. The ocean is an important part of this cycle. It influences the average temperature and precipitation of a place.

The overlay graph below uses two vertical scales to show monthly average precipitation and temperatures for Redding, California.

Use the graph for 10–13.

10. About how much precipitation falls in Redding, California, in February?

_____

11. What is the average temperature for Redding, California, in February?

_____

12. **Explain** how the overlay graph helps you relate precipitation and temperature for each month.

_____

_____

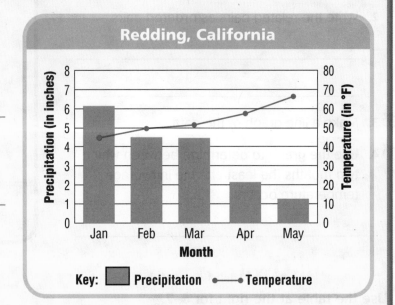

Redding, California

Key: ▬ Precipitation ●—● Temperature

13. **Write Math** ▶ **Describe** how the average temperature changes in the first 5 months of the year.

_____

14. **Test Prep** Which day had an increase of 3 feet of snow from the previous day?

Ⓐ Day 2

Ⓑ Day 3

Ⓒ Day 5

Ⓓ Day 6

Accumulated Snowfall

FOR MORE PRACTICE:
Standards Practice Book, pp. P189–P190

Name _____

▶ Vocabulary

Choose the best term from the box.

1. The _____ is the horizontal number line on the coordinate grid. (p. 373)

2. A _____ is a graph that uses line segments to show how data changes over time. (p. 381)

▶ Check Concepts

Use the line plot at the right for 3–5.

3. How many kittens weigh at least $\frac{3}{8}$ of a pound?

   _____

4. What is the combined weight of all the kittens?

   _____

5. What is the average weight of the kittens in the shelter?

   _____

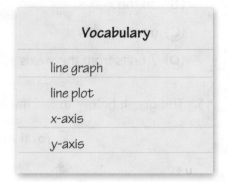

Weights of Kittens in the Animal Shelter (lb)

Use the coordinate grid at the right for 6–13.

Write an ordered pair for the given point.

6. A _____   7. B _____

8. C _____   9. D _____

Plot and label the point on the coordinate grid.

10. E (6, 2)   11. F (5, 0)

12. G (3, 4)   13. H (3, 1)

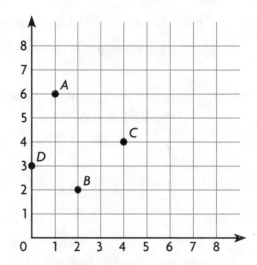

Fill in the bubble completely to show your answer.

14. The ordered pair (0, 7) is:

Ⓐ at the origin

Ⓑ on the *x*-axis

Ⓒ on the *y*-axis

Ⓓ 7 units from the *y*-axis

15. The graph below shows the amount of snowfall in a 6-hour period.

Total Amount of Snow

Based on the graph, which statement best describes the amount of snow that fell during that time period?

Ⓐ The greatest amount of snow fell between hour 1 and hour 2.

Ⓑ The greatest amount of snow fell between hour 5 and hour 6.

Ⓒ The least amount of snow fell between hour 2 and hour 4.

Ⓓ The least amount of snow fell between hour 4 and hour 5.

16. Joy recorded the distances she walked each day for five days. How far did she walk in 5 days?

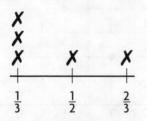

Distance Walked Each Day
(in miles)

Ⓐ $1\frac{1}{3}$ miles   Ⓒ 2 miles

Ⓑ $1\frac{2}{3}$ miles   Ⓓ $2\frac{1}{6}$ miles

Numerical Patterns

Essential Question How can you identify a relationship between two numerical patterns?

🔑 UNLOCK the Problem  REAL WORLD

On the first week of school, Joel purchases 2 movies and 6 songs from his favorite media website. If he purchases the same number of movies and songs each week, how does the number of songs purchased compare to the number of movies purchased from one week to the next?

• How many movies does Joel purchase each week?

_____

• How many songs does Joel purchase each week?

_____

STEP 1 Use the two rules given in the problem to generate the first 4 terms in the sequence for the number of movies and the sequence for number of songs.

• The sequence for the number of movies each week is:

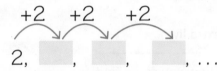

+2   +2   +2

2, ⬜ , ⬜ , ⬜ , . . .

• The sequence for the number of songs each week is:

+6   +6   +6

6, ⬜ , ⬜ , ⬜ , . . .

STEP 2 Write number pairs that relate the number of movies to the number of songs.

Week 1: ___2, 6___   Week 2: _____

Week 3: _____   Week 4: _____

STEP 3 For each number pair, compare the number of movies to the number of songs. Write a rule to describe this relationship.

Think: For each related number pair, the second number is _____ times as great as the first number.

Rule: _____

So, from one week to the next, the number of songs Joel purchased

is _____ times as many as the number of movies purchased.

🔓 Example

When Alice completes each level in her favorite video game, she wins 3 extra lives and 6 gold coins. What rule can you write to relate the number of gold coins to the number of extra lives she has won at any level? How many extra lives will Alice have won after she completes 8 levels?

Add _____.

Add _____.

Level	0	1	2	3	4	8
Extra Lives	0	3	6	9	12	
Gold Coins	0	6	12	18	24	48

Multiply by _____ or divide by _____.

STEP 1 To the left of the table, complete the rule for how you could find the number of extra lives won from one level to the next.

←— difference between consecutive terms

0,   3,   6,   9,   12

From one level to the next, Alice wins _____ more extra lives.

STEP 2 To the left of the table, complete the rule for how you could find the number of gold coins won from one level to the next.

←— difference between consecutive terms

0,   6,   12,   18,   24

From one level to the next, Alice wins _____ more gold coins.

STEP 3 Write number pairs that relate the number of gold coins to the number of extra lives won at each level.

Level 1: _6, 3_   Level 2: _____

Level 3: _____   Level 4: _____

STEP 4 Complete the rule to the right of the table that describes how the number pairs are related. Use your rule to find the number of extra lives at level 8.

Think: For each level, the number of extra lives is _____ as great as the number of gold coins.

Rule: _____

So, after 8 levels, Alice will have won _____ extra lives.

Math Talk
Explain how your rule would change if you were relating extra lives to gold coins instead of gold coins to extra lives.

Name _____

Share and Show

Use the given rules to complete each sequence. Then, complete the rule that describes how nickels are related to dimes.

1.

	Number of coins	1	2	3	4	5
Add 5.	Nickels (¢)	5	10	15	20	
Add 10.	Dimes (¢)	10	20	30	40	

Multiply by _____.

Complete the rule that describes how one sequence is related to the other. Use the rule to find the unknown term.

2. Multiply the number of books by _____ to find the amount spent.

Day	1	2	3	4	8
Number of Books	3	6	9	12	24
Amount Spent ($)	12	24	36	48	

3. Divide the weight of the bag by _____ to find the number of marbles.

Bags	1	2	3	4	12
Number of Marbles	10	20	30	40	
Weight of Bag (grams)	30	60	90	120	360

On Your Own

Complete the rule that describes how one sequence is related to the other. Use the rule to find the unknown term.

4. Multiply the number of eggs by _____ to find the number of muffins.

Batches	1	2	3	4	9
Number of Eggs	2	4	6	8	18
Muffins	12	24	36	48	

5. Divide the number of meters by _____ to find the number of laps.

Runners	1	2	3	4
Number of Laps	4	8	12	
Number of Meters	1,600	3,200	4,800	6,400

6. H.O.T. Suppose the number of eggs used in Exercise 4 is changed to 3 eggs for each batch of 12 muffins, and 48 eggs are used. How many batches and how many muffins will be made?

_____

Problem Solving REAL WORLD

7. Emily has a road map with a key that shows an inch on the map equals 5 miles of actual distance. If a distance measured on the map is 12 inches, what is the actual distance? Write the rule you used to find the actual distance.

_____

_____

8. To make a shade of lavender paint, Jon mixes 4 ounces of red tint and 28 ounces of blue tint into one gallon of white paint. If 20 gallons of white paint and 80 ounces of red tint are used, how much blue tint should be added? Write a rule that you can use to find the amount of blue tint needed.

_____

_____

_____

9. H.O.T. In the cafeteria, tables are arranged in groups of 4, with each table seating 8 students. How many students can sit at 10 groups of tables? Write the rule you used to find the number of students.

_____

_____

10. Test Prep What is the unknown number in Sequence 2 in the chart? What rule could you write that relates Sequence 1 to Sequence 2?

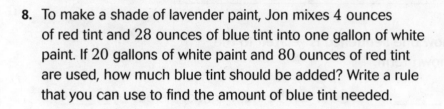

Sequence Number	1	2	3	5	7
Sequence 1	5	10	15	25	35
Sequence 2	15	30	45	75	?

(A) 70; Multiply by 2.

(B) 100; Add 25.

(C) 105; Multiply by 3.

(D) 150; Add 150.

Problem Solving • Find a Rule

Essential Question How can you use the strategy *solve a simpler problem* to help you solve a problem with patterns?

🔑 UNLOCK the Problem — REAL WORLD

On an archaeological dig, Gabriel separates his dig site into sections with areas of 15 square feet each. There are 3 archaeological members digging in every section. What is the area of the dig site if 21 members are digging at one time?

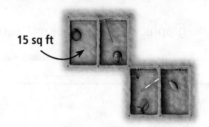

15 sq ft

Read the Problem

What do I need to find?	What information do I need to use?	How will I use the information?
I need to find the _____ _____ _____.	I can use the area of each section, which is _____, that there are _____ members in each section, and that there are 21 members digging.	I will use the information to search for patterns to solve a _____ problem.

Solve the Problem

Sections	1	2	3	4	5	6	7
Number of Members	3	6	9	12	15	18	21
Area (in square feet)	15	30	45	60	75	90	

Add 3.

Add 15.

Multiply by _____.

Multiply by _____.

So, the area of the dig site if 21 members are digging is _____ square feet.

Possible Rules:

• Multiply the number of sections by _____ to find the number of members.

• Multiply the number of members by _____ to find the total area. Complete the table.

Math Talk — MATHEMATICAL PRACTICES

Explain how you can use division to find the number of members if you know the dig site area is 135 square feet.

🔑 Try Another Problem

Casey is making a design with triangles and beads for a costume. In his design, each pattern unit adds 3 triangles and 18 beads. If Casey uses 72 triangles in his design, how many times does he repeat the pattern unit? How many beads does Casey use?

Use the graphic organizer below to solve the problem.

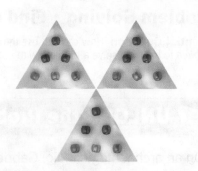

Read the Problem

What do I need to find?	What information do I need to use?	How will I use the information?

Solve the Problem

So, Casey repeats the pattern unit _____ times and

uses _____ beads.

- What rule could you use to find an unknown number of beads if you know the related number of triangles?

_____

_____

Share and Show

1. Max builds rail fences. For one style of fence, each section uses 3 vertical fence posts and 6 horizontal rails. How many posts and rails does he need for a fence that will be 9 sections long?

1 Section

2 Sections   3 Sections

First, think about what the problem is asking and what you know. As each section of fence is added, how does the number of posts and the number of rails change?

Next, make a table and look for a pattern. Use what you know about 1, 2, and 3 sections. Write a rule for the number of posts and rails needed for 9 sections of fence.

Number of Sections	1	2	3	9
Number of Posts	3	6	9	
Number of Rails	6	12	18	

Possible rule for posts: _____

_____

Possible rule for rails: _____

_____

Finally, use the rule to solve the problem.

2. 🔆 **H.O.T.** **What if** another style of rail fencing has 6 rails between each pair of posts? How many rails are needed for 9 sections of this fence?

Possible rule: _____

_____

Number of Sections	1	2	3	9
Number of Posts	3	6	9	
Number of Rails	12	24	36	

3. Leslie is buying a coat on layaway for $135. She will pay $15 each week until the coat is paid for. How much will she have left to pay after 8 weeks?

_____

Number of Weeks	1	2	3	8
Amount paid ($)	15	30	45	

On Your Own..........................

Choose a STRATEGY

Act It Out

Draw a Diagram

Make a Table

Solve a Simpler Problem

Work Backward

Guess, Check, and Revise

4. Jane works as a limousine driver. She earns $50 for every 2 hours that she works. How much does Jane earn in one week if she works 40 hours per week? Write a rule and complete the table.

Possible rule: _____

Hours Worked	2	4	6	40
Jane's Pay ($)	50	100	150	

5. Rosa joins a paperback book club. Members pay $8 to buy 2 tokens, and can trade 2 tokens for 4 paperback books. Rosa buys 30 tokens and trades them for 60 paperback books. How much money does she spend? Write a rule and complete the table.

Tokens	2	4	6	8	30
Cost ($)	8	16	24	32	
Books	4	8	12	16	60

Possible rule: _____

_____

_____

6. 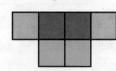 Paul is taking a taxicab to a museum. The taxi driver charges a $3 fee plus $2 for each mile traveled. How much does the ride to the museum cost if it is 8 miles away?

_____

7. Test Prep Which expression could describe the next figure in the pattern, Figure 4?

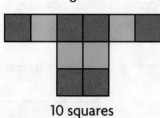

Figure 1   Figure 2   Figure 3

2 squares   6 squares   10 squares

Ⓐ 2 × 5

Ⓑ 2 + 4 + 4

Ⓒ 2 + 4 + 4 + 4

Ⓓ 16

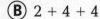

Graph and Analyze Relationships

Essential Question How can you write and graph ordered pairs
on a coordinate grid using two numerical patterns?

🔑 UNLOCK the Problem REAL WORLD

Sasha is making hot cocoa for a party. For each mug of cocoa,
he uses 3 tablespoons of cocoa mix and 6 fluid ounces of hot
water. If Sasha uses an entire 18-tablespoon container of cocoa
mix, how many fluid ounces of water will he use?

STEP 1 Use the two given rules in the problem to generate
the first four terms for the number of tablespoons
of cocoa mix and the number of fluid ounces of water.

Cocoa Mix (tbsp)	3				18
Water (fl oz)	6				

STEP 2 Write the number pairs as ordered pairs, relating the
number of tablespoons of cocoa mix to the number of
fluid ounces of water.

(3, 6) _____ _____ _____

STEP 3 Graph and label the ordered pairs. Then write a rule
to describe how the number pairs are related.

• What rule can you write that relates the amount
of cocoa mix to water?

_____

So, Sasha will use _____ fluid ounces of water if he uses
the entire container of cocoa mix.

• How many tablespoons of
cocoa mix does Sasha add for
each mug of cocoa?

_____

• How many fluid ounces of
water does Sasha add for each
mug of cocoa?

_____

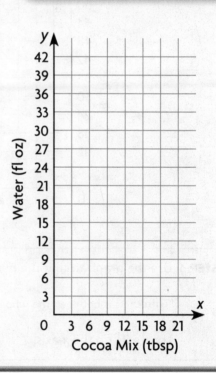

• Write the final number pair as an ordered pair. Then graph and label
it. Starting at the origin, connect the points with straight line segments.
What do the connected points form? **Explain** why this is formed.

_____

_____

_____

 **Example**

Jon is customizing an audio sound system. He needs to buy $3\frac{1}{2}$ feet of cable wire, but it is sold in inches. He knows there are 12 inches in 1 foot. How many inches of wire will he need?

Feet	1	2	3	4
Inches	12			

Rule: Multiply the number of feet by _____.

STEP 1 Write the number pairs as ordered pairs, relating the number of feet to the number of inches.

_____ _____ _____ _____

STEP 2 Graph the ordered pairs. Connect the points from the origin with straight line segments.

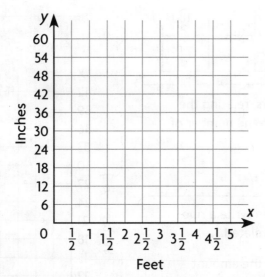

_____

STEP 3 Use the graph to find the number of inches in $3\frac{1}{2}$ feet.

Think: $3\frac{1}{2}$ is between the whole numbers _____ and _____.

Locate $3\frac{1}{2}$ on the x-axis.

_____

STEP 4 Draw a vertical line from $3\frac{1}{2}$ on the x-axis to the line that connects the ordered pairs. Then graph that point.

To find how many inches equal $3\frac{1}{2}$ feet, draw a horizontal line from that point left to the y-axis. What is the ordered pair for the point?

_____

So, Jon needs to buy _____ inches of cable wire.

Name _____

Graph and label the related number pairs as ordered pairs.
Then complete and use the rule to find the unknown term.

1. Multiply the number of tablespoons by _____ to find its weight in ounces.

Butter (tbsp)	1	2	3	4	5
Weight (oz)	2	4	6	8	

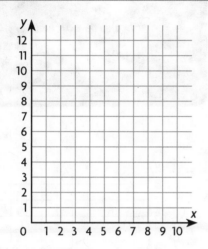

2. Multiply the number of hours by _____ to find the distance in miles.

Time (hr)	1	2	3	4
Distance walked (mi)	3	6	9	

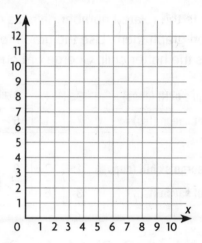

On Your Own

Graph and label the related number pairs as ordered pairs.
Then complete and use the rule to find the unknown term.

3. Multiply the number of inches by _____ to find the distance in miles.

Map (in.)	2	4	6	8	10
Miles	10	20	30	40	

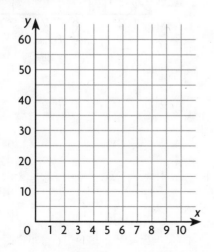

4. Multiply the number of centiliters by _____ to find the equivalent number of milliliters.

Centiliters	1	2	3	4	5
Milliliters	10	20	30	40	

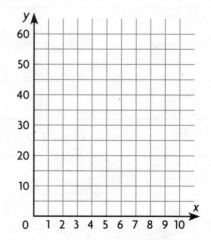

Problem Solving REAL WORLD

 Sense or Nonsense?

5. Elsa solved the following problem.

Lou and George are making chili for the Annual Firefighter's Ball. Lou uses 2 teaspoons of hot sauce for every 2 cups of chili that he makes, and George uses 3 teaspoons of the same hot sauce for every cup of chili in his recipe. Who has the hotter chili, George or Lou?

Write the related number pairs as ordered pairs and then graph them. Use the graph to compare who has the hotter chili, George or Lou.

Lou's chili (cups)	2	4	6	8
Hot sauce (tsp)	2	4	6	8

George's chili (cups)	1	2	3	4
Hot sauce (tsp)	3	6	9	12

Lou's chili: $(2, 2), (4, 4), (6, 6), (8, 8)$

George's chili: $(1, 3), (2, 6), (3, 9), (4, 12)$

Elsa said that George's chili was hotter than Lou's, because the graph showed that the amount of hot sauce in George's chili was always 3 times as great as the amount of hot sauce in Lou's chili. Does Elsa's answer make sense, or is it nonsense? **Explain.**

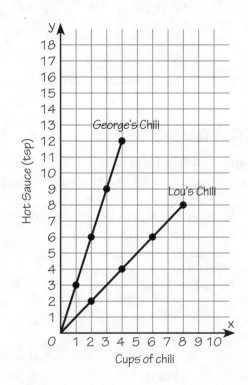

_____

_____

_____

_____

_____

FOR MORE PRACTICE:
Standards Practice Book, pp. P195–P196

Name _____

▶ Vocabulary

Choose the best term from the box.

1. The _____ is the point where the *x*-axis

   and *y*-axis meet. Its _____ is 0,

   and its _____ is 0. (p. 373)

2. A _____ uses line segments to show how
   data changes over time. (p. 381)

Vocabulary
line graph
line plot
origin
x-coordinate
y-coordinate

▶ Check Concepts

Use the table for 3–4.

Height of Seedling				
Weeks	1	2	3	4
Height (in cm)	2	6	14	16

3. Write related number pairs of data as ordered pairs.

   _____

4. Make a line graph of the data.

**Complete the rule that describes how one sequence
is related to the other. Use the rule to find the
unknown term.**

5. Multiply the number of eggs by _____ to find the
   number of cupcakes.

Batches	1	2	3	4	6
Number of Eggs	3	6	9	12	
Number of Cupcakes	18	36	54	72	

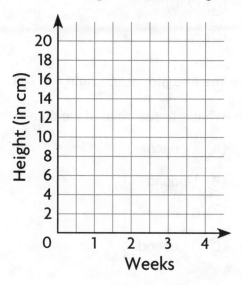

Height of Seedling

GO Online Assessment Options
Chapter Test

Fill in the bubble completely to show your answer.

6. The letters on the coordinate grid represent the locations of the first four holes on a golf course.

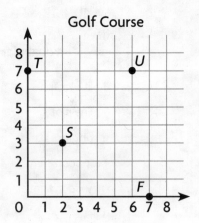

Golf Course

Which ordered pair describes the location of the hole labeled *T*?

Ⓐ (0, 7)

Ⓑ (1, 7)

Ⓒ (7, 0)

Ⓓ (7, 1)

Use the line plot at the right for 7–8.

7. What is the average of the data in the line plot?

Ⓐ $\frac{1}{2}$ pound

Ⓑ 1 pound

Ⓒ 6 pounds

Ⓓ $6\frac{3}{4}$ pounds

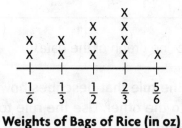

Weights of Bags of Rice (in oz)

8. How many bags of rice weigh at least $\frac{1}{2}$ pound?

Ⓐ 2

Ⓑ 3

Ⓒ 5

Ⓓ 8

Name _____

Fill in the bubble completely to show your answer.

Use the table for 9–10.

Week	1	2	3	4	10
Tori's savings	$20	$40	$60	$80	$200
Martin's savings	$5	$10	$15	$20	$50

9. Compare Tori's and Martin's savings. Which of the following statements is true?

   Ⓐ Tori saves 4 times as much per week as Martin.

   Ⓑ Tori will always have exactly $15 more in savings than Martin has.

   Ⓒ Tori will save 15 times as much as Martin will.

   Ⓓ On week 5, Martin will have $30 and Tori will have $90.

10. What rule could you use to find Tori's savings after 10 weeks?

   Ⓐ Add 10 from one week to the next.

   Ⓑ Multiply the week by 2.

   Ⓒ Multiply Martin's savings by 4.

   Ⓓ Divide Martin's savings by 4.

11. In an ordered pair, the *x*-coordinate represents the number of hexagons and the *y*-coordinate represents the total number of sides. If the x-coordinate is 7, what is the *y*-coordinate?

   Ⓐ 6

   Ⓑ 7

   Ⓒ 13

   Ⓓ 42

12. Point A is 2 units to the right and 4 units up from the origin. What ordered pair describes point A?

   Ⓐ (2, 0)

   Ⓑ (2, 4)

   Ⓒ (4, 2)

   Ⓓ (0, 4)

13. Mr. Stevens drives 110 miles in 2 hours, 165 miles in 3 hours, and 220 miles in 4 hours. How many miles will he drive in 5 hours?

_____

Explain how the number of hours he drives is related to the number of miles he drives.

_____

_____

► Performance Task

14. Tim opens the freezer door and measures the temperature of the air inside. He continues to measure the temperature every 2 minutes, as the door stays open, and records the data in the table.

Open Freezer Temperatures						
Time (in minutes)	0	2	4	6	8	10
Temperature (in °F)	0	6	12	14	16	18

Ⓐ On the grid below, make a line graph showing the data in the table.

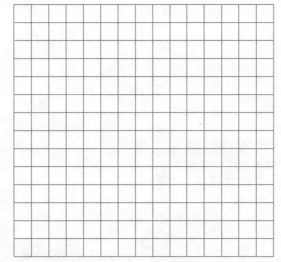

Ⓑ Use the graph to estimate the temperature at 7 minutes.

Estimate: _____

Ⓒ Write a question that can be answered by making a prediction. Then answer your question and explain how you made your prediction.

_____

_____

_____

Chapter 10

Convert Units of Measure

Show What You Know

Check your understanding of important skills.

Name _____

▶ **Measure Length to the Nearest Inch**
Use an inch ruler. Measure the length to the nearest inch.

1.    about _____ inches

2.    about _____ inches

▶ **Multiply and Divide by 10, 100, and 1,000** Use mental math.

3. $1 \times 5.98 = 5.98$
$10 \times 5.98 = 59.8$

$100 \times 5.98 =$ _____

$1,000 \times 5.98 =$ _____

4. $235 \div 1 = 235$
$235 \div 10 = 23.5$

$235 \div 100 =$ _____

$235 \div 1,000 =$ _____

▶ **Choose Customary Units** Write the appropriate unit to measure each.
Write *inch, foot, yard,* or *mile.*

5. length of a pencil _____

6. length of a football field _____

You can step out distances of 5 feet by using an estimate. Two steps or 2 paces is about 5 feet. Be a Math Detective and act out the directions on the map to find a treasure. About how many feet from start to finish is the path to the treasure?

Vocabulary Builder

▶ **Visualize It** •

Sort the review and preview words into the Venn diagram.

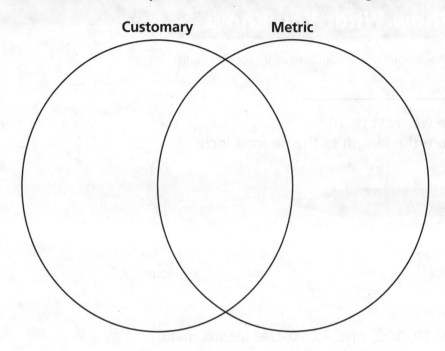

Customary   Metric

Review Words

decimeter

gallon

gram

length

liter

mass

meter

mile

milligram

milliliter

millimeter

pound

ton

weight

Preview Words

capacity

dekameter

▶ **Understand Vocabulary** •

Complete the sentences.

1. A metric unit of length that is equal to one tenth of a meter

   is a _____.

2. A metric unit of length that is equal to one thousandth

   of a meter is a _____.

3. A metric unit of capacity that is equal to one thousandth

   of a liter is a _____.

4. A metric unit of length that is equal to 10 meters

   is a _____.

5. A metric unit of mass that is equal to one thousandth

   of a gram is a _____.

GO Online • eStudent Edition • Multimedia eGlossary

Customary Length

Essential Question How can you compare and convert customary units of length?

 UNLOCK the Problem REAL WORLD

To build a new swing, Mr. Mattson needs 9 feet of rope for each side of the swing and 6 more feet for the monkey bar. The hardware store sells rope by the yard.

- How many feet of rope does Mr. Mattson

  need for the swing? _____

- How many feet does Mr. Mattson need for

  the swing and the monkey bar combined? _____

Mr. Mattson needs to find how many yards of rope he needs to buy. He will need to convert 24 feet to yards. How many groups of 3 feet are in 24 feet?

A 12-inch ruler is 1 foot.		

A yardstick is 1 yard.

_____ feet = 1 yard

🔑 **Use a bar model to write an equation.**

MODEL

3		3

24

RECORD

total feet	feet in 1 yard	total yards
↓	↓	↓
24	÷ _____	= _____

So, Mr. Mattson needs to buy _____ yards of rope.

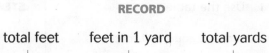 **Math Talk** MATHEMATICAL PRACTICES
What operation did you use when you found groups of 3 feet in 24 feet? Do you multiply or divide when you convert a smaller unit to a larger unit? Explain.

Example 1 Use the table to find the relationship between miles and feet.

The distance between the new high school and the football field is 2 miles. How does this distance compare to 10,000 feet?

When you convert larger units to smaller units, you need to multiply.

Customary Units of Length
1 foot (ft) = 12 inches (in.)
1 yard (yd) = 3 ft
1 mile (mi) = 5,280 ft
1 mile = 1,760 yd

STEP 1   Convert 2 miles to feet.

Think: 1 mile is equal to 5,280 feet.

I need to _____ the total

number of miles by _____ .

total miles		feet in 1 mile		total feet
↓		↓		↓
2	×	_____	=	_____

2 miles = _____ feet

STEP 2   Compare. Write <, >, or =.

_____ feet ◯ 10,000 feet

Since _____ is _____ than 10,000, the distance between the

new high school and the football field is _____ than 10,000 feet.

Example 2 Convert to mixed measures.

Mixed measures use more than one unit of measurement. You can convert a single unit of measurement to mixed measures.

Convert 62 inches into feet and inches.

STEP 1   Use the table.

Think: 12 inches is equal to 1 foot

I am changing from a smaller unit to

a larger unit, so I _____ .

STEP 2   Convert.

total inches		inches in 1 foot		feet		inches
↓		↓		↓		↓
62	÷	_____	is	_____	r	_____

So, 62 inches is equal to _____ feet _____ inches.

- **Explain** how to convert the mixed measures, 12 yards 2 feet, to a single unit of measurement in feet. How many feet is it?

Name _____

Share and Show . . . .

Convert.

1. 2 mi = _____ yd

✓ 2. 6 yd = _____ ft

✓ 3. 90 in. = _____ ft _____ in.

Math Talk MATHEMATICAL PRACTICES
Explain how you know when to multiply to convert a measurement.

On Your Own . . . .

Convert.

4. 57 ft = _____ yd

5. 13 ft = _____ in.

6. 240 in. = _____ ft

7. 6 mi = _____ ft

8. 96 ft = _____ yd

9. 75 in. = _____ ft _____ in.

Practice: Copy and Solve Convert.

10. 60 in. = ■ ft

11. ■ ft = 7 yd 1 ft

12. 4 mi = ■ yd

13. 125 in. = ■ ft ■ in.

14. 46 ft = ■ yd ■ ft

15. 42 yd 2 ft = ■ ft

Compare. Write <, >, or =.

16. 8 ft ◯ 3 yd

17. 2 mi ◯ 10,500 ft

18. 3 yd 2 ft ◯ 132 in.

Problem Solving REAL WORLD

19. H.O.T. Javon is helping his dad build a tree house. He has a piece of trim that is 13 feet long. How many pieces can Javon cut that are 1 yard long? How much of a yard will he have left over?

_____

20. **Test Prep** Katy's driveway is 120 feet long. How many yards long is Katy's driveway?

Ⓐ 60 yards

Ⓑ 40 yards

Ⓒ 20 yards

Ⓓ 10 yards

Connect to Reading

Compare and Contrast

When you compare and contrast, you tell how two or more things are alike and different. You can compare and contrast information in a table.

Complete the table below. Use the table to answer the questions.

Linear Units				
Yards	1	2	3	4
Feet	3	6	9	
Inches	36	72		

21. How are the items in the table alike? How are they different?

_____

_____

22. What do you notice about the relationship between the number of larger units and the number of smaller units as the length increases?

_____

_____

© Houghton Mifflin Harcourt Publishing Company

FOR MORE PRACTICE:
Standards Practice Book, pp. P201–P202

Customary Capacity

Essential Question How can you compare and convert customary units of capacity?

🔑 UNLOCK the Problem REAL WORLD

Mara has a can of paint with 3 cups of purple paint in it. She also has a bucket with a capacity of 26 fluid ounces. Will the bucket hold all of the paint Mara has?

The **capacity** of a container is the amount the container can hold.

- What capacity does Mara need to convert?

_____

- After Mara converts the units, what does she need to do next?

_____

_____

1 cup (c) = _____ fluid ounces (fl oz)

 Use a bar model to write an equation.

STEP 1 Convert 3 cups to fluid ounces.

MODEL		RECORD

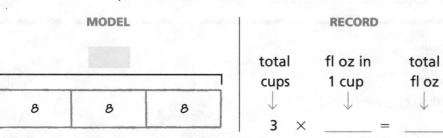

total cups	fl oz in 1 cup	total fl oz
↓	↓	↓
3	× _____	= _____

STEP 2 Compare. Write <, >, or =. _____ fl oz ◯ 26 fl oz

Since _____ fluid ounces is _____ than 26 fluid ounces,

Mara's bucket _____ hold all of the paint.

- **What if** Mara has 7 cups of green paint and a container filled with 64 fluid ounces of yellow paint? Which color paint does Mara have more of? **Explain** your reasoning.

_____

_____

🔑 Example

Coral made 32 pints of fruit punch for a party. She needs to transport the punch in 1-gallon containers. How many containers does Coral need?

Customary Units of Capacity
1 cup (c) = 8 fluid ounces (fl oz)
1 pint (pt) = 2 cups
1 quart (qt) = 2 pints
1 gallon (gal) = 4 quarts

To convert a smaller unit to a larger unit, you need to divide. Sometimes you may need to convert more than once.

Convert 32 pints to gallons.

STEP 1 Write an equation to convert pints to quarts.

total pints ↓   pints in 1 qt ↓   total quarts ↓

32 ◯ _____ ◯ _____

STEP 2 Write an equation to convert quarts to gallons.

total quarts ↓   quarts in 1 gal ↓   total gallons ↓

_____ ◯ _____ ◯ _____

So, Coral needs _____ 1-gallon containers to transport the punch.

Share and Show 📝 MATH BOARD ·····································

1. Use the picture to complete the statements and convert 3 quarts to pints.

   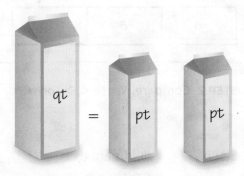

   a. 1 quart = _____ pints

   b. 1 quart is _____ than 1 pint.

   c. 3 qt ◯ _____ pt in 1 qt = _____ pt

Convert.

2. 3 gal = _____ pt

☑ **3.** 5 qt = _____ pt

☑ **4.** 6 qt = _____ c

Math Talk MATHEMATICAL PRACTICES
Explain how converting units of capacity is similar to converting units of length. How is it different?

Name _____

On Your Own ·

Convert.

5. 38 c = _____ pt

6. 36 qt = _____ gal

7. 104 fl oz = _____ c

8. 4 qt = _____ c

9. 7 gal = _____ pt

10. 96 fl oz = _____ pt

Practice: Copy and Solve Convert.

11. 200 c = ■ qt

12. 22 pt = ■ fl oz

13. 8 gal = ■ qt

14. 72 fl oz = ■ c

15. 2 gal = ■ pt

16. 48 pt = ■ gal

Compare. Write <, >, or =.

17. 28 c ◯ 14 pt

18. 25 pt ◯ 13 qt

19. 20 qt ◯ 80 c

20. 12 gal ◯ 50 qt

21. 320 fl oz ◯ 18 pt

22. 15 qt ◯ 63 c

23. **Write Math** ▶ Which of exercises 17–22 could you solve mentally?
Explain your answer for one exercise.

_____

_____

Problem Solving REAL WORLD

Show your work. For 24–26, use the table.

24. **H.O.T.** Complete the table, and make a graph showing
    the relationship between pints and quarts. Draw a line
    through the points to make the graph.

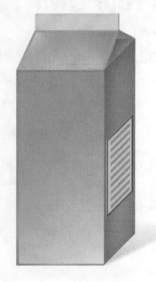

Pints	Quarts
0	0
2	
4	
6	
8	

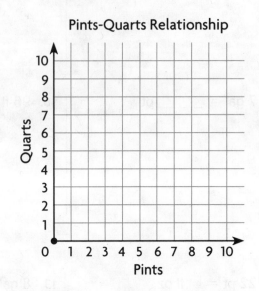

Pints-Quarts Relationship

25. **Describe** any pattern you notice in the pairs of numbers
    you graphed. Write a rule to describe the pattern.

    _____

    _____

    _____

    _____

26. **H.O.T.** **Explain** how you can use your graph to find the
    number of quarts equal to 5 pints.

    _____

    _____

    _____

27. **Test Prep** Shelby made 5 quarts of juice for a picnic. How
    many cups of juice did Shelby make?

    (A) 1 cup   (C) 10 cups

    (B) 5 cups   (D) 20 cups

FOR MORE PRACTICE:
Standards Practice Book, pp. P203–P204

Name _____

Weight

Essential Question How can you compare and convert customary units of weight?

🔑 UNLOCK the Problem REAL WORLD

Hector's school is having a model rocket competition. To qualify, each rocket must weigh 4 pounds or less. Hector's unpainted rocket weighs 62 ounces. What is the weight of the most paint he can use for his model rocket to qualify for entry?

- What weight does Hector need to convert?

  _____

- After Hector converts the weight, what does he need to do next?

  _____

1 pound = _____ ounces

 Use a bar model to write an equation.

STEP 1 Convert 4 pounds to ounces.

MODEL			
16	16	16	16

RECORD

total lb	oz in 1 lb	total oz
4 ⟶ ◯ _____ ◯ _____		

STEP 2 Subtract the rocket's weight from the total ounces a rocket can weigh to qualify.

_____ − 62 = _____

So, the weight of the paint can be at most _____ ounces for Hector's model rocket to qualify for entry.

Math Talk MATHEMATICAL PRACTICES
How did you choose which operation to use to change from pounds to ounces? Explain.

🔑 Example

The rocket boosters for a U.S. space shuttle weigh 1,292,000 pounds each when the shuttle is launched. How many tons does each rocket booster weigh?

Use mental math to convert pounds to tons.

STEP 1 Decide which operation to use.	Since pounds are smaller than tons, I need to _____ the number of pounds by _____.	**Units of Weight**

		Units of Weight
		1 pound (lb) = 16 ounces (oz)
		1 ton (T) = 2,000 lb

STEP 2 Break 2,000 into two factors that are easy to divide by mentally.

$2,000 =$ _____ $\times 2$

STEP 3 Divide 1,292,000 by the first factor. Then divide the quotient by the second factor.

$1,292,000 \div$ _____ $=$ _____

_____ $\div 2 =$ _____

So, each rocket booster weighs _____ tons when launched.

Share and Show

1. Use the picture to complete each equation.

   a. 1 pound = _____ ounces   b. 2 pounds = _____ ounces

   c. 3 pounds = _____ ounces   d. 4 pounds = _____ ounces

   e. 5 pounds = _____ ounces

Convert.

2. 15 lb = _____ oz

☑ 3. 3 T = _____ lb

☑ 4. 320 oz = _____ lb

MATHEMATICAL PRACTICES

Math Talk Explain how you can compare 11 pounds to 175 ounces mentally.

Name _____

On Your Own...

Convert.

5. 5 T = _____ lb

6. 19 T = _____ lb

7. 16,000 lb = _____ T

8. 192 oz = _____ lb

9. 416 oz = _____ lb

10. 24 lb = _____ oz

Practice: Copy and Solve Convert.

11. 23 lb = ▪ oz

12. 6 T = ▪ lb

13. 144 oz = ▪ lb

14. 15 T = ▪ lb

15. 352 oz = ▪ lb

16. 18 lb = ▪ oz

Compare. Write <, >, or =.

17. 130 oz ◯ 8 lb

18. 34 lb ◯ 544 oz

19. 14 lb ◯ 229 oz

20. 16 T ◯ 32,000 lb

21. 5 lb ◯ 79 oz

22. 85,000 lb ◯ 40 T

Problem Solving REAL WORLD

23. **Write Math** ▶ **Explain** how you can use mental math to compare 7 pounds to 120 ounces.

_____

_____

_____

_____

24. **Test Prep** Carlos used 32 ounces of walnuts in a muffin recipe. How many pounds of walnuts did Carlos use?

Ⓐ 8 pounds

Ⓑ 4 pounds

Ⓒ 2 pounds

Ⓓ 1 pound

Problem Solving REAL WORLD

H.O.T. Pose a Problem

25. Kia wants to have 4 pounds of munchies for her party. She has 36 ounces of popcorn and wants the rest to be pretzel sticks. How many ounces of pretzel sticks does she need to buy?

4 pounds = 64 ounces

36 ounces	_____ ounces

$64 - 36 =$ _____

So, Kia needs to buy _____ ounces of pretzel sticks.

Write a new problem using different amounts of snacks. Some weights should be in pounds and others in ounces. Make sure the amount of snacks given is less than the total amount of snacks needed.

Pose a Problem

_____

_____

_____

_____

_____

_____

Draw a bar model for your problem. Then solve.

- Write an expression you could use to solve your problem.
  Explain how the expression represents the problem.

_____

_____

_____

FOR MORE PRACTICE:
Standards Practice Book, pp. P205–P206

Name _____

Multistep Measurement Problems

Essential Question How can you solve multistep problems that include measurement conversions?

 UNLOCK the Problem REAL WORLD

A leaky faucet in Jarod's house drips 2 cups of water each day. After 2 weeks of dripping, the faucet is fixed. If it dripped the same amount each day, how many quarts of water dripped from Jarod's leaky faucet in 2 weeks?

 Use the steps to solve the multistep problem.

STEP 1

Record the information you are given.

The faucet drips _____ cups of water each day.

The faucet drips for _____ weeks.

STEP 2

Find the total amount of water dripped in 2 weeks.

Since you are given the amount of water dripped each day, you must convert 2 weeks into days and multiply.

Think: There are 7 days in 1 week.

cups each day	days in 2 weeks	total cups
↓	↓	↓
2	× _____	= _____

The faucet drips _____ cups in 2 weeks.

STEP 3

Convert from cups to quarts.

Think: There are 2 cups in 1 pint.

There are 2 pints in 1 quart.

_____ cups = _____ pints

_____ pints = _____ quarts

So, Jarod's leaky faucet drips _____ quarts of water in 2 weeks.

• **What if** the faucet dripped for 4 weeks before it was fixed? How many quarts of water would have leaked?

_____

🔑 Example

A carton of large, Grade A eggs weighs about 1.5 pounds. If a carton holds a dozen eggs, how many ounces does each egg weigh?

STEP 1

In ounces, find the weight of a carton of eggs.

Think: 1 pound = _____ ounces

Weight of a carton (in ounces):

total lb	oz in 1 lb	total oz
↓	↓	↓

1.5 × _____ = _____

The carton of eggs weighs about _____ ounces.

STEP 2

In ounces, find the weight of each egg in a carton.

Think: 1 carton (dozen eggs) = _____ eggs

Weight of each egg (in ounces):

total oz	eggs in 1 carton	oz of 1 egg
↓	↓	↓

24 ÷ _____ = _____

So, each egg weighs about _____ ounces.

Share and Show MATH BOARD ·

Solve.

1. After each soccer practice, Scott runs 4 sprints of 20 yards each. If he continues his routine, how many practices will it take for Scott to have sprinted a total of 2 miles combined?

   Scott sprints _____ yards each practice.

   Since there are _____ yards in 2 miles, he will need to continue his routine for

   _____ practices.

2. A worker at a mill is loading 5-lb bags of flour into boxes to deliver to a local warehouse. Each box holds 12 bags of flour. If the warehouse orders 3 Tons of flour, how many boxes are needed to fulfill the order?

   _____

3. Cory brings five 1-gallon jugs of juice to serve during parent night at his school. If the paper cups he is using for drinks can hold 8 fluid ounces, how many drinks can Cory serve for parent night?

   _____

Math Talk MATHEMATICAL PRACTICES
Explain the steps you took to solve Exercise 2.

Name _____

On Your Own ·

Solve.

4. A science teacher needs to collect lake water for a lab she is teaching. The lab requires each student to use 4 fluid ounces of lake water. If 68 students are participating, how many pints of lake water will the teacher need to collect?

5. A string of decorative lights is 28 feet long. The first light on the string is 16 inches from the plug. If the lights on the string are spaced 4 inches apart, how many lights are there on the string?

6. When Jamie's car moves forward such that each tire makes one full rotation, the car has traveled 72 inches. How many full rotations will the tires need to make for Jamie's car to travel 10 yards?

7. A male African elephant weighs 7 Tons. If a male African lion at the local zoo weighs $\frac{1}{40}$ of the weight of the male African elephant, how many pounds does the lion weigh?

8. An office supply company is shipping a case of wooden pencils to a store. There are 64 boxes of pencils in the case. If each box of pencils weighs 2.5 ounces, what is the weight, in pounds, of the case of wooden pencils?

9. **H.O.T.** A gallon of unleaded gasoline weighs about 6 pounds. About how many ounces does 1 quart of unleaded gasoline weigh? HINT: 1 quart $= \frac{1}{4}$ of a gallon

UNLOCK the Problem   REAL WORLD

10. At a local animal shelter there are 12 small-size dogs and 5 medium-size dogs. Every day, the small-size dogs are each given 12.5 ounces of dry food and the medium-size dogs are each given 18 ounces of the same dry food. How many pounds of dry food does the shelter serve in one day?

a. What are you asked to find? _____

_____

b. What information will you use? _____

_____

_____

c. What conversion will you need to do to solve the problem?

_____

_____

d. Show the steps you use to solve the problem.

e. Complete the sentences. The small-size dogs eat a total of _____ ounces of dry food each day.

The medium-size dogs eat a total of

_____ ounces of dry food each day.

The shelter serves _____ ounces,

or _____ pounds, of dry food each day.

11. Test Prep For a class assignment, students are asked to record the total amount of water they drink in one day. Melinda records that she drank four 8-fluid ounce glasses of water and two 1-pint bottles. How many quarts of water did Melinda drink during the day?

(A) 2 quarts   (C) 6 quarts

(B) 4 quarts   (D) 8 quarts

Name _____

 Mid-Chapter Checkpoint

▶ **Vocabulary**

Choose the best term from the box.

1. The _____ of an object is how heavy the object is. (p. 413)

2. The _____ of a container is the amount the container can hold. (p. 409)

▶ **Concepts and Skills**

Convert.

3. 5 mi = _____ yd

4. 48 qt = _____ gal

5. 9 T = _____ lb

6. 336 oz = _____ lb

7. 14 ft = _____ yd _____ ft

8. 11 pt = _____ fl oz

Compare. Write <, >, or =.

9. 96 fl oz ◯ 13 c

10. 25 lb ◯ 384 oz

11. 8 yd ◯ 288 in.

Solve.

12. A standard coffee mug has a capacity of 16 fluid ounces. If Annie needs to fill 26 mugs with coffee, how many total quarts of coffee does she need?

_____

Fill in the bubble completely to show your answer.

13. The length of a classroom is 34 feet. What is this measurement in yards and feet?

    (A) 17 yards 0 feet

    (B) 11 yards 1 foot

    (C) 8 yards 2 feet

    (D) 5 yards 4 feet

14. Charlie's puppy, Max, weighs 8 pounds. How many ounces does Max weigh?

    (A) 24 ounces

    (B) 88 ounces

    (C) 124 ounces

    (D) 128 ounces

15. Milton purchases a 5-gallon aquarium for his bedroom. To fill the aquarium with water, he uses a container with a capacity of 1 quart. How many times will Milton fill and empty the container before the aquarium is full?

    (A) 10

    (B) 15

    (C) 20

    (D) 25

16. Sarah uses a recipe to make 2 gallons of her favorite mixed-berry juice. The containers she plans to use to store the juice have a capacity of 1 pint. How many containers will Sarah need?

    (A) 4

    (B) 8

    (C) 10

    (D) 16

17. The average length of a female white-beaked dolphin is about 111 inches. What is this length in feet and inches?

    (A) 9 feet 2 inches   (C) 10 feet 0 inches

    (B) 9 feet 3 inches   (D) 10 feet 3 inches

Name _____

Metric Measures

Essential Question How can you compare and convert metric units?

🔑 UNLOCK the Problem › REAL WORLD

Using a map, Alex estimates the distance between his house and his grandparent's house to be about 15,000 meters. About how many kilometers away from his grandparent's house does Alex live?

- Underline the sentence that tells you what you are trying to find.
- Circle the measurement you need to convert.

The metric system is based on place value. Each unit is related to the next largest or next smallest unit by a power of 10.

🔓 One Way Convert 15,000 meters to kilometers.

kilo- (k)	hecto- (h)	deka- (da)	meter (m) liter (L) gram (g)	deci- (d)	centi- (c)	milli- (m)

Power of 10   Power of 10   Power of 10

STEP 1 Find the relationship between the units.

Meters are _____ powers of 10 smaller than kilometers.

There are _____ meters in 1 kilometer.

_____

STEP 2 Determine the operation to be used.

I am converting from a _____ unit to a

_____ unit, so I will _____.

_____

STEP 3 Convert.

number of meters	meters in 1 kilometer	number of kilometers
↓	↓	↓

15,000 ◯ _____ = _____

Math Talk MATHEMATICAL PRACTICES
Chose two units in the chart. **Explain** how you use powers of 10 to describe how the two units are related.

So, Alex's house is _____ kilometers from his grandparent's house.

🔑 Another Way Use a diagram.

Jamie made a bracelet 1.8 decimeters long.
How many millimeters long is Jamie's bracelet?

Convert 1.8 decimeters to millimeters.

kilo-	hecto-	deka-	meter liter gram	deci-	centi-	milli-
				1	8	

STEP 1 Show 1.8 decimeters.

Since the unit is decimeters, place the decimal point so that decimeters are the whole number unit.

STEP 2 Convert.

Cross out the decimal and rewrite it so that millimeters will be the whole number unit. Write zeros to the left of the decimal point as needed to complete the whole number.

STEP 3 Record the value with the new units.

1.8 dm = _____ mm

So, Jamie's bracelet is _____ millimeters long.

Try This! Complete the equation to show the conversion.

Ⓐ Convert 247 milligrams to centigrams, decigrams, and grams.

Are the units being converted to a larger unit or a smaller unit? _____

Should you multiply or divide by powers of 10 to convert? _____

247 mg ◯ 10 = _____ cg

247 mg ◯ 100 = _____ dg

247 mg ◯ 1,000 = _____ g

Ⓑ Convert 3.9 hectoliters to dekaliters, liters, and deciliters.

Are the units being converted to a larger unit or a smaller unit? _____

Should you multiply or divide by powers of 10 to convert? _____

3.9 hL ◯ 10 = _____ daL

3.9 hL ◯ 100 = _____ L

3.9 hL ◯ 1,000 = _____ dL

Name _____

Share and Show

Complete the equation to show the conversion.

1. 8.47 L $\bigcirc$ 10 = _____ dL

8.47 L $\bigcirc$ 100 = _____ cL

8.47 L $\bigcirc$ 1,000 = _____ mL

Think: Are the units being converted to a larger unit or a smaller unit?

2. 9,824 dg $\bigcirc$ 10 = _____ g

9,824 dg $\bigcirc$ 100 = _____ dag

9,824 dg $\bigcirc$ 1,000 = _____ hg

Convert.

3. 4,250 cm = _____ m

 4. 6,000 mL = _____ L

 5. 4 dg = _____ cg

MATHEMATICAL PRACTICES

Math Talk Explain how you can compare the lengths 4.25 dm and 4.25 cm without converting.

On Your Own

Convert.

6. 8 kg = _____ g

7. 5 km = _____ m

8. 40 mm = _____ cm

9. 7 g = _____ mg

10. 6,000 g = _____ kg

11. 1,521 mL = _____ dL

Compare. Write <, >, or =.

12. 32 hg $\bigcirc$ 3.2 kg

13. 6 km $\bigcirc$ 660 m

14. 525 mL $\bigcirc$ 525 cL

Problem Solving 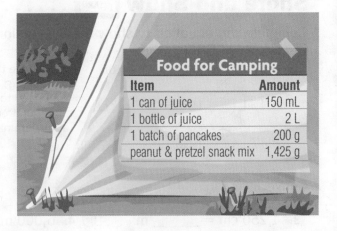 REAL WORLD

For 15–16, use the table.

15. Kelly made one batch of peanut and pretzel snack mix. How many grams does she need to add to the snack mix to make 2 kilograms?

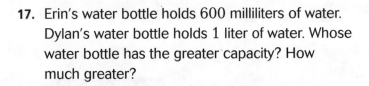

16. **H.O.T.** Kelly plans to take juice on her camping trip. Which will hold more juice, 8 cans or 2 bottles? How much more?

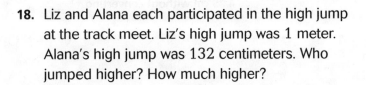

Food for Camping	
Item	**Amount**
1 can of juice	150 mL
1 bottle of juice	2 L
1 batch of pancakes	200 g
peanut & pretzel snack mix	1,425 g

17. Erin's water bottle holds 600 milliliters of water. Dylan's water bottle holds 1 liter of water. Whose water bottle has the greater capacity? How much greater?

SHOW YOUR WORK

18. Liz and Alana each participated in the high jump at the track meet. Liz's high jump was 1 meter. Alana's high jump was 132 centimeters. Who jumped higher? How much higher?

19. **H.O.T.** Are there less than 1 million, exactly 1 million, or greater than 1 million milligrams in 1 kilogram? **Explain** how you know.

20. **Test Prep** Monica has 426 millimeters of fabric. How many centimeters of fabric does Monica have?

   (A) 4,260 centimeters   (C) 4.26 centimeters

   (B) 42.6 centimeters   (D) 0.426 centimeters

FOR MORE PRACTICE:
Standards Practice Book, pp. P209–P210

Name _____

Problem Solving
Customary and Metric Conversions

Essential Question How can you use the strategy *make a table* to help you solve problems about customary and metric conversions?

? UNLOCK the Problem REAL WORLD

Aaron is making fruit punch for a family reunion. He needs to make 120 cups of punch. If he wants to store the fruit punch in gallon containers, how many gallon containers will Aaron need?

Use the graphic organizer below to help you solve the problem.

Conversion Table

	gal	qt	pt	c
1 gal	1	4	8	16
1 qt	$\frac{1}{4}$	1	2	4
1 pt	$\frac{1}{8}$	$\frac{1}{2}$	1	2
1 c	$\frac{1}{16}$	$\frac{1}{4}$	$\frac{1}{2}$	1

Read the Problem

What do I need to find?

I need to find _____

_____

_____ .

What information do I need to use?

I need to use _____

_____

_____ .

How will I use the information?

I will make a table to show the relationship between the

number of _____ and

the number of _____ .

Solve the Problem

There are _____ cups in 1 gallon. So, each cup is _____ of a gallon. Complete the table below.

c	1	2	3	4	120
gal	$\frac{1}{16}$	$\frac{1}{8}$	$\frac{3}{16}$	$\frac{1}{4}$	

Multiply by _____ .

So, Aaron needs _____ gallon containers to store the punch.

• Will all of the gallon containers Aaron uses be filled to capacity? **Explain.** _____

_____

🔒 Try Another Problem

Sharon is working on a project for art class. She needs to cut strips of wood that are each 1 decimeter long to complete the project. If Sharon has 7 strips of wood that are each 1 meter long, how many 1-decimeter strips can she cut?

Conversion Table

	m	dm	cm	mm
1 m	1	10	100	1,000
1 dm	$\frac{1}{10}$	1	10	100
1 cm	$\frac{1}{100}$	$\frac{1}{10}$	1	10
1 mm	$\frac{1}{1,000}$	$\frac{1}{100}$	$\frac{1}{10}$	1

Read the Problem

What do I need to find?	**What information do I need to use?**	**How will I use the information?**

Solve the Problem

So, Sharon can cut _____ 1-decimeter lengths to complete her project.

- What relationship did the table you made show? _____

_____

© Houghton Mifflin Harcourt Publishing Company

Math Talk MATHEMATICAL PRACTICES
Explain how you could use another strategy to solve this problem.

Share and Show

1. Edgardo has a drink cooler that holds 10 gallons of water. He is filling the cooler with a 1-quart container. How many times will he have to fill the quart container to fill the cooler?

   First, make a table to show the relationship between gallons and quarts. You can use a conversion table to find how many quarts are in a gallon.

gal	1	2	3	4	10
qt	4				

   Then, look for a rule to help you complete your table.

   number of gallons × _____ = number of quarts

   Finally, use the table to solve the problem.

   Edgardo will need to fill the quart container _____ times.

2. **H.O.T.** **What if** Edgardo only uses 32 quarts of water to fill the cooler. How can you use your table to find how many gallons that is?

   _____

   _____

   _____

   _____

3. If Edgardo uses a 1-cup container to fill the cooler, how will that affect the number of times he has to fill a container to fill the cooler? **Explain.**

   _____

   _____

   _____

   _____

SHOW YOUR WORK

On Your Own

4. Jeremy made a belt that was 6.4 decimeters long. How many centimeters long is the belt Jeremy made?

_____

5. Dan owns 9 DVDs. His brother Mark has 3 more DVDs than Dan has. Their sister, Marsha, has more DVDs than either of her brothers. Together, the three have 35 DVDs. How many DVDs does Marsha have?

_____

SHOW YOUR WORK

6. **H.O.T.** Kevin is making a picture frame. He has a piece of trim that is 4 feet long. How many 14-inch-long pieces can Kevin cut from the trim? How much of a foot will he have left over?

_____

_____

7. **Write Math** ▶ **Explain** how you could find the number of cups in five gallons of water.

_____

_____

_____

_____

8. Carla uses $2\frac{3}{4}$ cups of flour and $1\frac{3}{8}$ cups of sugar in her cookie recipe. How many cups does she use in all?

_____

9. Tony needs 16-inch-long pieces of gold chain to make each of 3 necklaces. He has a piece of chain that is $4\frac{1}{2}$ feet long. How much chain will he have left after making the necklaces?

(A) 6 inches   (C) 18 inches

(B) 12 inches   (D) 24 inches

Elapsed Time

Essential Question How can you solve elapsed time problems by converting units of time?

🔑 UNLOCK the Problem ⟩ REAL WORLD

A computer company claims its laptop has a battery that lasts 4 hours. The laptop actually ran for 200 minutes before the battery ran out. Did the battery last 4 hours?

1 hour = _____ minutes

Think: The minute hand moves from one number to the next in 5 minutes.

🔑 **Convert 200 minutes to hours and minutes.**

	total min	min in 1 hr	hr	min
STEP 1 Convert minutes into hours and minutes.

200 min = ____ hr ____ min   _____ ◯ _____ is _____ r _____

STEP 2 Compare. Write <, >, or =.   _____ hr _____ min ◯ 4 hr

Since _____ hours _____ minutes is _____ 4 hours, the

battery _____ last as long as the computer company claims.

Try This! Convert to mixed measures.

Jill spent much of her summer away from home. She spent 10 days with her grandparents, 9 days with her cousins, and 22 days at camp. How many weeks and days was she away from home?

STEP 1 Find the total number of days away.

10 days + 9 days + 22 days = _____ days

STEP 2 Convert the days into weeks and days.

_____ ÷ 7 is _____ r _____

So, Jill was away from home _____ weeks and _____ days.

Units of Time
60 seconds (s) = 1 minute (min)
60 minutes = 1 hour (hr)
24 hours = 1 day (d)
7 days = 1 week (wk)
52 weeks = 1 year (yr)
12 months (mo) = 1 year
365 days = 1 year

🔑 One Way Use a number line to find elapsed time.

Monica spent $2\frac{1}{2}$ hours working on her computer. If she started working at 10:30 A.M., what time did Monica stop working?

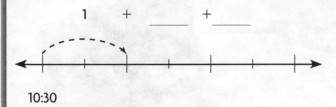

1   +   _____   +   _____

10:30   **Think:** $\frac{1}{2}$ hour = 30 minutes
_____   _____   _____

🔑 Another Way Use a clock to find elapsed time.

Start   End

So, Monica stopped working at _____.

Try This! Find a start time.

Robert's soccer team needs to be off the soccer field by 12:15 P.M. Each game is at most $1\frac{3}{4}$ hours long. What time should the game begin to be sure that the team finishes on time?

$\frac{1}{4}$ hour = 15 minutes, so $\frac{3}{4}$ hour = _____ minutes

STEP 1 Subtract the minutes first.

45 minutes earlier is _____.

STEP 2 Then subtract the hour.

1 hour and 45 minutes earlier is _____.

So, the game should begin at _____.

Math Talk MATHEMATICAL PRACTICES
Explain how you could convert 3 hours 45 minutes to minutes.

432

Name _____

Share and Show

Convert.

1. 540 min = _____ hr

2. 8 d = _____ hr

✓ **3.** 110 hr = _____ d _____ hr

Find the end time.

✓ **4.** Start time: 9:17 A.M.   Elapsed time: 5 hr 18 min

End time: _____

MATHEMATICAL PRACTICES

Math Talk Explain how to find how long a movie lasts if it starts at 1:35 P.M. and ends at 3:40 P.M.

On Your Own

Convert.

5. 3 min = _____ sec

6. 240 min = _____ hr

7. 1 hr = _____ sec

8. 3 yr = _____ d

9. 208 wk = _____ yr

10. 350 min = _____ hr _____ min

Find the start, elapsed, or end time.

11. Start time: 11:38 A.M.

Elapsed time: 3 hr 10 min

End time: _____

12. Start time: _____

Elapsed time: 2 hr 37 min

End time: 1:15 P.M.

13. Start time: _____

Elapsed time: $2\frac{1}{4}$ hr

End time: 5:30 P.M.

14. Start time: 7:41 P.M.

Elapsed time: _____

End time: 8:50 P.M.

Problem Solving 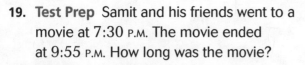 REAL WORLD

For 15–17, use the graph.

15. Which Internet services downloaded the podcast in less than 4 minutes?

_____

16. 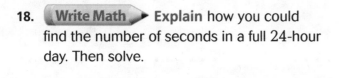 Which service took the longest to download the podcast? How much longer did it take than Red Fox in minutes and seconds?

_____

17. H.O.T. Which service was faster, Red Fox or Internet-C? How much faster in minutes and seconds?

_____

_____

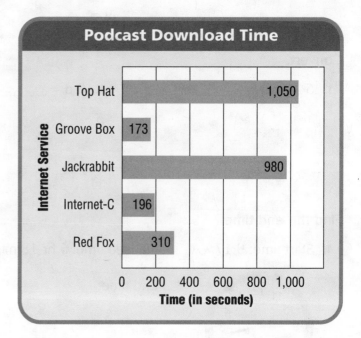

Podcast Download Time

Internet Service

- Top Hat — 1,050
- Groove Box — 173
- Jackrabbit — 980
- Internet-C — 196
- Red Fox — 310

0   200   400   600   800   1,000
Time (in seconds)

18. **Write Math** ▶ **Explain** how you could find the number of seconds in a full 24-hour day. Then solve.

_____

_____

_____

_____

_____

SHOW YOUR WORK

19. **Test Prep** Samit and his friends went to a movie at 7:30 P.M. The movie ended at 9:55 P.M. How long was the movie?

　Ⓐ 2 hours 25 minutes

　Ⓑ 2 hours 5 minutes

　Ⓒ 1 hour 25 minutes

　Ⓓ 1 hour 5 minutes

✓ 🏴 Chapter Review/Test

▶ **Vocabulary**

Choose the best term from the box.

1. A metric unit of mass that is equal to $\frac{1}{1,000}$ of a gram

   is called a _____. (p. 423)

2. A metric unit for measuring length that is equal to 10 meters

   is called a _____. (p. 423)

▶ **Concepts and Skills**

Convert.

3. 96 oz = _____ lb

4. 5 kg = _____ g

5. 500 min = _____ hr _____ min

6. 65 yd 2 feet = _____ ft

Compare. Write <, >, or =.

7. 7 wk ◯ 52 d

8. 4 L ◯ 3,000 mL

9. 72 in. ◯ 2 yd

Solve.

10. A girl walks 5,000 meters in one hour. If the girl walks at the same speed for 4 hours, how many kilometers will she have walked?

   _____

GO
Online

Assessment Options
Chapter Test

Fill in the bubble completely to show your answer.

11. Howard cuts 54 centimeters off a 1-meter board. How much of the board does Howard have left?

Ⓐ 53 centimeters

Ⓑ 53 meters

Ⓒ 46 meters

Ⓓ 46 centimeters

12. Joe's dog has a mass of 28,000 grams. What is the mass of Joe's dog in kilograms?

Ⓐ 2,800 kilograms

Ⓑ 280 kilograms

Ⓒ 28 kilograms

Ⓓ 2.8 kilograms

13. Cathy drank 600 milliliters of water at school and another 400 milliliters at home. How many liters of water did Cathy drink?

Ⓐ 1,000 liters

Ⓑ 100 liters

Ⓒ 10 liters

Ⓓ 1 liter

14. Mr. Banks left work at 5:15 P.M. It took him $1\frac{1}{4}$ hours to drive home. At what time did Mr. Banks arrive home?

Ⓐ 6:15 P.M.

Ⓑ 6:30 P.M.

Ⓒ 6:45 P.M.

Ⓓ 7:30 P.M.

Name _____

Fill in the bubble completely to show your answer.

15. A turtle walks 12 feet in one hour. How many inches does the turtle walk in one hour?

Ⓐ 12 inches

Ⓑ 24 inches

Ⓒ 124 inches

Ⓓ 144 inches

16. Jason and Doug competed in the long jump at a track meet. Jason's long jump was 98 inches. Doug's long jump was 3 yards. How much longer was Doug's jump than Jason's jump?

Ⓐ 1 inch

Ⓑ 10 inches

Ⓒ 12 inches

Ⓓ 20 inches

17. Sarita used 54 ounces of apples to make an apple pie. How many pounds and ounces of apples did Sarita use?

Ⓐ 2 pounds 6 ounces

Ⓑ 3 pounds 6 ounces

Ⓒ 4 pounds 6 ounces

Ⓓ 8 pounds 6 ounces

18. Morgan measures the capacity of a juice glass to be 12 fluid ounces. If she uses the glass to drink 4 glasses of water throughout the day, how many pints of water does Morgan drink?

Ⓐ 3 pints

Ⓑ 6 pints

Ⓒ 24 pints

Ⓓ 48 pints

▶ Constructed Response

19. Louisa needs 3 liters of lemonade and punch for a picnic. She has 1,800 milliliters of lemonade. How much punch does she need? **Explain** how you found your answer.

_____

_____

20. Maddie bought 10 quarts of ice cream. How many gallons and quarts of ice cream did Maddie buy? **Explain** how you found your answer.

_____

_____

_____

▶ Performance Task

21. The Drama Club is showing a video of their recent play. The first showing began at 2:30 P.M. The second showing was scheduled to start at 5:25 P.M. with a $\frac{1}{2}$-hour break between the showings.

Ⓐ How long is the video in hours and minutes? _____

Ⓑ **Explain** how you can use a number line to find the answer.

_____

_____

_____

_____

Ⓒ The second showing started 20 minutes late. Will the second showing be over by 7:45 P.M.? **Explain** why your answer is reasonable.

_____

_____

_____

11 Geometry and Volume

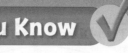

Show What You Know ✓

Check your understanding of important skills.

Name _____

▶ **Perimeter** Count the units to find the perimeter.

1.

Perimeter = _____ units

2.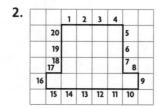

Perimeter = _____ units

▶ **Area** Write the area of each shape.

3.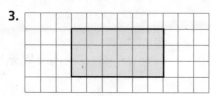

_____ square units

4.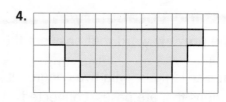

_____ square units

▶ **Multiply Three Factors** Write the product.

5. $3 \times 5 \times 4 =$ _____   6. $5 \times 5 \times 10 =$ _____   7. $7 \times 3 \times 20 =$ _____

Helen must find a certain polyhedron for a treasure hunt. Be a Math Detective by using the clues to help Helen identify the polyhedron.

Clues

- The polyhedron has 1 base.
- It has 4 lateral faces that meet at a common vertex.
- The edges of the base are all the same length.

 rectangular prism

 triangular prism

 hexagonal prism

 square pyramid

 triangular pyramid

 cube

Vocabulary Builder

© Houghton Mifflin Harcourt Publishing Company

▶ **Visualize It** ••••••••••••••••••••••••••••••••••••••

Sort the checked words into the circle map.

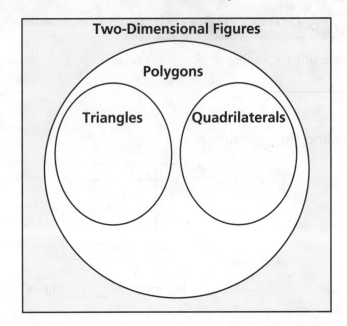

Review Words

✓ acute triangle

✓ decagon

equilateral triangle

✓ hexagon

isosceles triangle

✓ obtuse triangle

✓ octagon

✓ parallelogram

quadrilateral

✓ rectangle

✓ rhombus

✓ right triangle

scalene triangle

✓ trapezoid

Preview Words

base

congruent

heptagon

lateral face

nonagon

polygon

polyhedron

prism

pyramid

regular polygon

unit cube

volume

▶ **Understand Vocabulary** •••••••••••••••••••••••••••••••

Write the preview word that answers the riddle.

1. I am a solid figure with two congruent polygons that are bases, connected with lateral faces that are rectangles. _____

2. I am a polygon in which all sides are congruent and all angles are congruent. _____

3. I am a cube that has a length, width, and height of 1 unit. _____

4. I am a solid figure with faces that are polygons. _____

5. I am the measure of the amount of space a solid figure occupies. _____

6. I am a polygon that connects with the bases of a polyhedron. _____

GO
Online • eStudent Edition • Multimedia eGlossary

Name _____

Polygons

Essential Question How can you identify and classify polygons?

UNLOCK the Problem REAL WORLD

The Castel del Monte in Apulia, Italy, was built more than 750 years ago. The fortress has one central building with eight surrounding towers. Which polygon do you see repeated in the structure? How many sides, angles, and vertices does this polygon have?

A **polygon** is a closed plane figure formed by three or more line segments that meet at points called vertices. It is named by the number of sides and angles it has. To identify the repeated polygon in the fortress, complete the tables below.

Polygon	Triangle	Quadrilateral	Pentagon	Hexagon
Sides	3	4	5	
Angles				
Vertices				

Polygon	Heptagon	Octagon	Nonagon	Decagon
Sides	7	8		
Angles				
Vertices				

Math Idea
Sometimes the angles inside a polygon are greater than 180°.

275°

So, the _____ is the repeated polygon in the

Castel del Monte because it has _____ sides, _____ angles,

and _____ vertices.

Math Talk MATHEMATICAL PRACTICES
What pattern do you see among the number of sides, angles, and vertices a polygon has?

Chapter 11 441

Regular Polygons When line segments have the same length or when angles have the same measure, they are **congruent**. In a **regular polygon**, all sides are congruent and all angles are congruent.

regular polygon	not a regular polygon
All sides are congruent. 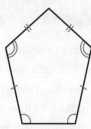 You can write measurements to show congruent sides and angles.	Not all sides are congruent. Not all angles are congruent. You can use the same markings to show the congruent sides and angles.
All angles are congruent.	

Try This! Label the Venn diagram to classify the polygons in each group. Then draw a polygon that belongs only to each group.

Congruent _____   Congruent _____

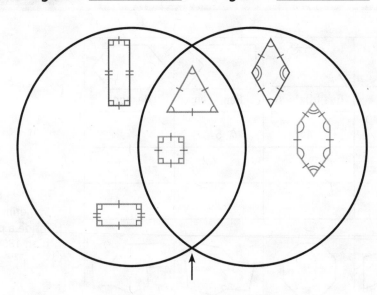

Regular _____

Share and Show

1. Name the polygon. Then use the markings on the figure to tell whether it is a *regular polygon* or *not a regular polygon*.

   a. Name the polygon. _____

   b. Are all the sides and all the angles congruent? _____

   c. Is the polygon a regular polygon? _____

Name _____

Triangles

Essential Question How can you classify triangles?

 UNLOCK the Problem REAL WORLD

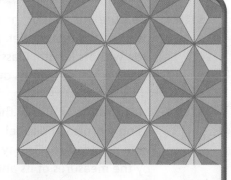

If you look closely at Epcot Center's Spaceship Earth building in Orlando, Florida, you may see a pattern of triangles. The triangle outlined in the pattern at the right has 3 congruent sides and 3 acute angles. What type of triangle is outlined?

🔑 **Complete the sentence that describes each type of triangle.**

Classify triangles by the lengths of their sides.	Classify triangles by the measures of their angles.
An **equilateral triangle** has _____ congruent sides. 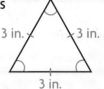	A **right triangle** has one 90°, or _____ angle. 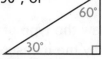
An **isosceles triangle** has _____ congruent sides. 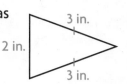	An **acute triangle** has 3 _____ angles. 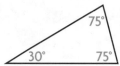
A **scalene triangle** has _____ congruent sides. 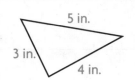	An **obtuse triangle** has 1 _____ angle.

The type of triangle outlined in the pattern can be classified by the length of its sides as an _____ triangle.

The triangle can also be classified by the measures of its angles as an _____ triangle.

Math Talk MATHEMATICAL PRACTICES
Is an equilateral triangle also a regular polygon? **Explain.**

Activity

Classify triangle ABC by the lengths of its sides and by the measures of its angles.

Materials ■ centimeter ruler ■ protractor

STEP 1 Measure the sides of the triangle using a centimeter ruler. Label each side with its length. Classify the triangle by the lengths of its sides.

STEP 2 Measure the angles of the triangle using a protractor. Label each angle with its measure. Classify the triangle by the measures of its angles.

- What type of triangle has 3 sides of different lengths?

  _____

- What is an angle called that is greater than 90° and less than 180°?

  _____

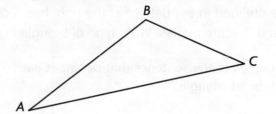

Triangle *ABC* is a _____ _____ triangle.

Try This! Draw the type of triangle described by the lengths of its sides and by the measures of its angles.

Triangle by Length of Sides		
	Scalene	**Isosceles**
Acute	**Think:** I need to draw a triangle that is acute and scalene.	
Obtuse		

Triangle by Angle Measure

Math Talk MATHEMATICAL PRACTICES
Can you draw a triangle that is right equilateral? **Explain.**

Share and Show

Classify each triangle. Write *isosceles*, *scalene*, or *equilateral*.
Then write *acute*, *obtuse*, or *right*.

1.

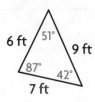

6 ft 51° 9 ft
87° 42°
7 ft

_____ _____

2.

_____ _____

3.

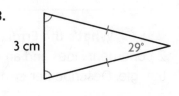

3 cm 29°

_____ _____

Math Talk Can you tell that a triangle is obtuse, right, or acute without measuring the angles? **Explain.**

MATHEMATICAL PRACTICES

On Your Own

Classify each triangle. Write *isosceles*, *scalene*, or *equilateral*.
Then write *acute*, *obtuse*, or *right*.

4.

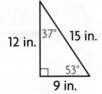

12 in. 37° 15 in.
53°
9 in.

_____ _____

5.

53°

_____ _____

6.

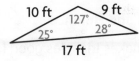

10 ft 127° 9 ft
25° 28°
17 ft

_____ _____

A triangle has sides with the lengths and angle measures given.
Classify each triangle. Write *isosceles*, *scalene*, or *equilateral*.
Then write *acute*, *obtuse*, or *right*.

7. sides: 3.5 cm, 6.2 cm, 3.5 cm

angles: 27°, 126°, 27°

_____ _____

8. sides: 2 in., 5 in., 3.8 in.

angles: 43°, 116°, 21°

_____ _____

9. Circle the figure that does not belong. **Explain.**

_____

Problem Solving REAL WORLD

10. Draw 2 equilateral triangles that are congruent and share a side.
What polygon is formed? Is it a regular polygon?

_____

11. **H.O.T.** **What's the Error?** Shannon said that a triangle with exactly
2 congruent sides and an obtuse angle is an equilateral obtuse
triangle. Describe her error.

_____

_____

12. **Test Prep** Which kind of triangle has exactly 2 congruent sides?

   (A) isosceles   (B) equilateral   (C) scalene   (D) right

Connect to Science

Forces and Balance

What makes triangles good for the construction of buildings or bridges?
The 3 fixed lengths of the sides of a triangle, when joined, can form no
other shape. So, when pushed, triangles don't bend or buckle.

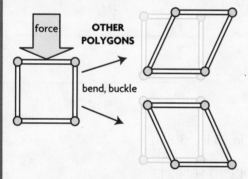

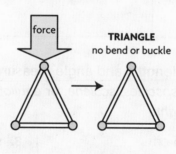

Classify the triangles in the structures below. Write _isosceles, scalene,_
or _equilateral._ Then write _acute, obtuse,_ or _right._

13.

14.

_____   _____

Name _____

Quadrilaterals

Essential Question How can you classify and compare quadrilaterals?

🔑 UNLOCK the Problem · REAL WORLD

A seating chart for a baseball field has many four-sided figures, or **quadrilaterals**. What types of quadrilaterals can you find in the seating chart?

There are five special types of quadrilaterals. You can classify quadrilaterals by their properties, such as parallel sides and perpendicular sides. Parallel lines are lines that are always the same distance apart. Perpendicular lines are lines that intersect to form four right angles.

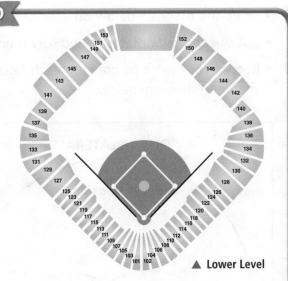

▲ Lower Level

🔒 **Complete the sentence that describes each type of quadrilateral.**

A general quadrilateral has 4 sides and 4 angles.

A **parallelogram** has

opposite _____

that are _____
and parallel.

A **rectangle** is a special

parallelogram with _____ right angles and 4 pairs of

_____ sides.

A **rhombus** is a special

parallelogram with _____ congruent sides.

A **square** is a special parallelogram with

_____ congruent sides

and _____ right angles.

A **trapezoid** is a quadrilateral with exactly

1 pair of _____ sides.

So, the types of quadrilaterals you can find in the seating chart of the

field are _____ .

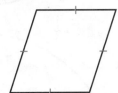

Math Talk · MATHEMATICAL PRACTICES
Explain how trapezoids and parallelograms are different.

 **Activity**

Materials ■ quadrilaterals ■ scissors

You can use a Venn diagram to sort quadrilaterals and find out how they are related.

• Draw the diagram below on your MathBoard.

• Cut out the quadrilaterals and sort them into the Venn diagram.

• Record your work by drawing each figure you have placed in the Venn diagram below.

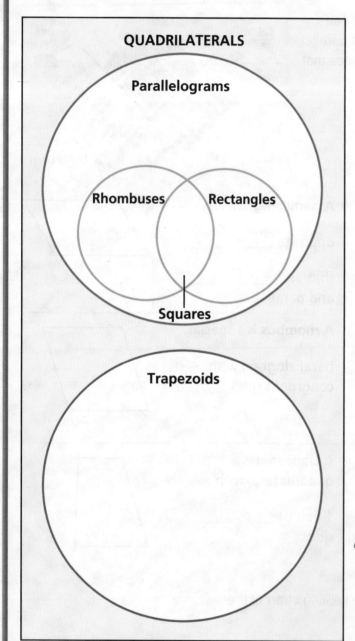

QUADRILATERALS

Parallelograms

Rhombuses   Rectangles

Squares

Trapezoids

Complete the sentences by writing *always, sometimes,* **or** *never.*

A rhombus is _____ a square.

A parallelogram is _____ a rectangle.

A rhombus is _____ a parallelogram.

A trapezoid is _____ a parallelogram.

A square is _____ a rhombus.

1. **Explain** why the circle for parallelograms does not intersect the circle for trapezoids.

_____

_____

_____

2. Draw a quadrilateral with four pairs of perpendicular sides and four congruent sides.

Name _____

Share and Show

1. Use quadrilateral *ABCD* to answer each question. Complete the sentence.

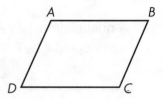

   a. Measure the sides. Are any of the sides congruent? _____
      Mark any congruent sides.

   b. How many right angles, if any, does the quadrilateral have? _____

   c. How many pairs of parallel sides, if any, does the quadrilateral have? _____

   So, quadrilateral *ABCD* is a _____.

Classify the quadrilateral in as many ways as possible. Write
quadrilateral, parallelogram, rectangle, rhombus, square, **or** *trapezoid.*

 2.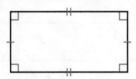

_____

3.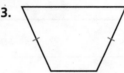

_____

MATHEMATICAL PRACTICES

Math Talk Can the parallel sides of a trapezoid be the same length? **Explain** your answer.

On Your Own

Classify the quadrilateral in as many ways as possible. Write
quadrilateral, parallelogram, rectangle, rhombus, square, **or** *trapezoid.*

4.

_____

5.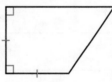

_____

6.

_____

7.

_____

Problem Solving ....

Solve the problems.

8. A quadrilateral has exactly 2 congruent sides. Which quadrilateral types could it be? Which quadrilaterals could it not be?

_____

_____

9. **H.O.T.** **What's the Error?** A quadrilateral has exactly 3 congruent sides. Davis claims that the figure must be a rectangle. Why is his claim incorrect? Use a diagram to **explain** your answer.

_____

_____

_____

10. The opposite corners of a quadrilateral are right angles. The quadrilateral is not a rhombus. What kind of quadrilateral is this figure? **Explain** how you know.

_____

_____

11. **Write Math** ▶ I am a figure with four sides. I can be placed in the following categories: quadrilateral, parallelogram, rectangle, rhombus, and square. Draw me. **Explain** why I fit into each category.

_____

_____

_____

12. **Test Prep** A quadrilateral has exactly 1 pair of parallel sides and no congruent sides. What type of quadrilateral is it?

Ⓐ rectangle Ⓒ parallelogram

Ⓑ rhombus Ⓓ trapezoid

Problem Solving

Properties of Two-Dimensional Figures

Essential Question How can you use the strategy *act it out* to approximate whether the sides of a figure are congruent?

🔑 UNLOCK the Problem — REAL WORLD

Lori has a quadrilateral with vertices *A*, *B*, *C*, and *D*. The quadrilateral has four right angles. She wants to show that quadrilateral *ABCD* is a square, but she does not have a ruler to measure the lengths of the sides. How can she show that the quadrilateral has four congruent sides and is a square?

Use the graphic organizer below to help you solve the problem.

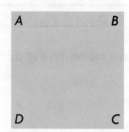

Read the Problem	Solve the Problem
What do I need to find? I need to determine whether the quadrilateral has 4 _____ sides and is a _____ .	I traced the quadrilateral and cut it out. I used *act it out* by folding to match each pair of sides. • I folded the quadrilateral to match side *AB* to side *CD*.
What information do I need to use? The quadrilateral has _____ angles. To be a square, it must also have _____ sides.	• I folded the quadrilateral to match side *AD* to side *BC*.
How will I use the information? I can trace the figure, cut it out, and then fold it to match each pair of sides to show that sides _____ are _____ .	• I folded the quadrilateral diagonally to match side *AD* to side *AB* and side *CD* to side *BC*.

1. What else do you need to do to solve the problem?

_____

_____

So, quadrilateral *ABCD* _____ a square.

Try Another Problem

Terrence has drawn a triangle with vertices *E, F,* and *G.* The triangle has three congruent angles. He wants to show that triangle *EFG* has three congruent sides, but he does not have a ruler to measure the lengths of the sides. How can he show that the triangle has three congruent sides?

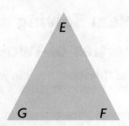

Read the Problem	Solve the Problem
What do I need to find?	Record your work by drawing your model after each fold. Label each drawing with the sides that you find are congruent.
What information do I need to use?	
How will I use the information?	

2. How can you use reasoning to show that all three sides of the triangle are congruent using just two folds? **Explain.**

_____

_____

_____

Name _____

Share and Show   . . . .

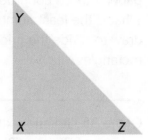

1. Erica thinks that triangle *XYZ*, at the right, has two congruent sides, but she does not have a ruler to measure the sides. Are two sides congruent?

    First, trace the triangle and cut out the tracing.

    Then, fold the triangle to match each pair of sides to determine if at least two of the sides are congruent. As you test the sides, record or draw the results for each pair to make sure that you have checked all pairs of sides. Possible drawings are shown.

    Finally, answer the question.

    _____

    _____

2. **What if** Erica also wants to show, without using a protractor, that the triangle has one right angle and two acute angles? Explain how she can show this.

    _____

    _____

    _____

    _____

3. December, January, and February were the coldest months in Kristen's town last year. February was the warmest of these months. December was not the coldest. What is the order of these months from coldest to warmest?

    _____

4. Jan enters a 20-foot by 30-foot rectangular room. The long sides face north and south. Jan enters the exact center of the south side and walks 10 feet north. Then she walks 8 feet east. How far is she from the east side of the room?

    _____

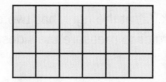
On Your Own

Choose a
STRATEGY

Act It Out
Draw a Diagram
Make a Table
Solve a Simpler Problem
Work Backward
Guess, Check, and Revise

5. **H.O.T.** Max drew a grid to divide a piece of paper into 18 congruent squares, as shown. What is the least number of lines Max can draw to divide the grid into 6 congruent rectangles?

_____

6. Of the 95 fifth and sixth graders going on a field trip, there are 27 more fifth graders than sixth graders. How many fifth graders are going on the field trip?

_____

Use the map to solve 7–8.

7. Sam's paper route begins and ends at the corner of Redwood Avenue and Oak Street. His route is made up of 4 streets, and he makes no 90° turns. What kind of polygon do the streets of Sam's paper route form? Name the streets in Sam's route.

_____

_____

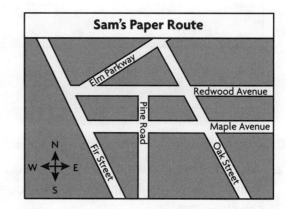

Sam's Paper Route

8. Sam's paper route includes all 32 houses on two pairs of parallel streets. If each street has the same number of houses, how many houses are on each street? Name the parallel streets.

_____

_____

9. **Test Prep** Which figure below is a quadrilateral that has opposite sides that are congruent and parallel?

Ⓐ

Ⓒ

Ⓑ

Ⓓ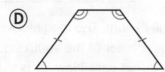

FOR MORE PRACTICE:
Standards Practice Book, pp. P225–P226

Name _____

Three-Dimensional Figures

Essential Question How can you identify, describe, and classify three-dimensional figures?

 UNLOCK the Problem

A solid figure has three dimensions: length, width, and height. **Polyhedrons**, such as prisms and pyramids, are three-dimensional figures with faces that are polygons.

A **prism** is a polyhedron that has two congruent polygons as **bases**.

A polyhedron's **lateral faces** are polygons that connect with the bases. The lateral faces of a prism are rectangles.

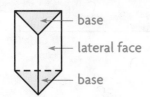
— base
— lateral face
— base

A prism's base shape is used to name the solid figure. The base shape of this prism is a triangle. The prism is a **triangular prism**.

 Identify the base shape of the prism. Use the terms in the box to correctly name the prism by its base shape.

Types of Prisms
decagonal prism
octagonal prism
hexagonal prism
pentagonal prism
rectangular prism
triangular prism

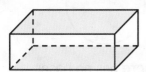

Base shape: _____
Name the solid figure.

_____

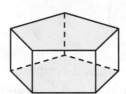

Base shape: _____
Name the solid figure.

_____

Base shape: _____
Name the solid figure.

_____

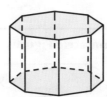

Base shape: _____
Name the solid figure.

_____

Math Talk MATHEMATICAL PRACTICES
What shapes make up a decagonal prism, and how many are there? **Explain.**

- What special prism has congruent squares for bases and lateral faces? _____

Pyramid A **pyramid** is a polyhedron with only one base. The lateral faces of a pyramid are triangles that meet at a common vertex.

Like a prism, a pyramid is named for the shape of its base.

 Identify the base shape of the pyramid. Use the terms in the box to correctly name the pyramid by its base shape.

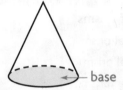

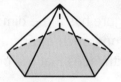

Base shape: _____

Name the solid figure.

_____

Base shape: _____

Name the solid figure.

_____

Base shape: _____

Name the solid figure.

_____

Non-polyhedrons Some three-dimensional figures have curved surfaces. These solid figures are *not* polyhedrons.

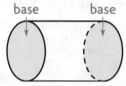

A **cone** has 1 circular base and 1 curved surface.

A **cylinder** has 2 congruent circular bases and 1 curved surface.

A **sphere** has no bases and 1 curved surface.

 Use the Venn diagram to sort the three-dimensional figures listed at the left.

Cones

Cylinders

Prisms

Pyramids

Spheres

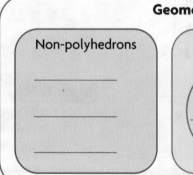

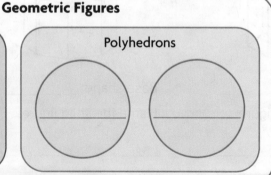

© Houghton Mifflin Harcourt Publishing Company

Name _____

Share and Show

Classify the solid figure. Write *prism, pyramid, cone, cylinder,* or *sphere*.

1.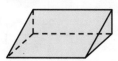

_____

2.

_____

✓3.

_____

Name the solid figure.

4.

_____

5.

_____

✓6.

_____

Math Talk MATHEMATICAL PRACTICES
Compare a prism and a pyramid. Tell how they are similar and how they are different.

On Your Own

Classify the solid figure. Write *prism, pyramid, cone, cylinder,* or *sphere*.

7.

_____

8.

_____

9.

_____

Name the solid figure.

10.

_____

11.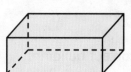

_____

12.

_____

13.

_____

14.

_____

15.

_____

Problem Solving · REAL WORLD

16. Mario is making a sculpture out of stone. He starts by carving a base with five sides. He then carves five triangular lateral faces that all meet at a point at the top. What three-dimensional figure does Mario make?

_____

17. H.O.T. What is another name for a cube? **Explain** your reasoning.

_____

_____

_____

Connect to Reading

Identify the Details

If you were given a description of a building and asked to identify which one of these three buildings is described, which details would you use to determine the building?

Word problems contain details that help you solve the problem. Some details are meaningful and are important to finding the solution and some details may not be. _Identify the details_ you need to solve the problem.

◄ **Flatiron Building, New York City, New York**

◄ **Nehru Science Center, Mumbai, India**

◄ **Luxor Hotel, Las Vegas, Nevada**

Example Read the description. **Underline the details you need to identify the solid figure that will name the correct building.**

This building is one of the most identifiable structures in its city's skyline. It has a square foundation and 28 floors. The building has four triangular exterior faces that meet at a point at the top of the structure.

Identify the solid figure and name the correct building.

18. Solve the problem in the Example.

Solid figure: _____

Building: _____

19. This building was completed in 1902. It has a triangular foundation and a triangular roof that are the same size and shape. The three sides of the building are rectangles.

Solid figure: _____

Building: _____

Name _____

 Mid-Chapter Checkpoint

▶ **Vocabulary**

Choose the best term from the box.

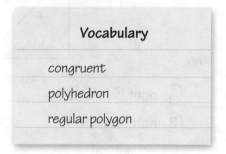

| Vocabulary |
| congruent |
| polyhedron |
| regular polygon |

1. A closed plane figure with all sides congruent and all angles

   congruent is called a _____. **(p. 442)**

2. Line segments that have the same length or angles that have

   the same measure are _____. **(p. 442)**

▶ **Concepts and Skills**

Name each polygon. Then tell whether it is a *regular polygon*
or *not a regular polygon*.

3.

_____

4.

_____

5.

_____

Classify each triangle. Write *isosceles, scalene,* or *equilateral*.
Then write *acute, obtuse,* or *right*.

6.

_____

7.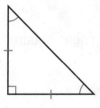

_____

8. 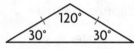 120° 30° 30°

_____

Classify the quadrilateral in as many ways as possible. Write *quadrilateral,
parallelogram, rectangle, rhombus, square,* or *trapezoid*.

9.

_____

_____

10.

_____

_____

11.

_____

_____

Fill in the bubble completely to show your answer.

12. What type of triangle is shown below?

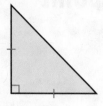

(A) right isosceles

(B) right scalene

(C) equilateral

(D) obtuse scalene

13. Classify the quadrilateral in as many ways as possible.

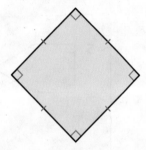

(A) quadrilateral, parallelogram, rhombus

(B) quadrilateral, parallelogram, rhombus, trapezoid

(C) quadrilateral, parallelogram, rhombus, rectangle, trapezoid, square

(D) quadrilateral, parallelogram, rhombus, rectangle, square

14. Classify the following figure.

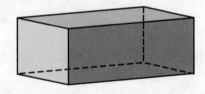

(A) cone

(B) cube

(C) rectangular prism

(D) rectangular pyramid

Name _____

Unit Cubes and Solid Figures

Essential Question What is a unit cube and how can you use it to build a solid figure?

Investigate

You can build rectangular prisms using unit cubes. How many different rectangular prisms can you build with a given number of unit cubes?

Materials ■ centimeter cubes

A **unit cube** is a cube that has a length, width, and height

of 1 unit. A cube has _____ square faces. All of its faces

are congruent. It has _____ edges. The lengths of all its edges are equal.

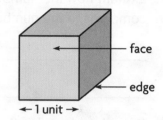

face

edge

← 1 unit →

A. Build a rectangular prism with 2 unit cubes.

Think: When the 2 cubes are pushed together, the faces and edges that are pushed together make 1 face and 1 edge.

- How many faces does the rectangular prism have? _____

- How many edges does the rectangular prism have? _____

B. Build as many different rectangular prisms as you can with 8 unit cubes.

C. Record in units the dimensions of each rectangular prism you built with 8 cubes.

Dimensions		

So, with 8 unit cubes, I can build _____ different rectangular prisms.

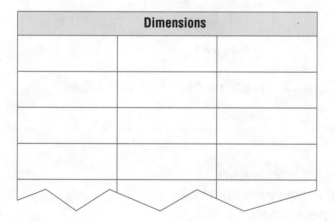

Math Talk MATHEMATICAL PRACTICES
Describe the different rectangular prisms that you can make with 4 unit cubes.

Draw Conclusions

1. **Explain** why a rectangular prism composed of 2 unit cubes has 6 faces. How do its dimensions compare to a unit cube?

_____

_____

_____

2. **Explain** how the number of edges for the rectangular prism compares to the number of edges for the unit cube.

_____

_____

3. **Describe** what all of the rectangular prisms you made in Step B have in common.

_____

_____

Make Connections

You can build other solid figures and compare the solid figures by counting the number of unit cubes.

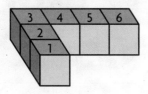

Figure 1

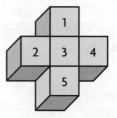

Figure 2

Figure 1 is made up of _____ unit cubes.   Figure 2 is made up of _____ unit cubes.

So, Figure _____ has more unit cubes than Figure _____.

- Use 12 unit cubes to build a solid figure that is not a rectangular prism. Share your model with a partner. Describe how your model is the same and how it is different from your partner's model.

_____

_____

Share and Show

Count the number of cubes used to build each solid figure.

1. The rectangular prism is made up of _____ unit cubes.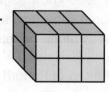

2.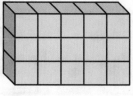

   _____ unit cubes

3.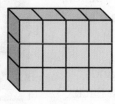

   _____ unit cubes

4.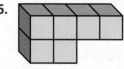

   _____ unit cubes

5.

   _____ unit cubes

6.

   _____ unit cubes

7.

   _____ unit cubes

8. **Write Math** ▶ How are the rectangular prisms in Exercises 3–4 related? Can you show a different rectangular prism with the same relationship? **Explain**.

   _____

   _____

Compare the number of unit cubes in each solid figure. Use <, > or =.

9.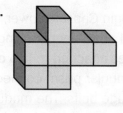

   _____ unit cubes ◯ _____ unit cubes

10.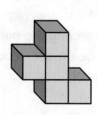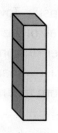

    _____ unit cubes ◯ _____ unit cubes

Architecture is the art and science of designing buildings and structures. An architect is a person who plans and designs the buildings.

Good architects are both artists and engineers. They must have a good knowledge of building construction, and they should know how to design buildings that meet the needs of the people who use them.

The Cube Houses of Rotterdam in the Netherlands, shown at the top right, were built in the 1970s. Each cube is a house, tilted and resting on a hexagon-shaped pylon, and is meant to represent an abstract tree. The village of Cube Houses creates a "forest".

The Nakagin Capsule Tower, shown at the right, is an office and apartment building in Tokyo, Japan, made up of modules attached to two central cores. Each module is a rectangular prism connected to a concrete core by four huge bolts. The modules are office and living spaces that can be removed or replaced.

Use the information to answer the questions.

11. There are 38 Cube Houses. Each house could hold 1,000 unit cubes that are 1 meter by 1 meter by 1 meter. Describe the dimensions of a cube house using unit cubes. Remember that the edges of a cube are all the same length.

_____

_____

12. **H.O.T.** The Nakagin Capsule Tower has 140 modules, and is 14 stories high. If all of the modules were divided evenly among the number of stories, how many modules would be on each floor? How many different rectangular prisms could be made from that number?

_____

Name _____

Understand Volume

Essential Question How can you use unit cubes to find the volume of a rectangular prism?

Investigate

CONNECT You can find the volume of a rectangular prism by counting unit cubes. **Volume** is the measure of the amount of space a solid figure occupies and is measured in **cubic units**. Each unit cube has a volume of 1 cubic unit.

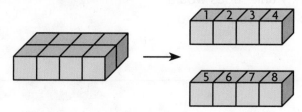

The rectangular prism above is made up of _____ unit cubes

and has a volume of _____ cubic units.

Materials ■ rectangular prism net A ■ centimeter cubes

A. Cut out, fold, and tape the net to form a rectangular prism.

B. Use centimeter cubes to fill the base of the rectangular prism without gaps or overlaps. Each centimeter cube has a length, width, and height of 1 centimeter and a volume of 1 cubic centimeter.

- How many centimeter cubes make up the length of the first layer? the width? the height?

  length: _____ width: _____ height: _____

- How many centimeter cubes are used to fill the base? _____

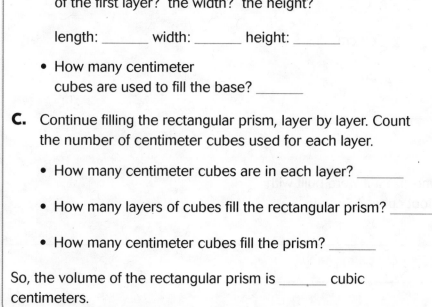

C. Continue filling the rectangular prism, layer by layer. Count the number of centimeter cubes used for each layer.

- How many centimeter cubes are in each layer? _____

- How many layers of cubes fill the rectangular prism? _____

- How many centimeter cubes fill the prism? _____

So, the volume of the rectangular prism is _____ cubic centimeters.

Draw Conclusions

1. **Describe** the relationship among the number of centimeter cubes you used to fill each layer, the number of layers, and the volume of the prism.

_____

_____

_____

2. **Apply** If you had a rectangular prism that had a length of 3 units, a width of 4 units, and a height of 2 units, how many unit cubes would you need for each layer? How many unit cubes would you need to fill the rectangular prism?

_____

Make Connections

To find the volume of three-dimensional figures, you measure in three directions. For a rectangular prism, you measure its length, width, and height. Volume is measured using cubic units, such as cu cm, cu in., or cu ft.

1 cu cm

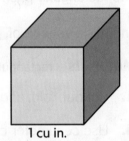
1 cu in.

- Which has a greater volume, 1 cu cm or 1 cu in.? **Explain**.

_____

_____

Find the volume of the prism if each cube represents 1 cu cm, 1 cu in., and 1 cu ft.

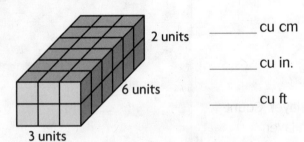
2 units
6 units
3 units

_____ cu cm

_____ cu in.

_____ cu ft

- Would the prism above be the same size if it were built with centimeter cubes, inch cubes, or foot cubes? **Explain**.

_____

_____

_____

Name _____

Share and Show ....

Use the unit given. Find the volume.

1.

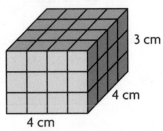

Each cube = 1 cu cm

Volume = _____ cu _____

2.

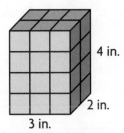

Each cube = 1 cu in.

Volume = _____ cu _____

3.

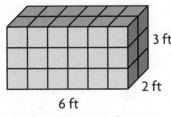

Each cube = 1 cu ft

Volume = _____ cu _____

4.

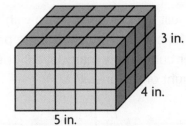

Each cube = 1 cu in.

Volume = _____ cu _____

Compare the volumes. Write <, >, or =.

5.

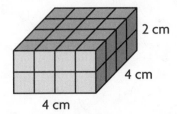

Each cube = 1 cu cm

6.

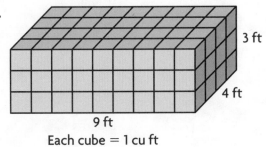

Each cube = 1 cu ft

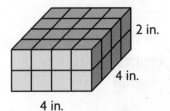

Each cube = 1 cu in.

_____ cu cm ◯ _____ cu in.

Each cube = 1 cu ft

_____ cu ft ◯ _____ cu ft

Problem Solving · REAL WORLD

7. What's the Error? Jerry says that a cube with edges that measure 10 centimeters has a volume that is twice as much as a cube with sides that measure 5 centimeters. **Explain** and correct Jerry's error.

_____

_____

_____

_____

_____

8. H.O.T. Pattie built a rectangular prism with cubes. The base of her prism has 12 centimeter cubes. If the prism was built with 108 centimeter cubes, how many layers does her prism have? What is the height of her prism?

_____

9. A packing company makes boxes with edges each measuring 3 feet. What is the volume of the boxes? If 10 boxes are put in a larger, rectangular shipping container and completely fill it with no gaps or overlaps, what is the volume of the shipping container?

_____

10. Test Prep Find the volume of the rectangular prism.

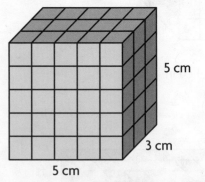

5 cm

3 cm

5 cm

Each cube = 1 cu cm

- Ⓐ 25 cubic feet
- Ⓑ 25 cubic meters
- Ⓒ 75 cubic meters
- Ⓓ 75 cubic centimeters

Name _____

Estimate Volume

Essential Question How can you use an everyday object to estimate the volume of a rectangular prism?

Investigate

Izzy is mailing 20 boxes of crayons to a children's-education organization overseas. She can pack them in one of two different-sized shipping boxes. Using crayon boxes as a cubic unit, about what volume is each shipping box, in crayon boxes? Which shipping box should Izzy use to mail the crayons?

Materials ■ rectangular prism net B ■ 2 boxes, different sizes

A. Cut out, fold, and tape the net to form a rectangular prism. Label the prism "Crayons." You can use this prism to estimate and compare the volume of the two boxes.

B. Using the crayon box that you made, count to find the number of boxes that make up the base of the shipping box. Estimate the length to the nearest whole unit.

Number of crayon boxes that fill the base:

Box 1: _____   Box 2: _____

C. Starting with the crayon box in the same position, count to find the number of crayon boxes that make up the height of the shipping box. Estimate the height to the nearest whole unit.

Number of layers:

Box 1: _____   Box 2: _____

Box 1 has a volume of _____ crayon boxes

and Box 2 has a volume of _____ crayon boxes.

So, Izzy should use Box _____ to ship the crayons.

Draw Conclusions ....

1. **Explain** how you estimated the volume of the shipping boxes.

   _____

   _____

   _____

   _____

2. **Analyze** If you had to estimate to the nearest whole unit to find the volume of a shipping box, how might you be able to ship a greater number of crayon boxes in the shipping box than you actually estimated? **Explain**.

   _____

   _____

   _____

Make Connections ....

The crayon box has a length of 3 inches, a width of 4 inches, and a height of 1 inch. The volume of the

crayon box is _____ cubic inches.

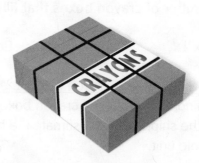

Using the crayon box, estimate the volume of the box at the right in cubic inches.

- The box to the right holds _____ crayon boxes in each

  of _____ layers, or _____ crayon boxes.

- Multiply the volume of 1 crayon box by the estimated number of crayon boxes that fit in the box at the right.

  _____ × _____ = _____

So, the volume of the shipping box at the right

is about _____ cubic inches.

Name _____

Share and Show

Estimate the volume.

1. Each tissue box has a volume of 125 cubic inches.

   There are _____ tissue boxes in the larger box.

   The estimated volume of the box holding the tissue

   boxes is _____ × 125 = _____ cu in.

2. Volume of chalk box: 16 cu in.

   Volume of large box: _____

3. Volume of small jewelry box: 30 cu cm

   Volume of large box: _____

On Your Own

Estimate the volume.

4. Volume of book: 80 cu in.

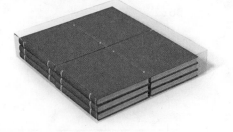

   Volume of large box: _____

5. Volume of spaghetti box: 750 cu cm

   Volume of large box: _____

6. Volume of cereal box: 324 cu in.

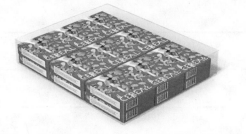

   Volume of large box: _____

7. Volume of pencil box: 4,500 cu cm

   Volume of large box: _____

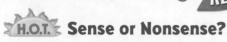
Problem Solving REAL WORLD

H.O.T. Sense or Nonsense?

8. Marcelle estimated the volume of the two boxes below, using one of his books. His book has a volume of 48 cubic inches. Box 1 holds about 7 layers of books, and Box 2 holds about 14 layers of books. Marcelle says that the volume of either box is about the same.

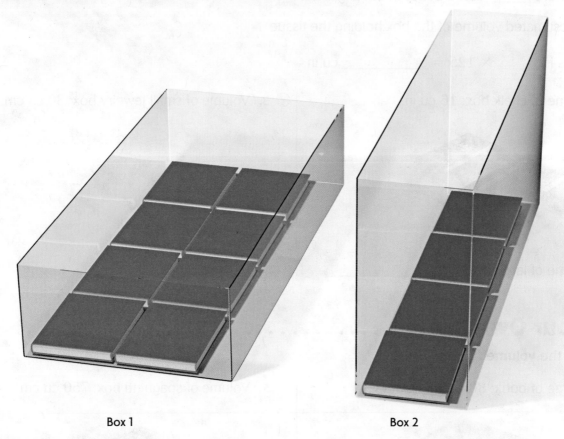

Box 1   Box 2

- Does Marcelle's statement make sense or is it nonsense? **Explain** your answer.

_____

_____

_____

_____

_____

_____

FOR MORE PRACTICE:
Standards Practice Book, pp. P233–P234

Name _____

Volume of Rectangular Prisms

Essential Question How can you find the volume of a rectangular prism?

CONNECT The base of a rectangular prism is a rectangle. You know that area is measured in square units, or units^2, and that the area of a rectangle can be found by multiplying the length and the width.

Volume is measured in cubic units, or units^3. When you build a prism and add each layer of cubes, you are adding a third dimension, height.

The area of the base

is _____ sq units.

🔑 UNLOCK the Problem REAL WORLD

Sid built the rectangular prism shown at the right, using 1-inch cubes. The prism has a base that is a rectangle and has a height of 4 cubes. What is the volume of the rectangular prism that Sid built?

You can find the volume of a prism in cubic units by multiplying the number of square units in the base shape by the number of layers, or its height.

Each layer of Sid's rectangular prism

is composed of _____ inch cubes.

+12

+12

+12

12

Height (in layers)	1	2	3	4
Volume (in cubic inches)	12	24		

Multiply the height by _____.

1. How does the volume change as each layer is added?

_____

_____

2. What does the number you multiply the height by represent?

_____

_____

So, the volume of Sid's rectangular prism is _____ in.^{3}

Relate Height to Volume

Toni stacks cube-shaped beads that measure 1 centimeter on each edge in a storage box. The box can hold 6 layers of 24 cubes with no gaps or overlaps. What is the volume of Toni's storage box?

- What are the dimensions of the base of the box?

  _____

- What operation can you use to find the area of the base shape?

  _____

🔒 One Way Use base and height.

The volume of each bead is _____ cm³.

The storage box has a base with an area of _____ cm².

The height of the storage box is _____ centimeters.

The volume of the storage box is

(_____ × _____), or _____ cm³.
  Base
  area

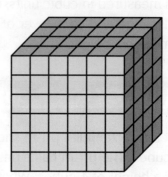

🔒 Another Way Use length, width, and height.

You know that the area of the base of the storage box is 24 cm².

The base has a length of _____ centimeters

and a width of _____ centimeters. The height

is _____ centimeters. The volume of the storage box is

(_____ × _____) × _____ , or _____ × _____ , or _____ cm³.
  Base area

So, the volume of the storage box is _____ cm³.

---

3. 🌟H.O.T.🌟 **What if** each cube-shaped bead measured 2 centimeters on each edge? How would the dimensions of the storage box change? How would the volume change?

_____

_____

_____

Name _____

Share and Show

Find the volume.

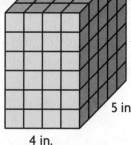

1. The length of the rectangular prism is _____.

   The width is _____. So, the area of the base is _____.

   The height is _____. So, the volume of the prism is _____.

6 in.

5 in.

4 in.

✓ 2.

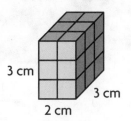

3 cm

3 cm

2 cm

Volume: _____

✓ 3.

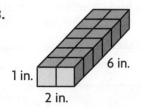

6 in.

1 in.

2 in.

Volume: _____

MATHEMATICAL PRACTICES

Math Talk Explain why the exponent 2 is used to express the measure of area and the exponent 3 is used to express the measure of volume.

On Your Own

Find the volume.

4.

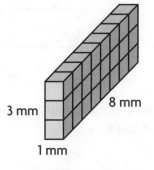

8 mm

3 mm

1 mm

Volume: _____

5.

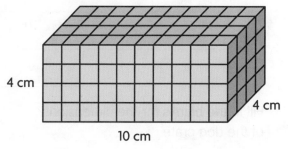

4 cm

4 cm

10 cm

Volume: _____

6.

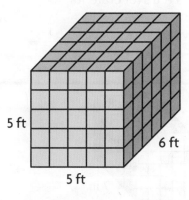

5 ft

6 ft

5 ft

Volume: _____

7.

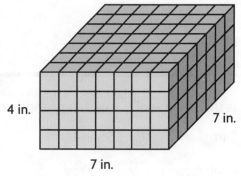

4 in.

7 in.

7 in.

Volume: _____

🔑 UNLOCK the Problem REAL WORLD

8. Rich is building a travel crate for his dog, Thomas, a beagle-
mix who is about 30 inches long, 12 inches wide, and 24
inches tall. For Thomas to travel safely, his crate needs to be
a rectangular prism that is about 12 inches greater than his
length and width, and 6 inches greater than his height. What
is the volume of the travel crate that Rich should build?

a. What do you need to find to solve the problem?

_____

_____

b. How can you use Thomas's size to help you solve the problem?

_____

_____

_____

c. What steps can you use to find the size of Thomas's crate?

_____

_____

_____

d. Fill in the blanks for the dimensions
of the dog crate.

length: _____

width: _____

height: _____

area of base: _____

e. Find the volume of the crate by multiplying
the base area and the height.

_____ × _____ = _____

So, Rich should build a travel crate for

Thomas that has a volume of _____.

9. What is the volume of the rectangular prism at the right?

Ⓐ 35 in.^{3} Ⓒ 155 in.^{3}

Ⓑ 125 in.^{3} Ⓓ 175 in.^{3}

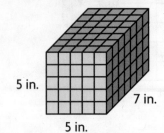

5 in.

7 in.

5 in.

© Houghton Mifflin Harcourt Publishing Company

Name _____

Apply Volume Formulas

Essential Question How can you use a formula to find the volume of a rectangular prism?

CONNECT Both prisms show the same dimensions and have the same volume.

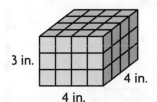

3 in.

4 in.

4 in.

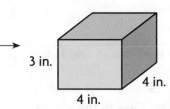

3 in.

4 in.

4 in.

The Earth

 UNLOCK the Problem REAL WORLD

Mike is making a box to hold his favorite DVDs. The length of the box is 7 inches, the width is 5 inches and the height is 3 inches. What is the volume of the box Mike is making?

- Underline what you are asked to find.
- Circle the numbers you need to use to solve the problem.

One Way Use length, width, and height.

You can use a formula to find the volume of a rectangular prism.

> Volume = length × width × height
>
> $V = l \times w \times h$

STEP 1 Identify the length, width, and height of the rectangular prism.

length = _____ in.

width = _____ in.

height = _____ in.

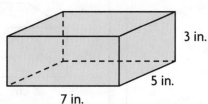

3 in.

5 in.

7 in.

STEP 2 Multiply the length by the width.

_____ × _____ = _____

STEP 3 Multiply the product of the length and width by the height.

35 × _____ = _____

So, the volume of Mike's DVD box is _____ cubic inches.

 MATHEMATICAL PRACTICES

Math Talk Explain how you can use the Associative Property to group the part of the formula that represents area.

You have learned one formula for finding the volume of a rectangular prism. You can also use another formula.

> Volume = Base area × height
> $V = B \times h$
> B = area of the base shape,
> h = height of the solid figure.

🔑 Another Way Use the area of the base shape and height.

Emilio's family has a sand castle kit. The kit includes molds for several solid figures that can be used to make sand castles. One of the molds is a rectangular prism like the one shown at the right. How much sand will it take to fill the mold?

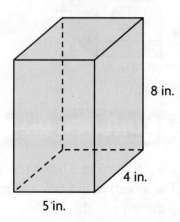

8 in.

4 in.

5 in.

V = _____ B _____ × h   Replace B with an expression for the area of the base shape. Replace h with the height of the solid figure.

V = (_____ × _____) × _____   Multiply.

V = _____ × _____

V = _____ cu in.

So, it will take _____ cubic inches of sand to fill the rectangular prism mold.

Try This!

Ⓐ Find the volume.

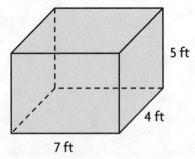

5 ft

4 ft

7 ft

V = l × w × h

V = _____ × _____ × _____

V = _____ × _____

V = _____ cu ft

Ⓑ Find the unknown measurement.

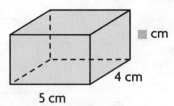

■ cm

4 cm

5 cm

V = l × w × h

60 = _____ × _____ × ■

60 = _____ × ■

Think: If I filled this prism with centimeter cubes, each layer would have 20 cubes. How many layers of 20 cubes are equal to 60?

So, the unknown measurement is _____ cm.

480

Name _____

Share and Show

Find the volume.

1.

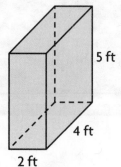

5 ft

4 ft

2 ft

V = _____

2.

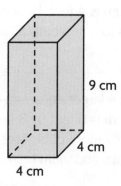

9 cm

4 cm

4 cm

V = _____

On Your Own

Find the volume.

3.

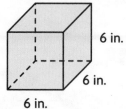

6 in.

6 in.

6 in.

V = _____

4.

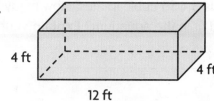

4 ft

4 ft

12 ft

V = _____

5.

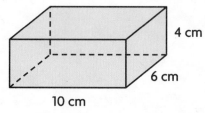

4 cm

6 cm

10 cm

V = _____

6.

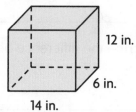

12 in.

6 in.

14 in.

V = _____

 Algebra Find the unknown measurement.

7.

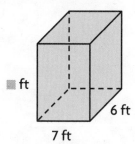

■ ft

6 ft

7 ft

V = 420 cu ft   ■ = _____ ft

8.

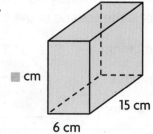

■ cm

15 cm

6 cm

V = 900 cu cm   ■ = _____ cm

Problem Solving REAL WORLD

9. The Jade Restaurant has a large aquarium on display in its lobby. The base of the aquarium is 5 feet by 2 feet. The height of the aquarium is 4 feet. How many cubic feet of water are needed to completely fill the aquarium?

_____

10. The Pearl Restaurant put a larger aquarium in its lobby. The base of their aquarium is 6 feet by 3 feet, and the height is 4 feet. How many more cubic feet of water does the Pearl Restaurant's aquarium hold than the Jade Restaurant's aquarium?

_____

11. **H.O.T.** Eddie measured his aquarium using a small fish food box. The box has a base area of 6 inches and a height of 4 inches. Eddie found that the volume of his aquarium is 3,456 cubic inches. How many boxes of fish food could fit in the aquarium? **Explain** your answer.

_____

_____

_____

_____

· · · · · SHOW YOUR WORK · · · · ·

12. **Write Math** ▶ **Describe** the difference between area and volume.

_____

_____

_____

13. **Test Prep** Adam stores his favorite CDs in a box like the one at the right. What is the volume of the box?

(A) 150 cubic centimeters

(B) 750 cubic centimeters

(C) 1,050 cubic centimeters

(D) 1,150 cubic centimeters

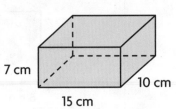

7 cm

10 cm

15 cm

FOR MORE PRACTICE:
Standards Practice Book, pp. P237–P238

Problem Solving • Compare Volumes

Essential Question: How can you use the strategy *make a table* to compare different rectangular prisms with the same volume?

🔑 UNLOCK the Problem › REAL WORLD

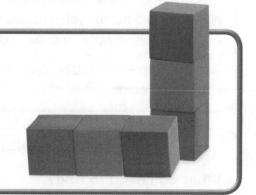

Adam has 50 one-inch cubes. The cubes measure 1 inch on each edge. Adam wonders how many rectangular prisms, each with a different-size base, that he could make with all of the one-inch cubes.

Use the graphic organizer below to help you solve the problem.

Read the Problem	Solve the Problem

Read the Problem

What do I need to find?

I need to find the number of _____,

each with a different-size _____, that have

a volume of _____.

What information do I need to use?

I can use the formula _____

_____ and the factors of _____.

How will I use the information?

I will use the formula and the factors of

50 in a _____ that shows all of the
possible combinations of dimensions with a

volume of _____ without repeating
the dimensions of the bases.

Solve the Problem

Complete the table.

Base (sq in.)	Height (in.)	Volume (cu in.)
(1 × 1)	50	(1 × 1) × 50 = 50
(1 × 2)	25	(1 × 2) × 25 = 50
(1 × 5)	10	(1 × 5) × 10 = 50
(1 × 10)	5	(1 × 10) × 5 = 50
(1 × 25)	2	(1 × 25) × 2 = 50
(1 × 50)	1	(1 × 50) × 1 = 50

1. What else do you need to do to solve the problem? _____

_____

2. How many rectangular prisms with different bases can Adam make

using fifty one-inch cubes? _____

🔑 Try Another Problem

Mrs. Wilton is planning a rectangular flower box for her front window. She wants the flower box to hold exactly 16 cubic feet of soil. How many different flower boxes, all with whole-number dimensions and a different-size base, will hold exactly 16 cubic feet of soil?

Use the graphic organizer below to help you solve the problem.

Read the Problem	Solve the Problem
What do I need to find?	
What information do I need to use?	
How will I use the information?	

Math Talk \text{MATHEMATICAL PRACTICES} Explain how a flower box with dimensions of $(1 \times 2) \times 8$ is different from a flower box with dimensions of $(2 \times 8) \times 1$.

3. How many flower boxes with different-size bases will hold exactly 16 cubic feet of soil, using whole-number dimensions?

_____

Name _____

Share and Show

🔓 UNLOCK the Problem   Tips

✓ Circle the question.
✓ Break the problem into easier steps.

1. Mr. Price makes cakes for special occasions. His most
popular-sized cakes have a volume of 360 cubic inches.
The cakes have a height, or thickness, of 3 inches, and
have different whole number lengths and widths. No cakes
have a length or width of 1 or 2 inches. How many different
cakes, each with a different-size base, have a volume of
360 cubic inches?

First, think about what the problem is asking you to solve,
and the information that you are given.

Next, make a table using the information from problem.

Finally, use the table to solve the problem.

_____

_____

_____

.... **SHOW YOUR WORK** ....

2. **H.O.T.** **What if** the 360 cubic-inch cakes are 4 inches thick
and any whole number length and width are possible? How
many different cakes could be made? Suppose that the cost
of a cake that size is $25, plus $1.99 for every 4 cubic inches
of cake. How much would the cake cost?

_____

_____

_____

3. One company makes inflatable swimming pools that come
in four sizes of rectangular prisms. The length of each pool is
twice the width and twice the depth. The depth of the pools
are each a whole number from 2 to 5 feet. If the pools are
filled all the way to the top, what is the volume of each pool?

_____

_____

_____

On Your Own.......

Choose a
STRATEGY

Act It Out
Draw a Diagram
Make a Table
Solve a Simpler Problem
Work Backward
Guess, Check, and Revise

4. Ray wants to buy the larger of two aquariums. One aquarium has a base that is 20 inches by 20 inches and a height that is 18 inches. The other aquarium has a base that is 40 inches by 12 inches and a height that is 12 inches. Which aquarium has a greater volume? By how much?

_____

5. Ken owns 13 CDs. His brother Keith has 7 more CDs than he does. Their brother, George, has more CDs than either of the younger brothers. Together, the three brothers have 58 CDs. How many CDs does George have?

_____

SHOW YOUR WORK

6. **H.O.T.** Kathy has ribbons that have lengths of 7 inches, 10 inches, and 12 inches. **Explain** how she can use these ribbons to measure a length of 15 inches.

_____

_____

_____

7. **H.O.T.** A park has a rectangular playground area that has a length of 66 feet and a width of 42 feet. The park department has 75 yards of fencing material. Is there enough fencing material to enclose the playground area? **Explain**.

_____

_____

_____

8. **Test Prep** John is making a chest that will have a volume of 1,200 cubic inches. The length is 20 inches and the width is 12 inches. How many inches tall will his chest be?

(A) 4 in.   (C) 6 in.

(B) 5 in.   (D) 7 in.

Find Volume of Composed Figures

Essential Question How can you find the volume of rectangular prisms that are combined?

 UNLOCK the Problem REAL WORLD

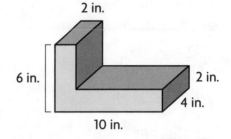

The shape at the right is a composite figure. It is made up of two rectangular prisms that are combined. How can you find the volume of the figure?

One Way Use addition.

STEP 1 Break apart the solid figure into two rectangular prisms.

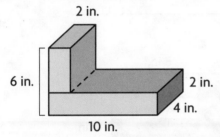

STEP 2 Find the length, width, and height of each prism.

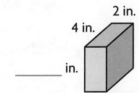 **Think:** The total height of both prisms is 6 inches. Subtract the given heights to find the unknown height. $6 - 2 = 4$

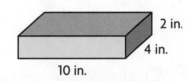

STEP 3 Find the volume of each prism.

$V = l \times w \times h$

$V = $ _____ $\times$ _____ $\times$ _____

$V = $ _____ in.^{3}

$V = l \times w \times h$

$V = $ _____ $\times$ _____ $\times$ _____

$V = $ _____ in.^{3}

STEP 4 Add the volumes of the rectangular prisms.

_____ $+$ _____ $=$ _____

So, the volume of the composite figure is _____ cubic inches.

- What is another way you could divide the composite figure into

  two rectangular prisms? _____

  _____

🔒 Another Way Use subtraction.

You can subtract the volumes of prisms formed in empty spaces from the greatest possible volume to find the volume of a composite figure.

STEP 1

Find the greatest possible volume.

length = _____ in.

width = _____ in.

height = _____ in.

V = _____ cubic inches

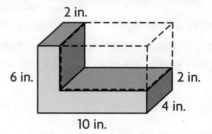

STEP 2

Find the volume of the prism in the empty space.

length = _____ in.   **Think:** 10 − 2 = 8

width = _____ in.

height = _____ in.   **Think:** 6 − 2 = 4

V = 8 × 4 × 4 = _____ cubic inches

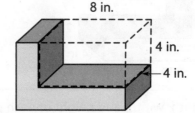

STEP 3

Subtract the volume of the empty space from the greatest possible volume.

_____ − _____ = _____ cubic inches

So, the volume of the composite figure is _____ cubic inches.

Try This!

Find the volume of a composite figure made by putting together three rectangular prisms.

V = _____ × _____ × _____ = _____ cu ft

V = _____ × _____ × _____ = _____ cu ft

V = _____ × _____ × _____ = _____ cu ft

Total volume = _____ + _____ + _____ = _____ cubic feet

© Houghton Mifflin Harcourt Publishing Company

Name _____

Share and Show

Find the volume of the composite figure.

1.

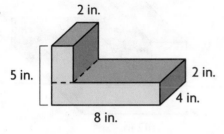

2 in.

5 in.   2 in.

4 in.

8 in.

V = _____

2.

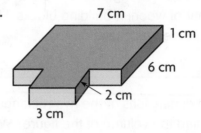

7 cm

1 cm

6 cm

2 cm

3 cm

V = _____

On Your Own

Find the volume of the composite figure.

3.

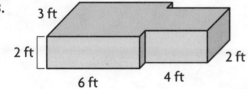

3 ft

2 ft   2 ft

6 ft   4 ft

V = _____

4.

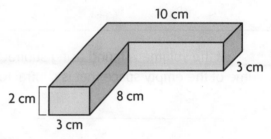

10 cm

3 cm

2 cm   8 cm

3 cm

V = _____

5.

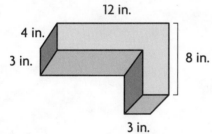

12 in.

4 in.

3 in.   8 in.

3 in.

V = _____

6.

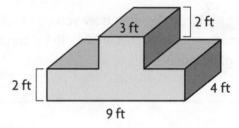

3 ft   2 ft

2 ft   4 ft

9 ft

V = _____

7.

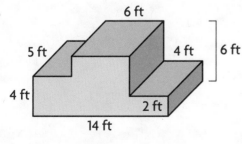

6 ft

5 ft   4 ft   6 ft

4 ft   2 ft

14 ft

V = _____

8.

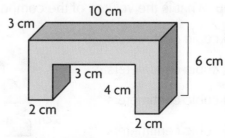

3 cm   10 cm

3 cm

4 cm

2 cm   6 cm

2 cm

V = _____

Problem Solving REAL WORLD

Use the composite figure at the right for 9–11.

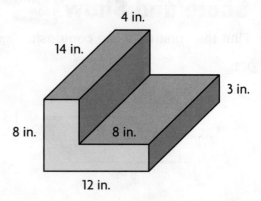

9. As part of a wood-working project, Jordan made the figure at the right out of wooden building blocks. How much space does the figure he made take up?

_____

10. What are the dimensions of the two rectangular prisms you used to find the volume of the figure? What other rectangular prisms could you have used?

_____

_____

_____

11. **H.O.T.** If the volume is found using subtraction, what is the volume of the empty space that is subtracted? **Explain**.

_____

_____

_____

12. **Write Math** ▶ **Explain** how you can find the volume of composite figures that are made by combining rectangular prisms.

_____

_____

_____

13. **Test Prep** What is the volume of the composite figure?

Ⓐ 126 cubic centimeters

Ⓑ 350 cubic centimeters

Ⓒ 450 cubic centimeters

Ⓓ 476 cubic centimeters

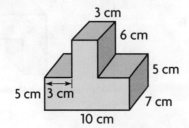

FOR MORE PRACTICE:
Standards Practice Book, pp. P241–P242

Name _____

Chapter Review/Test

▶ **Vocabulary**

Choose the best term from the box.

Vocabulary

polyhedron

prism

pyramid

1. A _____ has two congruent polygons as bases
   and rectangular lateral faces. (p. 457)

2. A _____ has only one base and triangular lateral faces. (p. 458)

▶ **Concepts and Skills**

**Name each polygon. Then tell whether it is a *regular polygon*
or *not a regular polygon*.**

3.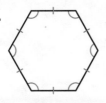

_____

4.

_____

5.

_____

6.

_____

Classify each figure in as many ways as possible.

7.

_____

8.

_____

Classify the solid figure. Write *prism, pyramid, cone, cylinder,* or *sphere*.

9.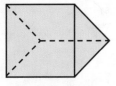

_____

10.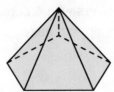

_____

Count the number of cubes used to build each solid figure.

11.

_____ unit cubes

12.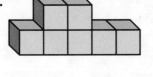

_____ unit cubes

13.

_____ unit cubes

GO Online — **Assessment Options** **Chapter Test**

Fill in the bubble completely to show your answer.

14. What type of triangle is shown below?

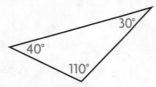

Ⓐ acute; isosceles

Ⓑ acute; scalene

Ⓒ obtuse; scalene

Ⓓ obtuse; isosceles

15. Angela buys a paperweight at the local gift shop. The paperweight is in the shape of a hexagonal pyramid.

Which of the following represents the correct number of faces, edges, and vertices in a hexagonal pyramid?

Ⓐ 6 faces, 12 edges, 18 vertices

Ⓑ 7 faces, 7 edges, 12 vertices

Ⓒ 7 faces, 12 edges, 7 vertices

Ⓓ 8 faces, 18 edges, 12 vertices

16. A manufacturing company constructs a shipping box to hold its cereal boxes. Each cereal box has a volume of 40 cubic inches. If the shipping box holds 8 layers with 4 cereal boxes in each layer, what is the volume of the shipping box?

Ⓐ 160 cu in.

Ⓑ 320 cu in.

Ⓒ 480 cu in.

Ⓓ 1,280 cu in.

Name _____

Fill in the bubble completely to show your answer.

17. Sharri packed away her old summer clothes in a storage tote
that had a length of 3 feet, a width of 4 feet, and a height of 3 feet.
What was the volume of the tote that Sharri used?

   Ⓐ 36 cu ft

   Ⓑ 24 cu ft

   Ⓒ 21 cu ft

   Ⓓ 10 cu ft

18. Which quadrilateral is NOT classified as a parallelogram?

   Ⓐ
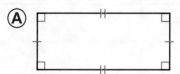

   Ⓑ

   Ⓒ
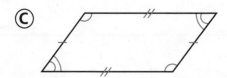

   Ⓓ
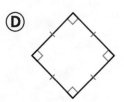

19. What is the volume of the composite figure below?

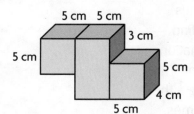

   Ⓐ 1,875 cm³   Ⓒ 360 cm³

   Ⓑ 480 cm³   Ⓓ 150 cm³

20. A video game store made a display of game console boxes shown at the right. The length, width, and height of each game console box is 2 feet.

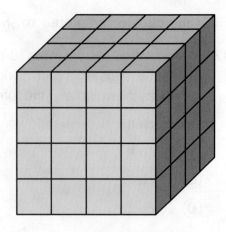

What is the volume of the display of game console boxes? Show your work and explain your answer.

_____

_____

_____

_____

On a busy Saturday, the video game store sold 22 game consoles. What is the volume of the game console boxes that are left?

_____

_____

▶ **Performance Task**

21. Look for two pictures of three-dimensional buildings in newspapers and magazines. The buildings should be rectangular prisms.

Ⓐ Paste the pictures on a large sheet of paper. Leave room to write information near the picture.

Ⓑ Label each building with their name and location.

Ⓒ Research the buildings, if the information is available. Find things that are interesting about the buildings or their location. Also find their length, width, and height to the nearest foot. If the information is not available, measure the buildings on the page in inches or centimeters, and make a good estimate of their width (such as $\frac{1}{2}$ the height, rounded to the nearest whole number). Find their volumes.

Ⓓ Make a class presentation, choosing one of the buildings you found.

Glossary

Pronunciation Key

a	add, map	ē	equal, tree	m	move, seem	ōō	pool, food	u̇	pull, book
ā	ace, rate	f	fit, half	n	nice, tin	p	pit, stop	û(r)	burn, term
â(r)	care, air	g	go, log	ng	ring, song	r	run, poor	yōō	fuse, few
ä	palm, father	h	hope, hate	o	odd, hot	s	see, pass	v	vain, eve
b	bat, rub	i	it, give	ō	open, so	sh	sure, rush	w	win, away
ch	check, catch	ī	ice, write	ô	order, jaw	t	talk, sit	y	yet, yearn
d	dog, rod	j	joy, ledge	oi	oil, boy	th	thin, both	z	zest, muse
e	end, pet	k	cool, take	ou	pout, now	th	this, bathe	zh	vision, pleasure
		l	look, rule	o͝o	took, full	u	up, done		

ə   the schwa, an unstressed vowel representing the sound spelled a in above, e in sicken, i in possible, o in melon, u in circus

Other symbols:
• separates words into syllables
′ indicates stress on a syllable

A

acute angle [ə•kyōōt′ ang′gəl] **ángulo agudo** An angle that has a measure less than a right angle (less than 90° and greater than 0°)
Example:

Word History

The Latin word for needle is *acus*. This means "pointed" or "sharp." You will recognize the root in the words *acid* (sharp taste), *acumen* (mental sharpness), and *acute*, which describes a sharp or pointed angle.

acute triangle [ə•kyōōt′ trī′ang•gəl] **triángulo acutángulo** A triangle that has three acute angles

addend [ad′end] **sumando** A number that is added to another in an addition problem

addition [ə•dish′ən] **suma** The process of finding the total number of items when two or more groups of items are joined; the opposite of subtraction

algebraic expression [al•jə•brā′ik ek•spresh′ən] **expresión algebraica** An expression that includes at least one variable
Examples: x + 5, 3a − 4

angle [ang′gəl] **ángulo** A shape formed by two line segments or rays that share the same endpoint
Example:

area [âr′ē•ə] **área** The measure of the number of unit squares needed to cover a surface

array [ə•rā′] **matriz** An arrangement of objects in rows and columns
Example:

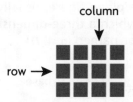

column

row →

Student Handbook   H1

Associative Property of Addition [ə•sō′shē•āt•iv präp′ ər•tē əv ə•dish′ən] **propiedad asociativa de la suma** The property that states that when the grouping of addends is changed, the sum is the same
Example: (5 + 8) + 4 = 5 + (8 + 4)

Associative Property of Multiplication [ə•sō′shē•āt•iv präp′ər•tē əv mul•tə•pli•kā′shən] **propiedad asociativa de la multiplicación** The property that states that factors can be grouped in different ways and still get the same product
Example: (2 × 3) × 4 = 2 × (3 × 4)

balance [bal′əns] **equilibrar** To equalize in weight or number

bar graph [bär graf] **gráfica de barras** A graph that uses horizontal or vertical bars to display countable data
Example:

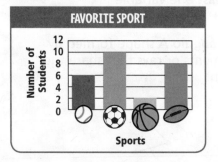

base (arithmetic) [bās] **base** A number used as a repeated factor (p. 17)
Example: $8^3 = 8 \times 8 \times 8$. The base is 8.

base (geometry) [bās] **base** In two dimensions, one side of a triangle or parallelogram that is used to help find the area. In three dimensions, a plane figure, usually a polygon or circle, by which a three-dimensional figure is measured or named (p. 457)
Examples:

benchmark [bench′märk] **punto de referencia** A familiar number used as a point of reference

capacity [kə•pas′i•tē] **capacidad** The amount a container can hold when filled (p. 409)

Celsius (°C) [sel′sē•əs] **Celsius (°C)** A metric scale for measuring temperature

centimeter (cm) [sen′tə•mēt•ər] **centímetro (cm)** A metric unit used to measure length or distance; 0.01 meter = 1 centimeter

closed figure [klōzd fig′yər] **figura cerrada** A figure that begins and ends at the same point

common denominator [käm′ən dē•näm′ə•nāt•ər] **denominador común** A common multiple of two or more denominators (p. 255)
Example: Some common denominators for $\frac{1}{4}$ and $\frac{5}{6}$ are 12, 24, and 36.

common factor [käm′ən fak′tər] **factor común** A number that is a factor of two or more numbers

common multiple [käm′ən mul′tə•pəl] **múltiplo común** A number that is a multiple of two or more numbers

Commutative Property of Addition [kə•myōōt′ə•tiv präp′ər•tē əv ə•dish′ən] **propiedad conmutativa de la suma** The property that states that when the order of two addends is changed, the sum is the same
Example: 4 + 5 = 5 + 4

Commutative Property of Multiplication [kə•myōōt′ə•tiv präp′ər•tē əv mul•tə•pli•kā′shən] **propiedad conmutativa de la multiplicación** The property that states that when the order of two factors is changed, the product is the same
Example: 4 × 5 = 5 × 4

compatible numbers [kəm•pat′ə•bəl num′bərz] **números compatibles** Numbers that are easy to compute with mentally

composite number [kəm•päz′it num′bər] **número compuesto** A number having more than two factors
Example: 6 is a composite number, since its factors are 1, 2, 3, and 6.

cone [kōn] **cono** A solid figure that has a flat, circular base and one vertex
Example:

congruent [kən·grŌŌ′ənt] **congruente** Having the same size and shape (p. 442)

coordinate grid [kō·ôrd′n·it grid] **cuadrícula de coordenadas** A grid formed by a horizontal line called the *x*-axis and á vertical line called the *y*-axis (p. 373)
Example:

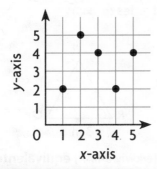

counting number [kount′ing num′bər] **número natural** A whole number that can be used to count a set of objects (1, 2, 3, 4, . . .)

cube [kyŌŌb] **cubo** A three-dimensional figure with six congruent square faces
Example:

cubic unit [kyŌŌ′bik yŌŌ′nit] **unidad cúbica** A unit used to measure volume such as cubic foot (ft³), cubic meter (m³), and so on (p. 467)

cup (c) [kup] **taza (t)** A customary unit used to measure capacity; 8 ounces = 1 cup

cylinder [sil′ən·dər] **cilindro** A solid figure that has two parallel bases that are congruent circles
Example:

data [dāt′ə] **datos** Information collected about people or things, often to draw conclusions about them

decagon [dek′ə·gän] **decágono** A polygon with ten sides and ten angles
Examples:

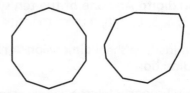

decagonal prism [dek·ag′ə·nəl priz′əm] **prisma decagonal** A three-dimensional figure with two decagonal bases and ten rectangular faces (p. 457)

decimal [des′ə·məl] **decimal** A number with one or more digits to the right of the decimal point

decimal point [des′ə·məl point] **punto decimal** A symbol used to separate dollars from cents in money, and to separate the ones place from the tenths place in a decimal

decimal system [des′ə·məl sis′təm] **sistema decimal** A system of computation based on the number 10

decimeter (dm) [des′i·mēt·ər] **decímetro (dm)** A metric unit used to measure length or distance; 10 decimeters = 1 meter

degree (°) [di·grē′] **grado (°)** A unit used for measuring angles and temperature

degree Celsius (°C) [di·grē′ sel′sē·əs] **grado Celcius** A metric unit for measuring temperature

degree Fahrenheit (°F) [di·grē′ fâr′ən·hīt] **grado Fahrenheit** A customary unit for measuring temperature

dekameter (dam) [dek′ə·mēt·ər] **decámetro** A metric unit used to measure length or distance; 10 meters = 1 dekameter (p. 423)

denominator [dē·näm′ə·nāt·ər] **denominador** The number below the bar in a fraction that tells how many equal parts are in the whole or in the group
Example: $\dfrac{3}{4}$ ← denominator

diagonal [dī·ag′ə·nəl] **diagonal** A line segment that connects two non-adjacent vertices of a polygon
Example:

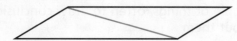

difference [dif′ər·əns] **diferencia** The answer to a subtraction problem

digit [dij′it] **dígito** Any one of the ten symbols 0, 1, 2, 3, 4, 5, 6, 7, 8, 9 used to write numbers

dimension [də·men′shən] **dimensión** A measure in one direction

Distributive Property [di·strib′yōō·tiv präp′ər·tē] **propiedad distributiva** The property that states that multiplying a sum by a number is the same as multiplying each addend in the sum by the number and then adding the products (p. 14)
Example: $3 \times (4 + 2) = (3 \times 4) + (3 \times 2)$
$$3 \times 6 = 12 + 6$$
$$18 = 18$$

divide [də·vīd′] **dividir** To separate into equal groups; the opposite operation of multiplication

dividend [div′ə·dend] **dividendo** The number that is to be divided in a division problem
Example: $36 \div 6$; $6\overline{)36}$ The dividend is 36.

division [də·vizh′ən] **división** The process of sharing a number of items to find how many equal groups can be made or how many items will be in each equal group; the opposite operation of multiplication

divisor [də·vī′zər] **divisor** The number that divides the dividend
Example: $15 \div 3$; $3\overline{)15}$ The divisor is 3.

edge [ej] **arista** The line segment made where two faces of a solid figure meet
Example:

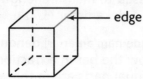

elapsed time [ē·lapst′ tīm] **tiempo transcurrido** The time that passes between the start of an activity and the end of that activity

endpoint [end′ point] **extremo** The point at either end of a line segment or the starting point of a ray

equal to (=) [ē′kwəl tōō] **igual a** Having the same value

equation [ē·kwā′zhən] **ecuación** An algebraic or numerical sentence that shows that two quantities are equal

equilateral triangle [ē·kwi·lat′ər·əl trī′ang·gəl] **triángulo equilátero** A triangle with three congruent sides (p. 445)
Example:

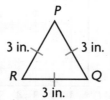

equivalent [ē·kwiv′ə·lənt] **equivalente** Having the same value

equivalent decimals [ē·kwiv′ə·lənt des′ə·məlz] **decimales equivalentes** Decimals that name the same amount
Example: $0.4 = 0.40 = 0.400$

equivalent fractions [ē·kwiv′ə·lənt frak′shənz] **fracciones equivalentes** Fractions that name the same amount or part
Example: $\frac{3}{4} = \frac{6}{8}$

estimate [es′tə·mit] *noun* **estimación (s)** A number close to an exact amount

estimate [es′tə·māt] *verb* **estimar (v)** To find a number that is close to an exact amount

evaluate [ē·val′yōō·āt] **evaluar** To find the value of a numerical or algebraic expression (p. 47)

even [ē′vən] **par** A whole number that has a 0, 2, 4, 6, or 8 in the ones place

expanded form [ek·span′did fôrm] **forma desarrollada** A way to write numbers by showing the value of each digit
Examples: $832 = 8 \times 100 + 3 \times 10 + 2 \times 1$
$$3.25 = (3 \times 1) + (2 \times \tfrac{1}{10}) + (5 \times \tfrac{1}{100})$$

exponent [eks′•pōn•ənt] **exponente** A number that shows how many times the base is used as a factor (p. 17)
Example: $10^3 = 10 \times 10 \times 10$.
3 is the exponent.

expression [ek•spresh′ən] **expresión** A mathematical phrase or the part of a number sentence that combines numbers, operation signs, and sometimes variables, but does not have an equal sign

face [fās] **cara** A polygon that is a flat surface of a solid figure
Example:

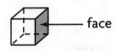
— face

fact family [fakt fam′ə•lē] **familia de operaciones** A set of related multiplication and division, or addition and subtraction, equations
Examples: $7 \times 8 = 56$; $8 \times 7 = 56$;
$56 \div 7 = 8$; $56 \div 8 = 7$

factor [fak′tər] **factor** A number multiplied by another number to find a product

Fahrenheit (°F) [fâr′ən•hīt] **Fahrenheit (°F)** A customary scale for measuring temperature

fluid ounce (fl oz) [floo′id ouns] **onza fluida** A customary unit used to measure liquid capacity; 1 cup = 8 fluid ounces

foot (ft) [foot] **pie (ft)** A customary unit used to measure length or distance; 1 foot = 12 inches

formula [fôr′myoo•lə] **fórmula** A set of symbols that expresses a mathematical rule
Example: $A = b \times h$

fraction [frak′shən] **fracción** A number that names a part of a whole or a part of a group

fraction greater than 1 [frak′shən grāt′ər than wun] **fracción mayor que 1** A number which has a numerator that is greater than its denominator
Example:

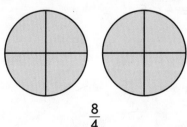

$\frac{8}{4}$

gallon (gal) [gal′ən] **galón (gal)** A customary unit used to measure capacity; 4 quarts = 1 gallon

general quadrilateral [jen′ər•əl kwä•dri•lat′ər•əl] **cuadrilátero en general** See *quadrilateral*.

gram (g) [gram] **gramo (g)** A metric unit used to measure mass; 1,000 grams = 1 kilogram

greater than (>) [grāt′ər than] **mayor que (>)** A symbol used to compare two numbers or two quantities when the greater number or greater quantity is given first
Example: $6 > 4$

greater than or equal to (≥) [grāt′ər than ôr ē′kwəl too] **mayor que o igual a** A symbol used to compare two numbers or quantities when the first is greater than or equal to the second

greatest common factor [grāt′əst käm′ən fak′tər] **máximo común divisor** The greatest factor that two or more numbers have in common
Example: 6 is the greatest common factor of 18 and 30.

grid [grid] **cuadrícula** Evenly divided and equally spaced squares on a figure or flat surface

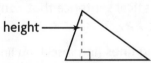

height [hīt] **altura** The length of a perpendicular from the base to the top of a two-dimensional or three-dimensional figure
Example:

height ⟶

heptagon [hep′tə•gän] **heptágono** A polygon with seven sides and seven angles (p. 441)

hexagon [hek′sə•gän] **hexágono** A polygon with six sides and six angles (p. 441)
Examples:

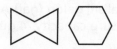

hexagonal prism [hek•sag′ə•nəl priz′əm] **prisma hexagonal** A three-dimensional figure with two hexagonal bases and six rectangular faces (p. 457)

horizontal [hôr•i•zänt′l] **horizontal** Extending left and right

hundredth [hun′drədth] **centésimo** One of 100 equal parts
Examples: 0.56, $\frac{56}{100}$, fifty-six hundredths

Identity Property of Addition [ī•den′tə•tē präp′ər•tē əv ə•dish′ən] **propiedad de identidad de la suma** The property that states that when you add zero to a number, the result is that number

Identity Property of Multiplication [ī•den′tə•tē präp′ər•tē əv mul•tə•pli•kā′shən] **propiedad de identidad de la multiplicación** The property that states that the product of any number and 1 is that number

inch (in.) [inch] **pulgada (pulg)** A customary unit used to measure length or distance; 12 inches = 1 foot

inequality [in•ē•kwôl′ə•tē] **desigualdad** A mathematical sentence that contains the symbol <, >, ≤, ≥, or ≠

intersecting lines [in•tər•sekt′ing līnz] **líneas secantes** Lines that cross each other at exactly one point
Example:

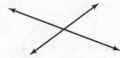

interval [in′tər•vəl] **intervalo** The difference between one number and the next on the scale of a graph (p. 381)

inverse operations [in′vûrs äp•ə•rā′shənz] **operaciones inversas** Opposite operations, or operations that undo each other, such as addition and subtraction or multiplication and division (p. 35)

isosceles triangle [ī•säs′ə•lēz trī′ang•gəl] **triángulo isósceles** A triangle with two congruent sides (p. 445)
Example:

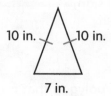

10 in.   10 in.

7 in.

key [kē] **clave** The part of a map or graph that explains the symbols

kilogram (kg) [kil′ō•gram] **kilogramo (kg)** A metric unit used to measure mass; 1,000 grams = 1 kilogram (p. 423)

kilometer (km) [kə•läm′ət•ər] **kilómetro (km)** A metric unit used to measure length or distance; 1,000 meters = 1 kilometer (p. 423)

lateral face [lat′ər•əl fās] **cara lateral** Any surface of a polyhedron other than a base (p. 457)

least common denominator [lēst käm′ən dē•näm′ə•nāt•ər] **mínimo común denominador** The least common multiple of two or more denominators
Example: The least common denominator for $\frac{1}{4}$ and $\frac{5}{6}$ is 12.

least common multiple [lēst käm′ən mul′tə•pəl] **mínimo común múltiplo** The least number that is a common multiple of two or more numbers

less than (<) [les <u>than</u>] **menor que (<)** A symbol used to compare two numbers or two quantities, with the lesser number given first
Example: 4 < 6

less than or equal to (≤) [les <u>than</u> ôr ē′kwəl tōō] **menor que o igual a** A symbol used to compare two numbers or two quantities, when the first is less than or equal to the second

line [līn] **línea** A straight path in a plane, extending in both directions with no endpoints
Example:

←————————————→

line graph [līn graf] **gráfica lineal** A graph that uses line segments to show how data change over time (p. 381)

line plot [līn plät] **diagrama de puntos** A graph that shows frequency of data along a number line
Example:

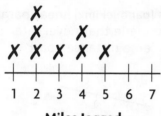

Miles Jogged

line segment [līn seg′mənt] **segmento** A part of a line that includes two points called endpoints and all the points between them

●————————————●

line symmetry [līn sim′ə•trē] **simetría axial** A figure has line symmetry if it can be folded about a line so that its two parts match exactly.

linear unit [lin′ē•ər yōō′nit] **unidad lineal** A measure of length, width, height, or distance

liquid volume [lik′wid väl′yōōm] **volumen de un líquido** The amount of liquid in a container

liter (L) [lēt′ər] **litro (L)** A metric unit used to measure capacity; 1 liter = 1,000 milliliters

mass [mas] **masa** The amount of matter in an object

meter (m) [mēt′ər] **metro (m)** A metric unit used to measure length or distance; 1 meter = 100 centimeters

mile (mi) [mīl] **milla (mi)** A customary unit used to measure length or distance; 5,280 feet = 1 mile

milligram (mg) [mil′i•gram] **milligramo** A metric unit used to measure mass; 1,000 milligrams = 1 gram (p. 423)

milliliter (mL) [mil′i•lēt•ər] **mililitro (mL)** A metric unit used to measure capacity; 1,000 milliliters = 1 liter

millimeter (mm) [mil′i•mēt•ər] **milímetro (mm)** A metric unit used to measure length or distance; 1,000 millimeters = 1 meter

million [mil′yən] **millón** 1,000 thousands; written as 1,000,000

mixed number [mikst num′bər] **número mixto** A number that is made up of a whole number and a fraction
Example: $1\frac{5}{8}$

multiple [mul′tə•pəl] **múltiplo** The product of two counting numbers is a multiple of each of those numbers

multiplication [mul•tə•pli•kā′shən] **multiplicación** A process to find the total number of items made up of equal-sized groups, or to find the total number of items in a given number of groups. It is the inverse operation of division

multiply [mul′tə•plī] **multiplicar** When you combine equal groups, you can multiply to find how many in all; the opposite operation of division

nonagon [nän′ə•gän] **eneágono** A polygon with nine sides and nine angles (p. 441)

not equal to (≠) [not ē′kwəl tōō] **no igual a** A symbol that indicates one quantity is not equal to another

number line [num′bər līn] **recta numérica** A line on which numbers can be located
Example:

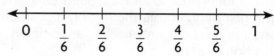

numerator [nōō′mər·āt·ər] **numerador** The number above the bar in a fraction that tells how many equal parts of the whole or group are being considered
Example: $\dfrac{3}{4}$ ← numerator

numerical expression [nōō·mer′i·kəl ek·spresh′ən] **expresión numérica** A mathematical phrase that uses only numbers and operation signs (p. 43)

O

obtuse angle [äb·tōōs′ ang′gəl] **ángulo obtuso** An angle whose measure is greater than 90° and less than 180°
Example:

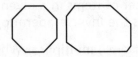

obtuse triangle [äb·tōōs′ trī′ang·gəl] **triángulo obtusángulo** A triangle that has one obtuse angle

octagon [äk′tə·gän] **octágono** A polygon with eight sides and eight angles (p. 441)
Examples:

octagonal prism [äk·tag′ə·nəl priz′əm] **prisma octagonal** A three-dimensional figure with two octagonal bases and eight rectangular faces (p. 457)

odd [od] **impar** A whole number that has a 1, 3, 5, 7, or 9 in the ones place

open figure [ō′pən fig′yər] **figura abierta** A figure that does not begin and end at the same point

order of operations [ôr′dər əv äp·ə·rā′shənz] **orden de las operaciones** A special set of rules which gives the order in which calculations are done in an expression (p. 47)

ordered pair [ôr′dərd pâr] **par ordenado** A pair of numbers used to locate a point on a grid. The first number tells the left-right position and the second number tells the up-down position (p. 373)

origin [ôr′ə·jin] **origen** The point where the two axes of a coordinate plane intersect; (0, 0) (p. 373)

ounce (oz) [ouns] **onza (oz)** A customary unit used to measure weight; 16 ounces = 1 pound

overestimate [ō′vər·es·tə·mit] **sobrestimar** An estimate that is greater than the exact answer

P

pan balance [pan bal′əns] **balanza de platillos** An instrument used to weigh objects and to compare the weights of objects

parallel lines [pâr′ə·lel līnz] **líneas paralelas** Lines in the same plane that never intersect and are always the same distance apart
Example:

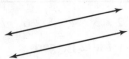

parallelogram [pâr·ə·lel′ə·gram] **paralelogramo** A quadrilateral whose opposite sides are parallel and have the same length, or are congruent
Example:

parentheses [pə·ren′thə·sēz] **paréntesis** The symbols used to show which operation or operations in an expression should be done first

partial product [pär′shəl präd′əkt] **producto parcial** A method of multiplying in which the ones, tens, hundreds, and so on are multiplied separately and then the products are added together

partial quotient [pär′shəl kwō′shənt] **cociente parcial** A method of dividing in which multiples of the divisor are subtracted from the dividend and then the quotients are added together

pattern [pat′ərn] **patrón** An ordered set of numbers or objects; the order helps you predict what will come next
Examples: 2, 4, 6, 8, 10

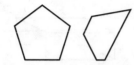

pentagon [pen′tə•gän] **pentágono** A polygon with five sides and five angles
Examples:

pentagonal prism [pen•tag′ə•nəl priz′əm] **prisma pentagonal** A three-dimensional figure with two pentagonal bases and five rectangular faces (p. 457)

pentagonal pyramid [pen•tag′ə•nəl pir′ə•mid] **pirámide pentagonal** A pyramid with a pentagonal base and five triangular faces (p. 458)

perimeter [pə•rim′ə•tər] **perímetro** The distance around a closed plane figure

period [pir′ē•əd] **período** Each group of three digits separated by commas in a multi-digit number (p. 9)
Example: 85,643,900 has three periods.

perpendicular lines [pər•pən•dik′yoo•lər līnz] **líneas perpendiculares** Two lines that intersect to form four right angles
Example:

picture graph [pik′chər graf] **gráfica con dibujos** A graph that displays countable data with symbols or pictures
Example:

HOW WE GET TO SCHOOL	
Walk	✹ ✹ ✹
Ride a Bike	✹ ✹ ✹ ✹
Ride a Bus	✹ ✹ ✹ ✹ ✹ ◗
Ride in a Car	✹ ✹

Key: Each ✹ = 10 students.

pint (pt) [pīnt] **pinta** A customary unit used to measure capacity; 2 cups = 1 pint

place value [plās val′yoo] **valor posicional** The value of each digit in a number based on the location of the digit

plane [plān] **plano** A flat surface that extends without end in all directions
Example:

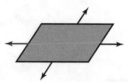

plane figure [plān fig′yər] **figura plana** See *two-dimensional figure*

point [point] **punto** An exact location in space

polygon [päl′i•gän] **polígono** A closed plane figure formed by three or more line segments (p. 441)
Examples:

Polygons   Not Polygons

polyhedron [päl•i•hē′drən] **poliedro** A solid figure with faces that are polygons (p. 457)
Examples:

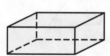

pound (lb) [pound] **libra (lb)** A customary unit used to measure weight;
1 pound = 16 ounces

prime number [prīm num′bər] **número primo** A number that has exactly two factors: 1 and itself
Examples: 2, 3, 5, 7, 11, 13, 17, and 19 are prime numbers. 1 is not a prime number.

prism [priz′əm] **prisma** A solid figure that has two congruent, polygon-shaped bases, and other faces that are all rectangles (p. 457)
Examples:

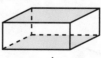

rectangular prism   triangular prism

product [präd′əkt] **producto** The answer to a multiplication problem

protractor [prō′trak·tər] **transportador** A tool used for measuring or drawing angles

pyramid [pir′ə·mid] **pirámide** A solid figure with a polygon base and all other faces are triangles that meet at a common vertex (p. 458)
Example:

Word History

A fire is sometimes in the shape of a pyramid, with a point at the top and a wider base. This may be how *pyramid* got its name. The Greek word for fire was *pura,* which may have been combined with the Egyptian word for pyramid, *pimar.*

 Q

quadrilateral [kwä·dri·lat′ər·əl] **cuadrilátero** A polygon with four sides and four angles
Example:

quart (qt) [kwôrt] **cuarto (ct)** A customary unit used to measure capacity; 2 pints = 1 quart

quotient [kwō′shənt] **cociente** The number, not including the remainder, that results from dividing
Example: 8 ÷ 4 = 2. The quotient is 2.

 R

range [rānj] **rango** The difference between the greatest and least numbers in a group

ray [rā] **semirrecta** A part of a line; it has one endpoint and continues without end in one direction
Example:

rectangle [rek′tang·gəl] **rectángulo** A parallelogram with four right angles
Example:

rectangular prism [rek·tang′gyə·lər priz′əm] **prisma rectangular** A three-dimensional figure in which all six faces are rectangles (p. 457)
Example:

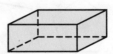

rectangular pyramid [rek·tang′gyə·lər pir′ə·mid] **pirámide rectangular** A pyramid with a rectangular base and four triangular faces (p. 458)

regroup [rē·grōōp′] **reagrupar** To exchange amounts of equal value to rename a number
Example: 5 + 8 = 13 ones or 1 ten 3 ones

regular polygon [reg′yə·lər päl′i·gän] **polígono regular** A polygon in which all sides are congruent and all angles are congruent (p. 442)

related facts [ri·lāt′id fakts] **operaciones relacionadas** A set of related addition and subtraction, or multiplication and division, number sentences
Examples: 4 × 7 = 28   28 ÷ 4 = 7
7 × 4 = 28   28 ÷ 7 = 4

remainder [ri·mān′dər] **residuo** The amount left over when a number cannot be divided equally

rhombus [räm′bəs] **rombo** A parallelogram with four equal, or congruent, sides
Example:

Word History

Rhombus is almost identical to its Greek origin, *rhombos.* The original meaning was "spinning top" or "magic wheel," which is easy to imagine when you look at a rhombus, an equilateral parallelogram.

right angle [rīt ang′gəl] **ángulo recto** An angle that forms a square corner and has a measure of 90°
Example:

right triangle [rīt trī′ang•gəl] **triángulo rectángulo** A triangle that has a right angle
Example:

round [round] **redondear** To replace a number with one that is simpler and is approximately the same size as the original number
Example: 114.6 rounded to the nearest ten is 110 and to the nearest one is 115.

scale [skāl] **escala** A series of numbers placed at fixed distances on a graph to help label the graph (p. 381)

scalene triangle [skā′lēn trī′ang•gəl] **triángulo escaleno** A triangle with no congruent sides (p. 445)
Example:

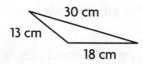

second (sec) [sek′ənd] **segundo (seg)** A small unit of time; 60 seconds = 1 minute

sequence [sē′kwəns] **sucesión** an ordered list of numbers (p. 143)

simplest form [sim′pləst fôrm] **mínima expresión** A fraction is in simplest form when the numerator and denominator have only 1 as a common factor

skip count [skip kount] **contar salteado** A pattern of counting forward or backward
Example: 5, 10, 15, 20, 25, 30, . . .

solid figure [sä′lid fig′yər] **cuerpo geométrico** See *three-dimensional figure*

solution [sə•lōō′shən] **solución** A value that makes an equation true

sphere [sfir] **esfera** A solid figure whose curved surface is the same distance from the center to all its points (p. 458)
Example:

square [skwâr] **cuadrado** A polygon with four equal, or congruent, sides and four right angles

square pyramid [skwâr pir′ə•mid] **pirámide cuadrada** A solid figure with a square base and with four triangular faces that have a common vertex (p. 458)
Example:

square unit [skwâr yōō′nit] **unidad cuadrada** A unit used to measure area such as square foot (ft²), square meter (m²), and so on

standard form [stan′dərd fôrm] **forma normal** A way to write numbers by using the digits 0–9, with each digit having a place value
Example: 456 ← standard form

straight angle [strāt ang′gəl] **ángulo llano** An angle whose measure is 180°
Example:

subtraction [səb•trak′shən] **resta** The process of finding how many are left when a number of items are taken away from a group of items; the process of finding the difference when two groups are compared; the opposite of addition

sum [sum] **suma o total** The answer to an addition problem

tablespoon (tbsp) [tā′bəl•spŏŏn] **cucharada (cda)** A customary unit used to measure capacity; 3 teaspoons = 1 tablespoon

tally table [tal′ē tā′bəl] **tabla de conteo** A table that uses tally marks to record data

teaspoon (tsp) [tē′spŏŏn] **cucharadita (cdta)** A customary unit used to measure capacity; 1 tablespoon = 3 teaspoons

tenth [tenth] **décimo** One of ten equal parts
Example: 0.7 = seven tenths

term [tûrm] **termino** a number in a sequence (p. 143)

thousandth [thou′zəndth] **milésimo** One of one thousand equal parts (p. 105)
Example: 0.006 = six thousandths

three-dimensional [thrē də•men′shə•nəl] **tridimensional** Measured in three directions, such as length, width, and height

three-dimensional figure [thrē də•men′shə•nəl fig′yər] **figura tridimensional** A figure having length, width, and height
Example:

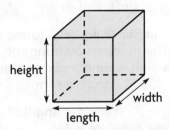

ton (T) [tun] **tonelada** A customary unit used to measure weight; 2,000 pounds = 1 ton

trapezoid [trap′i•zoid] **trapecio** A quadrilateral with exactly one pair of parallel sides
Examples:

triangle [trī′ang•gəl] **triángulo** A polygon with three sides and three angles
Examples:

triangular prism [trī•ang′gyə•lər priz′əm] **prisma triangular** A solid figure that has two triangular bases and three rectangular faces (p. 457)

triangular pyramid [trī•ang′gyə•lər pir′ə•mid] **pirámide triangular** A pyramid that has a triangular base and three triangular faces (p. 458)

two-dimensional [tŏŏ də•men′shə•nəl] **bidimensional** Measured in two directions, such as length and width

two-dimensional figure [tŏŏ də•men′shə•nəl fig′yər] **figura bidimensional** A figure that lies in a plane; a figure having length and width

underestimate [un•dər•es′tə•mit] **subestimar** An estimate that is less than the exact answer

unit cube [yŏŏ′nit kyŏŏb] **cubo unitaria** A cube that has a length, width, and height of 1 unit (p. 463)

unit fraction [yŏŏ′nit frak′shən] **fracción unitaria** A fraction that has 1 as a numerator

variable [vâr′ē•ə•bəl] **variable** A letter or symbol that stands for an unknown number or numbers

Venn diagram [ven dī′ə•gram] **diagrama de Venn** A diagram that shows relationships among sets of things
Example:

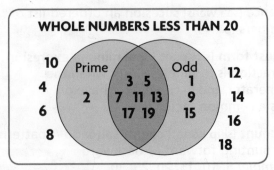

WHOLE NUMBERS LESS THAN 20

10   Prime   Odd   12
4   2   3   5   1   14
  7   11   13   9
6   17   19   15   16
8   18

H12 Glossary

vertex [vûr′teks] **vértice** The point where two or more rays meet; the point of intersection of two sides of a polygon; the point of intersection of three (or more) edges of a solid figure; the top point of a cone; the plural of vertex is vertices
Examples:

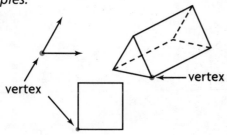

vertex

vertex

Word History

The Latin word *vertere* means "to turn" and also relates to "highest." You can turn a figure around a point, or *vertex*.

vertical [vûr′ti·kəl] **vertical** Extending up and down

volume [väl′yo͞om] **volumen** The measure of the space a solid figure occupies (p. 467)

weight [wāt] **peso** How heavy an object is

whole [hōl] **entero** All of the parts of a shape or group

whole number [hōl num′bər] **número entero** One of the numbers 0, 1, 2, 3, 4, . . . ; the set of whole numbers goes on without end

word form [wûrd fôrm] **en palabras** A way to write numbers in standard English
Example: 4,829 = four thousand, eight hundred twenty-nine

x*-axis** [eks ak′sis] **eje de la *x The horizontal number line on a coordinate plane (p. 373)

x*-coordinate** [eks kō·ôrd′n·it] **coordenada *x The first number in an ordered pair; tells the distance to move right or left from (0, 0) (p. 373)

yard (yd) [yärd] **yarda (yd)** A customary unit used to measure length or distance; 3 feet = 1 yard

y*-axis** [wī ak′sis] **eje de la *y The vertical number line on a coordinate plane (p. 373)

y*-coordinate** [wī kō·ôrd′n·it] **coordenada *y The second number in an ordered pair; tells the distance to move up or down from (0, 0) (p. 373)

Zero Property of Multiplication [zē′rō präp′ər·tē əv mul·tə·pli·kā′shən] **propiedad del cero de la multiplicación** The property that states that when you multiply by zero, the product is zero

Table of Measures

METRIC	CUSTOMARY
Length	
1 centimeter (cm) = 10 millimeters (mm) 1 meter (m) = 1,000 millimeters 1 meter = 100 centimeters 1 meter = 10 decimeters (dm) 1 kilometer (km) = 1,000 meters	1 foot (ft) = 12 inches (in.) 1 yard (yd) = 3 feet, or 36 inches 1 mile (mi) = 1,760 yards, or 5,280 feet
Capacity	
1 liter (L) = 1,000 milliliters (mL) 1 metric cup = 250 milliliters 1 liter = 4 metric cups 1 kiloliter (kL) = 1,000 liters	1 cup (c) = 8 fluid ounces (fl oz) 1 pint (pt) = 2 cups 1 quart (qt) = 2 pints, or 4 cups 1 gallon (gal) = 4 quarts
Mass/Weight	
1 gram (g) = 1,000 milligrams (mg) 1 gram = 100 centigrams (cg) 1 kilogram (kg) = 1,000 grams	1 pound (lb) = 16 ounces (oz) 1 ton (T) = 2,000 pounds

TIME

1 minute (min) = 60 seconds (sec)

1 half hour = 30 minutes

1 hour (hr) = 60 minutes

1 day = 24 hours

1 week (wk) = 7 days

1 year (yr) = 12 months (mo), or
about 52 weeks

1 year = 365 days

1 leap year = 366 days

1 decade = 10 years

1 century = 100 years

1 millennium = 1,000 years

SYMBOLS

$=$	is equal to	$\overleftrightarrow{AB}$	line AB
$\neq$	is not equal to	$\overrightarrow{AB}$	ray AB
$>$	is greater than	$\overline{AB}$	line segment AB
$<$	is less than	$\angle ABC$	angle ABC, or angle B
$(2, 3)$	ordered pair (x, y)	$\triangle ABC$	triangle ABC
$\perp$	is perpendicular to	$°$	degree
$\parallel$	is parallel to	$°C$	degrees Celsius
		$°F$	degrees Fahrenheit

FORMULAS

Perimeter		**Area**	
Polygon	P = sum of the lengths of sides	Rectangle	$A = b \times h$, or $A = bh$
Rectangle	$P = (2 \times l) + (2 \times w)$, or $P = 2l + 2w$		
Square	$P = 4 \times s$, or $P = 4s$		

Volume

Rectangular prism $V = B \times h$, or $V = l \times w \times h$

B = area of base shape, h = height of prism